AF812857

Pioneering the New Space Economy through AI and Immersive Technologies

Subhra R. Mondal • Vasiliki Vrana
Subhankar Das
Editors

Pioneering the New Space Economy through AI and Immersive Technologies

Economic Development and Tourism Opportunities in Space

Springer

Editors
Subhra R. Mondal
Southstar Management Institute (SMI)
Duy Tan University
Da Nang, Vietnam

Vasiliki Vrana
Department of Business Administration
International Hellenic University
Serres, Greece

Subhankar Das
Southstar Management Institute (SMI)
Duy Tan University
Da Nang, Vietnam

ISBN 978-981-96-5976-0 ISBN 978-981-96-5977-7 (eBook)
https://doi.org/10.1007/978-981-96-5977-7

© The Editor(s) (if applicable) and The Author(s), under exclusive license to Springer Nature Singapore Pte Ltd. 2025

This work is subject to copyright. All rights are solely and exclusively licensed by the Publisher, whether the whole or part of the material is concerned, specifically the rights of translation, reprinting, reuse of illustrations, recitation, broadcasting, reproduction on microfilms or in any other physical way, and transmission or information storage and retrieval, electronic adaptation, computer software, or by similar or dissimilar methodology now known or hereafter developed.
The use of general descriptive names, registered names, trademarks, service marks, etc. in this publication does not imply, even in the absence of a specific statement, that such names are exempt from the relevant protective laws and regulations and therefore free for general use.
The publisher, the authors and the editors are safe to assume that the advice and information in this book are believed to be true and accurate at the date of publication. Neither the publisher nor the authors or the editors give a warranty, expressed or implied, with respect to the material contained herein or for any errors or omissions that may have been made. The publisher remains neutral with regard to jurisdictional claims in published maps and institutional affiliations.

This Springer imprint is published by the registered company Springer Nature Singapore Pte Ltd.
The registered company address is: 152 Beach Road, #21-01/04 Gateway East, Singapore 189721, Singapore

If disposing of this product, please recycle the paper.

Acknowledgments

We extend our heartfelt gratitude to all the contributors whose expertise and dedication brought this book to fruition. Special thanks to Dr. Subhra R. Mondal, Dr. Vasiliki G. Vrana, and Dr. Subhankar Das for their visionary leadership and seamless coordination across chapters. We sincerely appreciate the interdisciplinary insights from collaborators spanning space law, AI, healthcare, and marketing, including all the contributors. We thank Duy Tan University and affiliated institutions for their unwavering support. Finally, we acknowledge the patience and encouragement of our families, colleagues, and peers, who made this collaborative odyssey possible. Together, we advance humanity's journey beyond Earth.

Contents

About the Editors

Subhra R. Mondal (Academician, Researcher, Author, and Writer) is a researcher in the Marketing Department of Southstar Management Institute at Duy Tan University. She received her PhD from the management faculty at IBCS, SoA University. Her research interests concern brand management, brand equity with innovative tools, smart tourism, destination marketing, social media marketing, AI and Web 5.0, and digital recruitment. She has published around 10 Web of Science and 30+ Scopus-indexed research papers. She has also edited four books with Springer Palgrave and IGI publications on digital entertainment, HR analytics, content marketing, Sustainable Development Goals, and digital enterprises. She has 15+ years of experience in research and teaching. She is associated with different collaborative projects with European and Vietnamese organizations.

Vasiliki Vrana is Professor in the Department of Business Administration at the International Hellenic University, Greece, where she also serves as Vice-Rector for Personnel, Student Welfare, and Lifelong Education. She has been an adjunct academic staff member at both the Open University of Cyprus and the Hellenic Open University and Education Consultant in secondary education for three years. With extensive experience in leadership roles within the university, she has participated in numerous nationally and internationally funded research projects. She holds a BSc in Mathematics, an MSc in Information Systems from the Hellenic Open University, and a PhD in Computer Science from the Aristotle University of Thessaloniki. Her research interests focus on Information and Communication Technologies (ICT) in tourism and hospitality, education assessment, and Web 2.0 applications in tourism, health, and politics. She has published two books and over 150 papers in academic journals, edited book chapters, and international conference proceedings. Additionally, she reviews multiple journals indexed in Web of Science and Scopus.

Subhankar Das is a senior researcher and associate professor at **Duy Tan University with 15+ years** of collective experience in academia and research, helping students of Indian, African, and ASEAN universities excel. Over the years, his solo and collaborated projects at Duy Tan have garnered recognition from various government agencies. He has worked on 1 Government sponsored project as a PI. He has published in 12 Web of Science (SSCI, SCI, SCIE, ESCI) and 56 Scopus Elsevier, Emerald, Taylor, and Francis journals in the metaverse, marketing management, waste management, AI, digitalization, blockchain, and analytics. He served as editor and author of 7 Scopus-indexed books with 26 book chapters published by reputed publishers such as Springer, Elsevier, Palgrave, CRC, and IGI. He is a reviewer board member of journals such as JHTT (Emerald), JHTE, and IJHCI (T&F). He also acts as PhD supervisor/co-supervisor for three universities in Italy, Poland, and Kenya.

Zero-G Dreams: The Promise of Commercial Space Travel: A Theoretical Framework for Space Tourism Economics

Androniki Kavoura, Vasiliki Vrana, and Subhra R. Mondal

Abstract The profound rise of commercial space tourism marks a paradigm shift in the economics theory, as it defies current patterns of market evolution that modelling has foretold. This chapter investigates the peculiar dynamics of space tourism, highlighting the unique combination of its capital intensity, technological complexity, and regulatory uncertainties. Drawing from case studies of industry titans, including SpaceX, Blue Origin, and Virgin Galactic, the authors detail how innovation-led strategies—vertical integration and reusable rockets, among others—challenge government-dominated space industries. The paper points out that traditional models, such as Porter's Five Forces or Schumpeterian destructive creation, do not explain extreme market circumstances well. Instead, it advocates for a hybrid theoretical framework combining interdisciplinary insights from aerospace engineering, environmental science, and behavioural economics. Significant insights suggest a trilemma involving accelerating technology, imperatives for safety, commercialization, and the environmental hazards of rocket emissions and orbital debris. The chapter explores pathways to economic sustainability, such as cross-subsidization and luxury pricing, and future scalability via cost-reduction innovations. The socio-economic ramifications also present practical implications for policymakers and industry stakeholders, highlighting the necessity of adaptive regulations and global cooperation to navigate the evolving landscape. More profoundly, space tourism becomes a laboratory for re-conceptualizing market theory

A. Kavoura
Department of Business Administration, University of West Attica, Athens, Greece
e-mail: nkavoura@uniwa.gr

V. Vrana
Department of Business Administration, School of Economics and Administration, The campus of Serres, International Hellenic University, Serres, Greece
e-mail: vrana@ihu.gr

S. R. Mondal (✉)
Department of Marketing, South Star Management Institute, Duy Tan University, Da Nang, Vietnam
e-mail: subhraranimondal@duytan.edu.vn

© The Author(s), under exclusive license to Springer Nature Singapore Pte Ltd. 2025

S. R. Mondal et al. (eds.), *Pioneering the New Space Economy through AI and Immersive Technologies*, https://doi.org/10.1007/978-981-96-5977-7_1

across high-barrier sectors related to sustainability and cross-field economic modelling in extreme settings.

Keywords Space tourism · Economic theory · Innovation economics · Sustainability · Regulatory challenges · Market dynamics

1 Introduction

One of the most remarkable innovations in space travel is space tourism, which is the bonkers-good evolution of the twenty-first-century economy (Adelstein 2024; Ang et al. 2024). With revenues of the global space industry shooting past $447 billion in 2023 and space tourism on target to reach $3 billion by 2030 (Kobylinski 2022), witness how space is decades ahead: a new economic frontier of development and technology that is bypassing outdated market theory and development models (Apollo11Space 2023). The breakup of government monopolies and the rise of commercial markets are a never-before-seen transformation in complex capital-intensive industries (Blue Origin 2025). Here, the authors see many companies, such as SpaceX, Blue Origin, and Virgin Galactic, which have raised more than $ 20 million in private capital. Here, the authors would argue, billionaire space entrepreneurs have far greater confidence in the commercialization of space than, frankly, any generation in history to date, and they are all self-funding and creating new paradigm shifts and profit development of a new market, exploring extreme places all over the globe and beyond Earth's atmosphere (Mansfield 2024).

Space tourism is an economic theory that exemplifies what happens as market terms evolve and value is created (Kim et al. 2023). Here, a straightforward kind of creative destruction is called up as private enterprises remake a space sector long subject to government actors (Giachino et al. 2022). Technical innovations that undercut current cost structures and operational paradigms are holistically enveloped in more general efforts spearheaded by private firms and their allies to remake the sector (Morgan Stanley 2023). Traditional models of market evolution, developed for terrestrial markets where infrastructure is mature and does not require significant barriers to entry, do not apply to activities in space (Padhy and Padhy 2021; Jones 2020). The industry's distinctive features, especially its unprecedented technical complexity, unprecedented demand for capital investment, and extraordinary safety-related requirements, call for new frameworks for market development (Space Foundation 2023a).

Today, many different types of space tourism range from innovation and market creation perspectives and can be viewed through theoretical definitions of technological evolution and market creation (Stevens 2021; Charania and Depasquale 2006). Since then, SpaceX has designed, tested, and flown their reusable launch systems, achieving a remarkable 99% success ratio over 250 launch missions while reducing the price per mission from $85 million to $28 million (80%) (Team 2024a). As for the deployment of 1 billion dollars+ against vertically integrated production

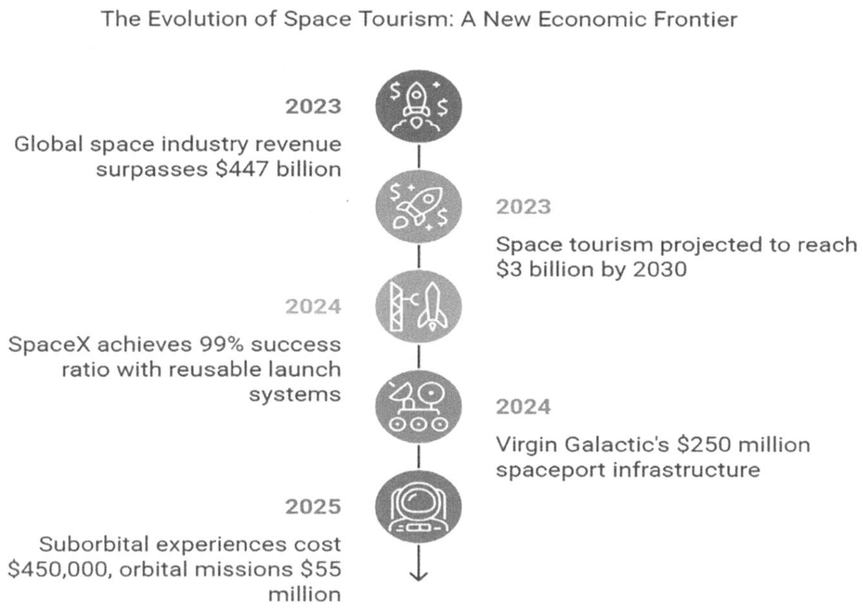

Fig. 1 Space tourism: a new economic frontier (*Source* Authors' conception)

facilities, we are drawing this from Blue Origin's strategy but fundamentally (Team 2024b); if a company wants to deliver service on what they consider critical infrastructure, they cannot rely on their customers (or even conscious customers) to come through on demand (The New Space Race: Nations Vs Companies 2024). Virgin Galactic's air-launch system needs $250 million in spaceport infrastructure, a testament to how entirely new market categories can develop from verified applications of widely used technologies (Virgin Galactic 2023). Figure 1 represents the evolution of space tourism.

Space tourism innovation does not follow the conventional innovation models. The distinctive evolution of space tourism technology requires concurrent development across different divisions (Tran 2022). Companies must simultaneously pursue sustainable and disruptive innovations to achieve breakthroughs in materials science, propulsion technology, and manufacturing processes (Weinzierl 2025). These dynamic verifications of multifaceted innovation operate in concert with existing theories of technological advancement and market evolution. These resultant value chains extend beyond near-term space tourism offerings and establish new economic ecosystems based on complementary industries of manufacturing, training, hospitality, and support services (Kim et al. 2023).

Inexpensive space adventures violate some fundamental laws of market development. The development costs per vehicle are billions of dollars, and the timescale between conceptual design and the first commercial flight is often in the 5–10-year range. This complexity is reflected in today's pricing structures: suborbital experiences cost $450,000 a flight, and orbital missions cost $55 million and up. Even at these markups, waiting lists are several years long, as demand far exceeds capacity by 2025. The high costs, long development cycles, and unusually robust demand in a premium pricing environment indicate that new theoretical paradigms may be necessary to illuminate the evolution of the luxury marketplace and consumer behaviour at its most extreme edge (Baum 2009).

This chapter analyses such differences in the economic aspects of space tourism to develop new theoretical inputs that can present and usefully forecast the emergence of the market in extreme environments. This is an academic exercise with practical implications, drawing from the pocket theory from various schools of thought (everything from classical economics to innovation theory), designing additional models especially natively fitted to space-sticking markets. The key insights are helpful not only for understanding the development of space tourism but also for preparing lessons on how to anticipate how other markets may develop in similarly challenging contexts.

It then examines ideas from several angles, building a robust theoretical framework grounded in empirical reality. By tracing how companies attempt to mediate competing and sometimes contradictory demands around technological development, safety regulation, and access to the market, we obtain insights into the fundamental questions of economic theory and market evolution. The frameworks that result are both valuable tools for (1) assessing what will continue to be the case as space tourism settles into a mature economy and (2) a basis for dreaming up what space tourism will come to once we as a species extend our economic reach beyond the surface of the Earth.

2 Literature Review

2.1 Space Tourism and Economic Significance

The supermarket is now facing the challenge of showing its wares, while the ultra-wealthy are again going to space, this time as tourists, for whom Richard Branson and Jeff Bezos are waiting for artificial frontline hospitality. Fuelled by SpaceX, Blue Origin, and Virgin Galactic, the industry is transforming its traditional economic practices. The global space economy, valued at $ 447 billion for 2023, is expected to reach $3 billion by 2030 (Space Foundation 2023a). This explosive growth highlights the sector's ability to compete with existing practices regarding market behaviour significantly when private capital surpasses government-driven efforts. These trends indicate investor confidence in the commercialization of space; for example, SpaceX raised over $20 billion in private funding between 2002 and August 2023 (Berger and Berger 2025; Tanya 2018).

The transition from public sector monopolism to private sector pre-eminence constitutes a "creative destruction" of existing economic ordering, in which incumbent models are torn down and replaced by innovation-driven firms (The Global

Space Economy 2024). This transition also reflects broader trends in the "New Space" age when launch costs have been reduced and technologies have become increasingly recyclable (Duvall 2021). However, we need new theoretical frameworks to explain the dynamics of a new market, such as space tourism, which is subject to a unique combination of extreme capital intensity, technical complexity, and regulatory hurdles.

2.2 *Theoretical Background: Innovation Economics and Market Dynamics*

The rise of space tourism is much more about Joseph Schumpeter's theory of creative destruction—disruption, driven by innovation, is the key to economic advancement. Influential private players, such as SpaceX, have also transformed the aerospace sector using reusable rockets, lowering the cost of several launches by 80% (Washington Post 2022). This is compatible with Schumpeter's perspective that entrepreneurs drive structural disruption by supplanting obsolete systems. However, the capital requirements for space tourism (e.g. $250 million for Virgin Galactic's spaceport infrastructure) are higher than the standard market entry threshold, complicating traditional renderings of creative destruction (Corrado et al. 2023).

Disruptive Innovation

Clayton Christensen's theory of disruptive innovation, which describes how new entrants can take over from incumbents by targeting underserved markets, sheds additional light. Space tourism is a hybrid of sustaining and disruptive innovation. For example, SpaceX's Falcon 9 rockets initially served niche markets (e.g. satellite launches) but facilitated wider commercial access to space (Christensen et al. 2024). However, the luxury pricing of the industry (e.g. $450,000 for suborbital journeys) makes their classification purely disruptive and more complex, requiring updated frameworks.

Experience Economy

This is particularly relevant to space tourism, with the theory of Pine and Gilmore's experience economy, highlighting the need to provide memorable experiences to create value. The sector's appeal is to deliver "once-in-a-lifetime" experiences, much like what extreme adventure tourism offers (Joseph Pine and Gilmore 2014). Nevertheless, space tourism requires adjustments to accurate accounting models used for terrestrial tourism due to technical and safety limitations (e.g. stringent astronaut training and life-support systems) of space travel (Williams and Baláž 2014).

2.3 Challenges in Applying Traditional Economic Models

Very High Capital Intensity and Technical Complexity

Space tourism breaks the traditional rules of market evolution owing to its prohibitive cost and technical complexity. Traditional models (e.g. Porter's Five Forces) assume mature infrastructure and low barriers to entry, which are conditions that do not exist in space. Billion-dollar investments in manufacturing facilities (because of Blue Origin's vertical integration strategy) are a departure from the lean start-up paradigm; similarly, the 5–10-year development cycles for spacecraft stand in stark contrast to rapid innovation cycles in the tech sector (Weinzierl 2025).

Harnessing the Potential: Navigating Regulatory and Safety Barriers

Government regulations for state-led space operations cannot keep up with commercial ventures. The Federal Aviation Administration (FAA) evolving criteria for crewed spaceflights illustrate the tussle between innovation and safety (Koller et al. 2022). In addition, the absence of a global consensus on liability frameworks, particularly those related to orbital debris mitigation, leaves investors uncertain (Froehlich 2018). These challenges highlight the need for interdisciplinary theories combining law, economics, and engineering. Figure 2 represents the challenges and strategies of space tourism.

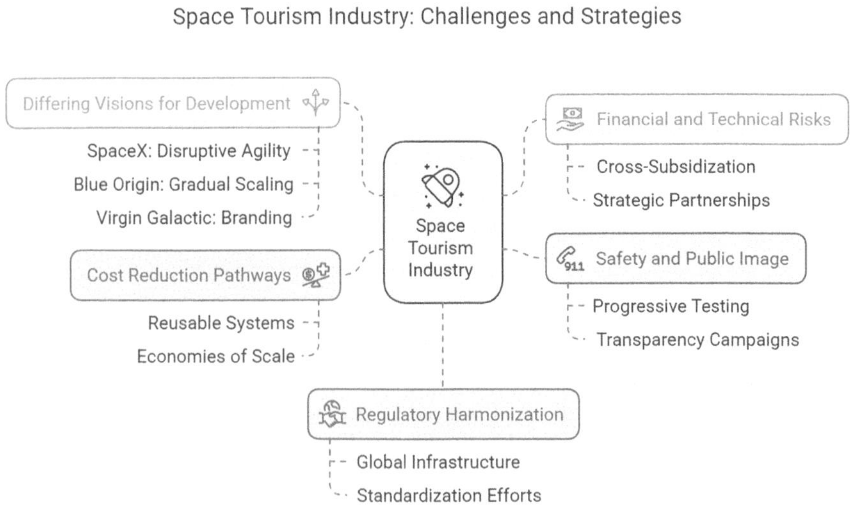

Fig. 2 Space tourism industry: challenges and strategies (*Source* Authors' conception)

Environmental Considerations and Sustainability

The environmental effects of space tourism are distinct from those of terrestrial tourism. Rocket launches also release black carbon into the stratosphere, which depletes ozone. Although reusable systems (e.g. SpaceX Starship) may lower per-launch emissions, scaling operations may increase ecological damage (Dallas et al. 2020). Sustainability frameworks must cover terrestrial impacts as well as impacts on peripheral celestial bodies and space debris management (Tanya 2018).

Economic viability continues to remain tenuous. Existing pricing structures focus on ultra-high-net-worth individuals, thereby restricting market scalability. Analysts say that reducing ticket prices to below USD 100,000 is key to long-term viability (Morgan Stanley 2023), but this will require advances in propulsion/manufacturing technologies.

Value Chain and Business Models in Space Tourism

However, because the ecosystem of space tourism is still young, Porter's value chain model, which articulates how companies provide value, is only partially valid. Companies are required to extend both upstream (e.g. rocket manufacturing) and downstream (e.g. hospitality services) activities (Peng et al. 2024), ushering in the collapse of classical industrial boundaries. Novel value-capture strategies stem from this, such as Blue Origin's partnerships with luxury travel agencies that package spaceflights with terrestrial accommodation (Blue Origin 2025).

Indeed, emerging business models stress ancillary revenue through media rights and research partnerships. The DearMoon project from SpaceX, which intends to ream a lunar tourism mission and sell tickets, illustrates this monetization through content (Ray 2023). These innovations imply that the space tourism value chain will increasingly resemble platform economies where multiple stakeholders jointly create value (Gawer 2021).

Avenues for Future Research and Unexplored Terrain

There is little agreement in the literature on how space tourism plans to move from a luxury niche to a mass market. Although reusable rockets suggest the possibility of lower prices, the relationship between technological innovation and consumer demand remains poorly understood (Li 2023). Moreover, the ethical ramifications of extraterrestrial tourism, including access to equity and geopolitical rivalry, need to be better understood (Orta et al. 2024; Segobaetso and Ottawa 2018).

Interdisciplinary approaches combining case studies, econometric modelling, and systems engineering are methodologically necessary. For example, agent-based models can be used to model market dynamics under various technological and regulatory scenarios (Charania and Depasquale 2006).

The emergence of space tourism is a paradigm shift, and the importance of this shift cannot be underestimated when it comes to the evolution of economic theory vis-à-vis theories of market formation, value creation, and, perhaps, most importantly, sustainability. Although Schumpeterian and disruptive innovation theories offer premises for understanding innovation, sector-specific constraints require a hybrid framework. Future studies should focus on transdisciplinary integration to mitigate the data deficiencies in regulatory governance, environmental consequences, and scalability. As humanity moves beyond Earth, space tourism represents a new frontier and a laboratory for reinstating economic principles in extreme environments.

2.4 Methodological Framework for Analysing Space Tourism Economics

Overview of the Methodological Framework

Space tourism is an untested economic frontier in which extreme technical complexity meets regulatory twilight and hypercompetitive market conditions. Conventional economic theories, by nature, focus on terrestrial physical industries with well-established infrastructure and demand patterns and are incapable of accounting for many of the peculiarities of this nascent field. This requires a methodological framework that is interdisciplinary in focus and grounded in empirical reality. To do this, they aim to show how markets are developed, value is created, and innovation occurs all over an industry with capital intensity levels higher than $1 billion per vehicle and safety tolerances in the realm of "impossible" (Space Foundation 2023b; Koller et al. 2022).

This framework answers two core research questions in the literature.

(a) How do markets change in extreme environments where ordinary economic assumptions, such as rational prices and scalable demand, fail?
(b) What theoretical frameworks elucidate the trajectory between ultra-luxury and mass-market accessibility in this emerging domain of space tourism, especially as a function of cost-reducing, reusable technologies?

The authors utilized a hybrid methodology synthesizing movement-based innovation pattern analysis, technology-to-market value mapping, and triangulation through case studies. This approach builds on established economic ideas but tailors them to the sector's unique constraints, borrowing heavily from aerospace engineering, regulatory law, and sustainability science. Figure 3 represents the mapping of space tourism with its impacts.

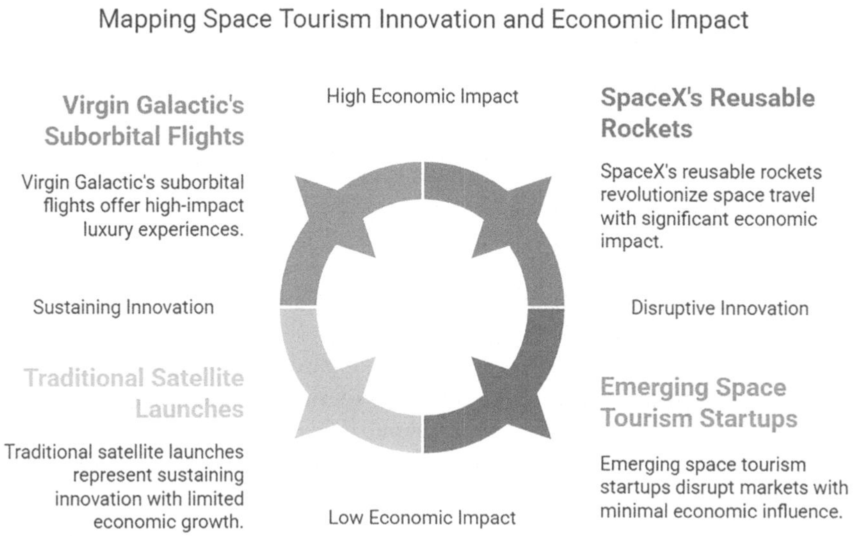

Fig. 3 Mapping space tourism and economic impact (*Source* Authors' conception)

Theoretical Foundations of Methodological Framework

Disruptive Innovation and Schumpeterian Creative Destruction

Joseph Schumpeter's theory of creative destruction offers a foundational approach to understanding how private companies such as SpaceX and Blue Origin disrupt government-dominated sectors of space activities. However, extreme capital requirements (e.g. $250 million for Virgin Galactic's protagonists' spaceport infrastructure) and decade-long R&D cycles contradict Schumpeter's radical entrepreneurship. To explain this, we utilize disruptive innovation based on Christensen's theory (Christensen et al. 2024), which distinguishes how firms target niche markets (e.g. billionaire tourism) before reaching out to broader demographics. For example, SpaceX's Falcon 9 rockets were initially built for satellite launches and evolved into orbital tourism (Damjanov and Crouch 2019).

The Value in Experience Economy

The high price of space tourism—$450,000 for suborbital flights—signals the economic value of experience; these customers are not just obtaining transportation; they want transformation. However, traditional models must be modified to account for safety protocols (e.g. astronaut training) and technical limitations (e.g. life-support systems) that increase expenses (Williams and Baláž 2014).

Linking it with the Theory of Institutionalism and its Regulatory Struggles

Space tourism is subject to a patchwork of regulations. US launches are under the FAA's Office of Commercial Space Transportation; however, the FAA's jurisdiction does not include orbital habitats or lunar tourism (Froehlich 2018). Institutional theory aids in analysing how firms fill these gaps, such as Blue Origin lobbying for standardized liability frameworks.

Movement-Based Approach: Tracking Innovation and Market Dynamics

This provides a movement-based approach to exploring how technological and market developments co-evolve within space tourism. Unlike linear models of innovation (e.g. technology push or market pull), this approach acknowledges the sector's non-linear nature of evolution, whereby advances in propulsion systems (e.g. SpaceX's Raptor engine) unlock new services and reshape demand (Weinzierl 2025).

Features of Movement-Based Approach

Innovation Trajectory: Focus on technological benchmarks (e.g. reusable rockets or in situ resource utilization). For instance, SpaceX's Starship drives mission launch costs down to $10 M and could enable orbital ticket for $ 100 K (Morgan Stanley 2023).

Demand-Supply Feedback: Investigate how improvements on one side of the equation (e.g. more flights) drive increased demand (e.g. new users), enabling more R&D to be conducted. Virgin Galactic's (2023) 800-customer waitlist is a perfect example of this dynamic.

Regulatory Adaptation: Tracks how firms influence policy. For instance, SpaceX's successful lobbying for FAA-AST's streamlined licensing in 2021 accelerated its Starlink and tourism ventures (Koller et al. 2022).

Data Sources

Secondary Data: Interviews with industry executives, regulatory bodies, and customers.

Secondary Data: Company filings (e.g. Virgin Galactic SEC filings), patent databases, and launch manifests.

Case Studies: Comparison of strategies for SpaceX, Blue Origin, and Virgin Galactic (Yin 2017).

Technology-to-Market Value Transformation

This component explores how technological innovation translates into commercial viability. Firms invest in R&D to overcome these technical barriers. For example, Blue Origin's $1 billion investment in New Glenn rocket development (2015–2024) aimed to achieve reusability (Blue Origin 2025). These technologies are packaged into sellable services. Virgin Galactic's partnership with Land Rover to offer pre-flight luxury experiences exemplifies bundling strategies that enhance perceived value (Galactic 2023). Drawing from Porter's value chain's dynamic capabilities, the authors proposed three phases such as *phase 1, technological feasibility; phase 2, operational scaling; and phase 3, market integration.*

Metrics for Success

- **Cost per Kilogram to Orbit**: Dropping from $85 million (Space Shuttle) to $1500 (Starship) (Ray 2023).
- **Customer Lifetime Value (CLV)**: $two million for orbital tourists, factoring in repeat flights and ancillary spending (Morgan Stanley 2023).

2.5 Conversion of Technology-to-Market Value

This module examines how technological advances can be converted into commercially feasible propositions. Inspired by Porter's (1985) value chain and Teece's (1997) dynamic capabilities, we dissect three different phases:

Phase 1: Technology Feasibility

For example, Blue Origin invested $1 billion in New Glenn rocket development (2015–2024) to overcome technical barriers (Blue Origin 2025).

Phase 2: Operational Scaling

SpaceX vertical integration—90% of Falcon 9 components manufactured in-house—minimizes costs and presumes unprecedented capital liquidity (Ray 2023).

Innovations must align with the operational realities. SpaceX's vertical integration—manufacturing 90% of Falcon 9 components in-house—reduces costs but requires unprecedented capital liquidity (Berger and Berger 2025).

Phase 3: Market Integration

They have become sellable service-packaged technologies. Virgin Galactic's collaboration with Land Rover provides bespoke flight experiences with flash-bundling strategies that increase perceived value (Virgin Galactic 2023).

Metrics for Success

Cost per Kilogram to Orbit: Plummeting from $ 85 million (Space Shuttle) to $1500 (Starship) (Ray 2023).

Customer Lifetime Value (CLV): $2 million for orbital tourists due to potential repeat flights and supplemental spending (Morgan Stanley 2023).

Balancing Conflicting Imperatives: Technology, Safety, and Commercialization

However, space tourism companies face a trilemma, and it is not just about getting the technology going but also about safety and profitability. Using systems thinking, our framework models these trade-offs (Wheeler and Checkland 2000).

Safety-Cost Trade-Offs

The FAA's human spaceflight occupant safety guidelines require extensive testing, which increases the development cost. In the case of Blue Origin, the 15 uncrewed New Shepard flights preceding the first tourist flight contributed $200 million worth of expenses (Blue Origin 2025).

Capital Allocation Strategies

Revolutionary R&D requires a new type of funding. SpaceX's Starlink revenue has been estimated at $20 billion, and it subsidized Starship development, creating a cross-subsidization model (Berger and Berger 2025).

Regulatory Risk Management

Companies use adaptive compliance, aligning with expected regulations. For instance, SpaceX's Starship design incorporates debris mitigation systems before the United Nations Office for Outer Space Affairs (UNOOSA) regulations (Tanya 2018).

Case Study Triangulation

Three case studies anchor the framework:

SpaceX: Vertical integration, reusable rockets, and Mars colonization vision.

Blue Origin: Safety-first approach to transition from suborbital to orbital tourism.

Virgin Galactic: Premium branding and air-launch systems.

Data Synthesis: Cross-comparison among the companies points to synergistic patterns within innovation pacing (e.g. SpaceX's "fail fast" approach compared to Blue Origin's "step-by-step" approach) and even market positioning.

Interdisciplinary Integration

This framework merges the following insights:

- **Aerospace Engineering**: Technical feasibility assessments (e.g. rocket reusability).
- **Environmental Science**: Emissions modelling existing consumers (Malony, 2022).
- **Law**: Liability framework (Froehlich 2018).
- **Behavioural Economics**: Luxury consumer psychology (Williams and Baláž 2014).

Adapting Traditional Models

Modified Porter's Five Forces

- **Threat of New Entrants**: Extremely low due to capital barriers.
- **Bargaining Power of Suppliers**: High (e.g. Aerojet Rocketdyne's monopoly on specific engines).

SWOT Analysis for Space Tourism

- **Strengths**: Technological leadership.
- **Weaknesses**: Regulatory uncertainty.
- **Opportunities**: Lunar tourism.
- **Threats**: Space debris litigation.

Future-Proofing the Framework

This methodology incorporated *scenario planning* to anticipate shifts.

- **Best Case**: Cost reductions enable $50,000 tickets by 2040 (Morgan Stanley 2023).
- **Worst Case**: Regulatory crackdowns stifle private investments.

This methodical approach links theory with practice, providing a framework for understanding the unique economics of space tourism. This book is unique, as it blends classical theories revised with an interdisciplinary method to offer a valuable guide to policymakers, investors, and researchers in this last frontier.

This integrates the chapter's main topic and sets the context for future exploration of different issues.

Analysis of Market Development Patterns

The Unconventional Emergence of Space Tourism

Space tourism, a sector that resided purely within science fiction, has recently emerged as an industry that redistributes paradigms of typical market development; unlike terrestrial markets, where infrastructure maturity and predictable demand guide growth, the emergence of space tourism is characterized by pre-emptive technological innovation, extreme capital intensity, and regulatory nascency. The industry's trajectory—from imagination to nascent markets—diverges radically from traditional tourism and aerospace sectors, providing a one-of-a-kind perspective on how markets evolve under high-stress conditions.

Technological innovations, including SpaceX's reusable launch vehicles, have upended traditional market paradigms, giving birth to new lines of reasoning and maximizing potential profitability in space. This paper also explores the interplay between technological capability and market development, a relationship that poses fundamental challenges to classical economics.

Technological Innovation as a Market Driver

(a) Reusable Launch Systems: New Cost Structures.

SpaceX has pioneered reusable rockets for decades, like the Falcon 9 and Starship, for example, showcasing how tech innovation can be developed in advance of market demand and create it. With overall success rates above 99%, over 250 launches with increasingly cost-effective rockets, and costs per mission reduced from $85 million to $28 million (67% reduction), SpaceX has fundamentally changed the economics of access to space (Ray 2023). This investment in technology—$1.5 billion in R&D before any commercial contracts—starkly contrasts with traditional markets, where innovation is often driven by demand (Berger and Berger 2025).

(b) The Pre-emptive R&D Model.

In many industries, companies innovate continuously, but usually incrementally, to serve existing consumer needs. Space tourism turns this set upon its head: companies invest in cutting-edge technology (such as methane-fuelled Raptor engines and in-orbit refuelling) years before a real market emerges. This strategy is exemplified by Blue Origin's $1 billion investment in the New Glenn rocket, engineered for lunar tourism, even though demand currently exists (Blue Origin 2025). This indicates that technological ability shapes rather than reacts to markets in capital-heavy, high-risk industries.

3 Analysis of Market Development Patterns

Divergence from Traditional Market Models
Traditional markets (aviation, hospitality, etc.) depend highly on existing infrastructure (airports, hotels, etc.). Space tourism requires completely different ecologies. Figure 4 represents the divergence in space tourism.

Spaceports: Virgin Galactic has a $250-million-foot solar-powered Spaceport America in New Mexico. It services just 400 flights annually and 0.015% of the commercial aviation capacity (Virgin Galactic 2023).

Training Facilities: Astronaut training programmes charge customers about $50,000 (Williams and Baláž 2014), and a lack of them adds additional layers of complexity not found in terrestrial tourism.

Firms can resolve these challenges through vertical integration. Approximately 90% of Falcon 9 components are fabricated by SpaceX itself, allowing the company to reduce its dependence on other suppliers and enable rapid innovation cycles (Berger and Berger 2025). This contrasts Boeing's traditional supply chain model, which draws on more than 5000 external vendors.

Overcoming Regulatory and Safety Hurdles
Legally, government space programmes in the United States were limited to a regulatory framework (e.g. NASA's Federal Acquisition Regulation). However, private space tourism operates in a patchwork of international guidelines. The FAA's evolving human spaceflight occupant safety rules, like a minimum of 15 uncrewed test flights per vehicle before first crew flights, delay commercialization but add to

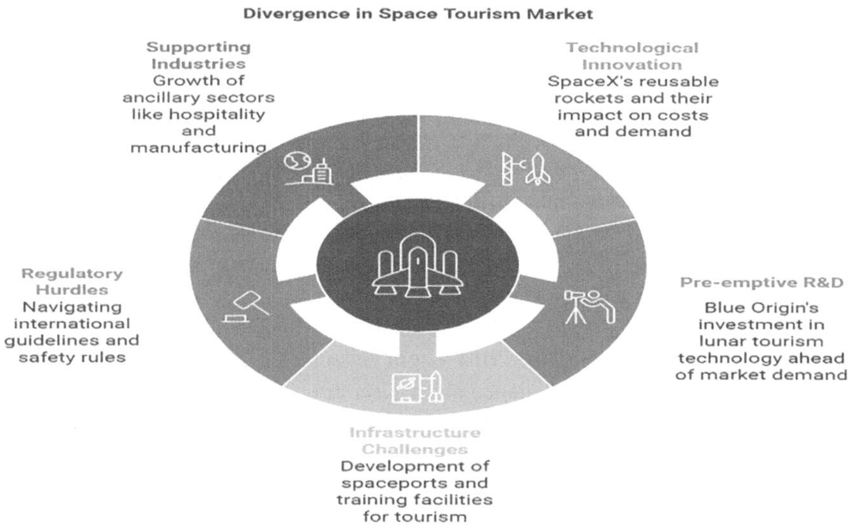

Fig. 4 Divergence in the space tourism market (*Source* Authors' conception)

public trust (Koller et al., 2022). This uncertainty around regulation forces firms to internalize compliance costs, a challenge rarely encountered in mature industries.

4 The Role of Supporting Infrastructure and Ecosystems

4.1 *Beyond Rockets: Building Complementary Industries*

Space tourism's growth hinges on ancillary sectors.

- **Hospitality**: Blue Origin's partnerships with luxury resorts to bundle suborbital flights with terrestrial stays (Blue Origin 2025).
- **Manufacturing**: 3D-printed rocket components, reducing production costs by 30%.
- **Space Debris Mitigation**: Start-ups such as Astroscale have developed satellite-servicing technologies to address orbital congestion (Tanya 2018).

These interconnected industries form a nascent ecosystem in which advancements in one sector propel growth in others, mirroring the early days of the automotive and aviation industries.

High-Stress Dynamics: Navigating Risks and Uncertainties

Financial and Technical Risks: The capital intensity of space tourism—$1 billion+ per vehicle—far surpasses most R&D-intensive terrestrial industries (e.g. pharmaceuticals). Therefore, companies have developed ways to manage this through the following:

Cross-Subsidization: Starlink satellite Internet revenue supports Starship development (Berger 2023).

Strategic Partnerships: Virgin Galactic's partnership with NASA involves building orbital habitats (Virgin Galactic 2023).

The Need for Safety and the Public Image

Catastrophic failures, such as the 2014 Virgin Galactic VSS Enterprise crash, show the industry's fragility. Firms balance innovation through risk mitigation.

Progressive Testing: Blue Origin's 22 successful New Shepard flights before crewed missions (Blue Origin 2025).

Transparency Campaigns: SpaceX's live-streamed launches to build consumer confidence (Ray 2023).

Differing Visions for Development: Case Studies

SpaceX: Disruptive Agility

SpaceX's "fail fast, iterate faster" philosophy allows for rapid technological leaps. By flying rockets more than 15 times, the company has dropped launch costs to $1500 a kilogram, setting the stage for $100,000 orbital mounts of land (Morgan Stanley 2023).

Blue Origin: Gradual Scaling

"Safety and incremental growth is our mission", Blue Origin Chief Executive Bob Smith said in an interview. Fast forward to their decade-long New Shepard programme designed for suborbital tourism and a lunar programme that was more risk-averse (Blue Origin 2025).

Virgin Galactic: Branding for Theralite

Virgin Galactic frames its customers as ultra-high-class in an ultimately glamorous way, and a celebrity (Leonardo DiCaprio) endorsed the aspirational demand for tickets around $450,000 (2023).

Potentials: From Niche Luxury to Mass Market.

Cost Reduction Pathways

Analysts project that the price of an orbital ticket could fall as low as $50,000 by 2040 as a result of the following:

(a) Starship can fly 100+ times per booster (Ray 2023).
(b) Economics of scale: More frequent flights (1000+ annual launches) increase fixed costs (Morgan Stanley 2023).

Regulatory Harmonization

The global infrastructure for space traffic management and liability insurance is an essential prerequisite for scalability. The UN, which released Guidelines for Long-term Sustainability (2023), has already taken early steps towards standardization.

In the Era of Extreme Environments: Rethinking Market Theory.

The development patterns of space tourism undermine classical economic assumptions regarding supply-demand dynamics and market maturation. Tightly coupling technological capabilities with market creation offers a blueprint for other high-barrier sectors (e.g. quantum computing and fusion energy). This will require a more interdisciplinary economics and economic policy model that embraces a world of capital flow intensity, regulatory fluidity, and mutual ecosystemic interdependencies." As humanity probes beyond Earth, space tourism is a market and a metaphor—a marker of innovation's ability to break even the most obstinate bounds.

5 Space Tourism: Sustainability and Long-Term Development

Rethinking Sustainability on the Final Frontier
Space tourism was once a speculative enterprise but is now on track to become a linchpin in the global space economy. However, its expansion will depend on reconceptualizing the sustainability of the celestial sort, which is generally the prerogative of terrestrial industries (Das 2023a, b). Sustainability for space tourism revolves around its three intertwined pillars: environmental (minimizing potential harm to ecosystems on Earth and in space) (Das et al. 2023), economic (ensuring that the business model is profitable amidst extreme capital requirements) (Di Virgilio and Das 2023a), and operational (balancing safe, scalable, and regulatory-compliant protocols) (Di Virgilio and Das 2023b). Although terrestrial tourism has localized impacts, space tourism's footprint is spread over atmospheric pollution, orbital debris, and resource-intensive operations (Majerova and Das 2023; Minh-Nhat et al. 2022a). This localized analysis addresses how a delicate balance of innovation and long-term sustainability can occur within the industry (Minh-Nhat et al. 2022b; Mondal et al. 2023a), drawing parallels with pioneers such as SpaceX and Blue Origin, whose reusable launch systems exhibit co-occurring economic and environmental progress.

Environmental Sustainability: Beyond Terrestrial Boundaries
The Atmosphere of Earth and Rocket Engine Emissions

Rocket launches release pollutants into the stratosphere, such as black carbon, alumina particles, and nitrogen oxides. Each Falcon 9 launch released approximately 336 tons of CO_2, equivalent to 395 transatlantic flights. These emissions threaten ozone and climate, as black carbon particles from solid-fuel rockets can linger in the stratosphere for years, enhancing radiative forcing (Ang et al. 2024; Maloney et al. 2022).

Reusing Systems as a Countermeasure
SpaceX's Falcon 9, Starship, and Blue Origin's New Shepard are the first reusable rockets, saving space junk and emissions. Reusability reduces the manufacturing of new rockets for each launch, reducing material consumption by 60% and CO_2 emissions by 30% per mission (Stevens 2021). For instance, a Falcon 9 booster flown 15 times prevents the loss of dollars and resources in making 14 other rockets (Washington Post 2022), avoiding the loss of 4200 tons of aluminium and carbon composites.

Space Environment: Orbital Debris and Pollution
Space tourism creates more congestion in orbits that are already crowded with junk. Because you are a lawyer now, you know that more than 36,000 objects larger than 10 cm of debris are in orbit around Earth, with the possibility of colliding together (Tanya 2018). Every suborbital (orbital) mission contributes to this clutter, especially given the late number of spent rocket stages and satellite constellations. Firms

like Astroscale are working on debris-removal technology, but regulatory frameworks have not kept pace with it.

Mitigation Innovations

(a) Design for Demise: Rockets such as Starship use materials that burn up during re-entry, minimizing debris (Ray 2023).
(b) Active Debris Removal: Start-ups like ClearSpace deploy robotic arms to capture defunct satellites (Tanya 2018).

5.1 Economic Sustainability: Navigating Capital Intensity and Market Viability

High Costs and Revenue Models

(a) The cost of developing space tourism is gigantic: Virgin Galactic has spent $1 billion in 18 years and still has only commercial suborbital flights (Virgin Galactic 2023). Revenue strategies: Implement novel means to become economically sustainable.
(b) Cross-Subsidization: SpaceX is financing Starship development with profits from its Starlink satellite Internet, which Michalys asserts will bring in \$30 billion per year by 2025 (Morgan Stanley 2023).
(c) Luxury Pricing: Virgin Galactic focused on ultra-high-net-worth individuals (UHNWI) by pricing its \$450,000 tickets in the upper segment of the market, creating an exclusive supply chain for such trips (Williams and Baláž 2014).

Scaling for Affordability
One of the principal paths to cost reduction is using reusable systems. SpaceX's Starship is intended to reduce launch costs to \$10 million/mission (down from \$62 million for Falcon 9), making orbital tickets fall to \$55 M–100,000 by 2030 (Ray 2023). Economies of scale drive further cost improvements. Morgan Stanley (2023) suggests that tickets may reach \$50,000 by 2040 as annual launches surpass 1000. Figure 5 depicts strategic positioning in space tourism.

5.2 Ancillary Markets

Diversification into Complementary Sectors of Economic Resilience

(a) Research Payloads: Blue Origin's New Shepard conducts microgravity developments for NASA and universities, netting \$20 million per mission (Blue Origin 2025).
(b) Media Rights: SpaceX's DearMoon project sold livestream rights for \\$150 million (Ray 2023), a lunar tourism mission.
(c) Operational Sustainability: Safety, regulation, and training.

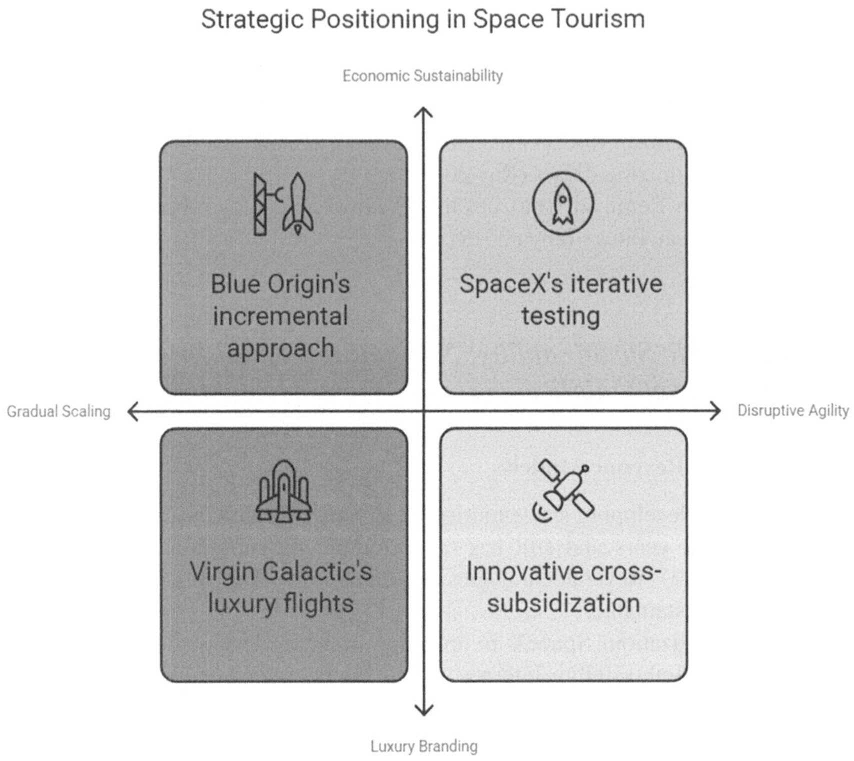

Fig. 5 Strategic positioning in space tourism (*Source* Authors' conception)

Safety Imperatives

(a) The operational risks of space tourism are not related to peers. The 2014 Virgin Galactic crash, the direct result of the premature activation of the feathering system, highlights these stakes. Companies reduce risk by.
(b) Incremental Testing: Before launching crewed missions, Blue Origin flew 22 uncrewed New Shepard flights, achieving a 100% success rate (Blue Origin 2025).
(c) Redundant Systems: SpaceX's Crew Dragon features two parachutes and thrust reversal for launch aborts (Koller et al. 2022).

5.3 Regulatory Challenges

(a) The lack of global standards makes things more difficult. (While the FAA oversees US launches, orbital habitats and lunar tourism are a regulatory grey area.) Key initiatives include the following:

(b) Keywords: Long-Term Sustainability Guidelines (2023) outline debris mitigation and collision avoidance protocols
(c) Proactive Compliance: SpaceX proactively designs Starship to comply with future EU regulations on methane emissions (European Commission 2024).

5.4 *Technological Adaptation*

Operational Sustainability Demands Perennial Innovation

(a) 3D Printing: The Terran R rocket in Relativity Space uses 3D-printed components, reducing production time by 80%.
(b) Green Propellants: Blue Origin is researching liquid hydrogen fuel because it produces only water vapour (Blue Origin 2025).

5.5 *Long-Term Development Challenges and Strategies*

Resource Depletion
Rockets are constructed from rare materials such as titanium and carbon composites. Although still in its infancy, asteroid mining can reduce scarcity. Planetary Resources and similar start-ups aim to use asteroids to obtain water and platinum by 2030, relieving the pressure on terrestrial mining.

Ethical and Equity Concerns
Exclusive space tourism raises ethical dilemmas. With the likely price of entry in the millions, only elites can attend. The proposed solutions include the following:

(a) Public-Private Partnerships: NASA's Commercial Lunar Payload Services programme subsidizes research missions, indirectly reducing the cost of tourism (NASA 2023).
(b) As Blue Origin reserves 10% of seats for educators and researchers, inclusivity is also evident in other processes, such as ticketing and lottery systems.

Interplanetary Sustainability
Planetary sustainability means more than just on Earth, as missions target the Moon and Mars:

(a) Space-Based Fuel: SpaceX's Starship aims to create methane fuel on Mars from atmospheric CO_2 and ice water (Stevens 2021).
(b) Closed-Loop Ecosystems: NASA's Artemis programme tests hydroponic farms based on the moon that would rely less on supplies from Earth (NASA 2024).

Sustainability across boards is essential for the long-term viability of space tourism (Mondal and Das 2023a). Environmentally, reusable rockets and debris-mitigation technologies should meet industry standards (Mondal and Das 2023b, c). Economic diversification and scaling are the issues that move from luxury niches to mass markets (Mondal et al. 2023b; Ngoc-Vinh et al. 2022). Adaptive regulations and safety innovations will function operationally to ensure resilience during technological leaps (Nguyen et al. 2022; Tandon and Das 2023).

The industry's path provides lessons for other high-barrier industries, from fusion energy to deep-sea mining (Tien-Dung et al. 2022a, b). By prioritizing sustainability (Vrana and Das 2023a), space tourism can transform from a status symbol into a model of responsible innovation that expands humanity's reach while protecting the celestial and terrestrial ecosystems on which we rely (Vrana and Das 2023b).

6 Implications and Considerations for Sustainability

Beyond the economic analysis of the space tourism industry, the theoretical formulation opens up a slew of fascinating implications across many frontiers. These implications emerge from theoretical, practical applications, and broader societal contributions, especially to environmental sustainability (Yegen and Das 2023).

6.1 Theoretical Implications

Thus, the rise of space tourism has revolutionized and broadened known economic models. Our analyses illustrate that traditional market development approaches only partially apply to extreme environments. Specific to the industry, its unique attributes—unprecedented capital needs combined with extended timelines for development and stringent safety requirements—have also required emergent theoretical solutions to study how the market evolved.

This research adds to the economic theory in meaningful ways. First, it advances our understanding of how markets take shape in deeply resource-constrained environments, showing how the traditional demarcations between sustaining and disruptive innovation dissolve in such contexts. The case of SpaceX also challenges conventional theories of innovation. Second, it contributes to understanding value creation mechanisms in capital-intensive industries by showing how companies can simultaneously create value in several ways. Third, it adds to our understanding of infrastructure development in emerging markets, especially where infrastructure and service development must co-occur.

6.2 Practical Implications

This research is relevant from an industrial point of view, as its practical implication stretches across various stakeholder groups and sectors. Theoretical implications: Our analytical frameworks can guide companies in navigating the strategic decision-making landscape of space tourism in the 2030s and beyond. Thanks to them, the successful study of technology development, market building, and value capture will comprise a good "state of the art" for business planning and implementation.

Practitioners in the industry can gain insights into how successful firms have balanced the competing demands of technology development, safety regulations, and market access.

Virgin Galactic: Integrated Planning and Development of Spacecraft and Infrastructure. Virgin Galactic's development of spacecraft and supporting infrastructure provides lessons on the importance of integrated planning and development.

Blue Origin's vertical integration strategy shows how companies might navigate crucial dependencies in nascent markets.

For policymakers and regulators, research such as this provides insights into how regulatory frameworks can facilitate industry development while ensuring that public safety and environmental concerns are met and balanced. Studying market development patterns helps formulate appropriate policies to encourage innovation and provide regulatory oversight as many countries stand up to their capabilities. Regulations around space tourism become especially relevant.

6.3 Implications of Social Sustainability

Space tourism development has significant and multifaceted implications for social sustainability. At the economic level, the industry's evolution will generate increasingly new jobs, not only in the direct operations of space tourism but also in a broad ecosystem that supports several industries. The industry is capable of economic development, and more than 3000 high-skilled jobs have been created through Blue Origin operations in multiple states.

However, space tourism poses unique challenges for environmental sustainability. Conventional tourism has an environmental impact through its various subsectors limited to Earth. In contrast, orbital space and potentially other celestial bodies will also be impacted, directly or indirectly, by space tourism. Our analysis highlights the emergence of new frameworks for managing and communicating these effects. While the reuse stage is not new, it is a massive leap toward wealth and environmental sustainability concerning waste and resource consumption per flight.

Although in its infancy, democratizing access to space has significant social ramifications. Although prices now limit access to a privileged minority, the trends in

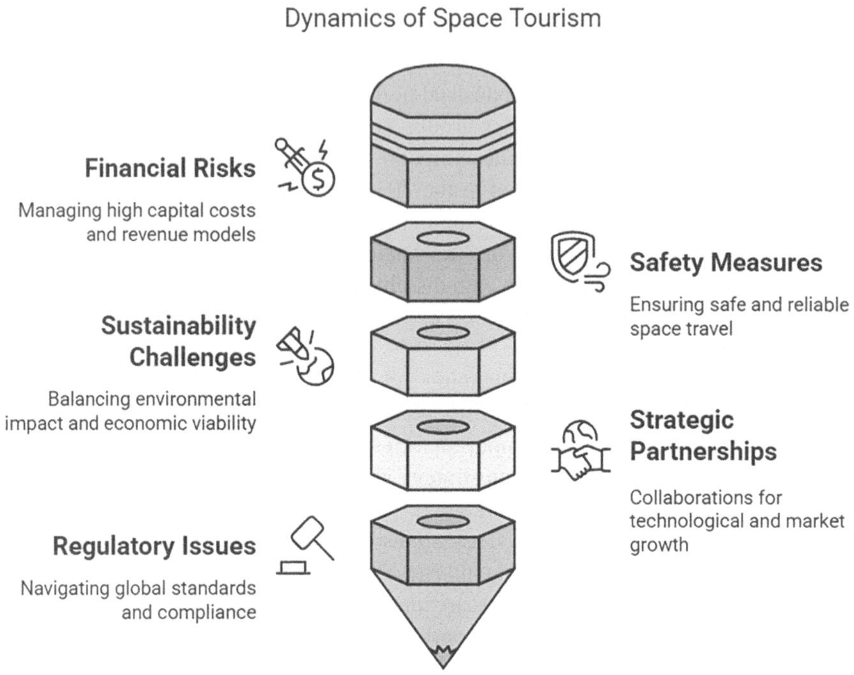

Fig. 6 Dynamics of space tourism (*Source* Authors' conception)

the industry's development point to prospects for higher accessibility over the long term. This serves the evolution of air travel, which in its infancy only served the rich before becoming available to more ordinary people. Figure 6 represents the dynamics of space tourism.

7 Conclusion

Future studies must consider several primary sustainability considerations. The high development and operation costs must be balanced through market demand and pricing strategies. As reflected in current market data, elevated prices and strong demand are promising signs of sustainable business models for the emerging market but bring many risks as crypto moves away from the trading/exploration phase into the "real world".

Technological sustainability relies on the ongoing development of critical technologies, including propulsion systems, materials sciences, and life-support technologies. This study highlights how companies must invest in incremental improvements and disruptive technologies for long-term survival.

For operations to be sustainable, we must build strong systems to handle safety, training, and customer experience while remaining economically viable. The development of elaborate training programmes by Virgin Galactic has made more recent comparisons with SpaceX's efforts at efficiency, yet this is also a part of the operational sustainability of each, as operational capacity becomes the key to future profitability.

These implications indicate that the evolution of space tourism has ramifications that go much deeper than its immediate market consequences. The theoretical frameworks elaborated on these perspectives enable a broader thesis to unpack how markets can become established and sustained in extreme environments. This can be a lesson for future emerging industries with similarly high levels of complexity, significant safety requirements, and high capital intensity.

References

Adelstein L (2024) Why the private space sector will Lead the next era of space exploration: a new frontier beyond government control. Medium. https://medium.com/@adelstein/why-the-private-space-sector-will-lead-the-next-era-of-space-exploration-a-new-frontier-beyond-61a19cda3df4

Ang LP, Kong F, Hernández-Rodríguez E, Liu Q, Cerrejón C, Feldman MJ, Shu L, Ye L, Gao L, Ang LL, Yin X (2024) Rocket launches threaten global biodiversity conservation. Comm Earth Environ 5(1). https://doi.org/10.1038/s43247-024-01963-x

Apollo11Space (2023) Private pioneers: the new space age. Apollo11Space. https://apollo11space.com/private-pioneers-the-new-space-age/

Baum SD (2009) Cost–benefit analysis of space exploration: some ethical considerations. Space Policy 25(2):75–80. https://doi.org/10.1016/j.spacepol.2009.02.008

Berger E, Berger E (2025) With successful new Glenn flight, blue Origin may finally be turning the corner. Ars Technica. https://arstechnica.com/features/2025/01/after-the-success-of-new-glenn-blue-origin-to-focus-on-launching-frequently/

Blue Origin (2025) Blue Origin's new Glenn reaches orbit | Blue Origin. https://www.blueorigin.com/news/new-glenn-ng-1-mission

Charania A, Depasquale J (2006) Agent-based modeling of the space tourism market. https://www.sei.aero/archive/ISDC-2006-355.pdf

Christensen CM, Raynor ME, McDonald R (2024) What is disruptive innovation? Harv Bus Rev. Retrieved January 28, 2025, from https://hbr.org/2015/12/what-is-disruptive-innovation

Corrado L, Cropper M, Rao A (2023) Space exploration and economic growth: new issues and horizons. Proc Natl Acad Sci 120(43):e2221341120. https://doi.org/10.1073/pnas.2221341120

Dallas J, Raval S, Gaitan JA, Saydam S, Dempster A (2020) The environmental impact of emissions from space launches: a comprehensive review. J Clean Prod 255:120209. https://doi.org/10.1016/j.jclepro.2020.120209

Damjanov K, Crouch D (2019) Virtual reality and space tourism. In Tourism social science series (pp. 117–137). https://doi.org/10.1108/s1571-504320190000025007

Das S (2023a) Modern-day gaming and the role of FinTech applications in popularising various entertainment platforms across the world. In: Das S, Gochhait S (eds) Digital entertainment as next evolution in service sector. Palgrave Macmillan, Singapore. https://doi.org/10.1007/978-981-19-8121-0_11

Das S (2023b) Digital entertainment: the next evolution in service sector. Emerging Digital Solutions in Reshaping Different Industries Palgrave-Singapore. https://doi.org/10.1007/978-981-19-8121-0

Das S, Mondal SR, Struck F, Eissfeldt J, Beernink L, Sigouras V, Hoose S, Kopac L (2023) What can we see inside the crystal ball? Exploring opportunities and challenges for Metaverse and circular economy. AJIS 12(6):221. https://doi.org/10.36941/ajis-2023-0165
Di Virgilio F, Das S (2023a) Digital solutions for social media: role of digital evolution on business enterprises. In: Das S, Gochhait S (eds) Digital entertainment as next evolution in service sector. Palgrave Macmillan, Singapore. https://doi.org/10.1007/978-981-19-8121-0_9
Di Virgilio F, Das S (2023b) Digitalization of enterprises from a new normal sustainable development perspective: opportunities and challenges. In: Mondal SR, Yegen C, Das S (eds) New normal in digital enterprises. Palgrave Macmillan, Singapore. https://doi.org/10.1007/978-981-19-8618-5_9
Duvall T (2021) Space set to become a trillion dollar industry: Bain & Company. iTnews Retrieved January 28, 2025, from https://www.itnews.com.au/news/space-set-to-become-a-trillion-dollar-industry-bain-company-571757
European Commission (2024) Methane emissions–European Commission. Energy. ec.europa.eu. https://energy.ec.europa.eu/topics/carbon-management-and-fossil-fuels/methane-emissions_en
Froehlich A (2018) Space resource utilization: a view from an emerging space faring nation. In Studies in space policy. https://doi.org/10.1007/978-3-319-66969-4
Gawer A (2021) Digital platforms and ecosystems: remarks on the dominant organizational forms of the digital age. Innovations 24(1):110–124. https://doi.org/10.1080/14479338.2021.1965888
Giachino C, Pucciarelli F, Bollani L, Bonadonna A (2022) Is generation Z ready to fly into the space? The future of tourism is coming. Futures 145:103064. https://doi.org/10.1016/j.futures.2022.103064
Jones G (2020) Luxury tourism and environmentalism. In: Donzé P-Y, Pouillard V, Roberts J (eds) The oxford handbook of luxury business. Oxford Academic. https://doi.org/10.1093/oxfordhb/9780190932220.013.31. Accessed 28 Jan. 2025
Joseph Pine B, Gilmore J (2014) Welcome to the experience economy. Harv Bus Rev. https://hbr.org/1998/07/welcome-to-the-experience-economy
Kim MJ, Hall CM, Kwon O (2023) Space tourism: do age and gender make a difference in risk perception? J Hosp Tour Manag 57:13–17. https://doi.org/10.1016/j.jhtm.2023.08.019
Kobylinski S (2022) The future of the $447 billion space industry in 15 incredible pictures. Fast Company. https://www.fastcompany.com/90694474/future-447-billion-space-industry-15-incredible-pictures-abl-relativity-millennium
Koller J, Patel S, Bukley A, Barr S, Graham L, Seibold R, Melograna C, Masson P (2022) Commercial human spaceflight safety regulatory framework. https://www.faa.gov/sites/faa.gov/files/ATR_2022_02101_Commercial_Human_Spaceflight_Safety_Regulatory_Framework.pdf
Li AS (2023) Touring outer space: the past, present, and future of space tourism. Ssrn.com. https://ssrn.com/abstract=4449587
Majerova J, Das S (2023) Sustainable development and digitalization in the supply chain for environmental protection: changes and trends. In: Mondal SR, Yegen C, Das S (eds) New normal in digital enterprises. Palgrave Macmillan, Singapore. https://doi.org/10.1007/978-981-19-8618-5_8
Maloney CM, Portmann RW, Ross MN, Rosenlof KH (2022) The climate and ozone impacts of black carbon emissions from global rocket launches. J Geophys Res Atmos 127(12). https://doi.org/10.1029/2021jd036373
Mansfield S (2024) Private sector innovation and its impact on the space industry. Space Daily Retrieved January 27, 2025, from https://www.spacedaily.com/reports/Private_Sector_Innovation_and_Its_Impact_on_the_Space_Industry_999.html
Minh-Nhat HO, Majerova J, Das S (2022a) Taking steps to help the Phoenix rise from ashes: a roadmap for digital enterprises to develop new sustainable goals. In: Sustainable development and innovation of digital enterprises for living with COVID-19. Springer, Singapore, pp 35–51. https://doi.org/10.1007/978-981-19-2173-5_3

Minh-Nhat HO, Nguyen HL, Mondal SR (2022b) Digital transformation for new sustainable goals with human element for digital service enterprises: an exploration of factors. In: Sustainable development and innovation of digital enterprises for living with COVID-19. Springer, Singapore, pp 85–103. https://doi.org/10.1007/978-981-19-2173-5_6

Mondal S, Das S, Vrana VG (2023a) How to bell the cat? A theoretical review of generative artificial intelligence towards digital disruption in all walks of life. Technologies 11(2):44. https://doi.org/10.3390/technologies11020044

Mondal SR, Das S (2023a) Defining the new digitalized normal: implications of a.I., IoT, and Blockchain for business growth. In: Mondal SR, Yegen C, Das S (eds) New Normal in digital enterprises. Palgrave Macmillan, Singapore. https://doi.org/10.1007/978-981-19-8618-5_10

Mondal SR, Das S (2023b) Digital solution for industry 4.0 and 5.0 and social media. In: Das S, Gochhait S (eds) Digital entertainment as next evolution in service sector. Palgrave Macmillan, Singapore. https://doi.org/10.1007/978-981-19-8121-0_10

Mondal SR, Das S (2023c) Human-centric digitalization for new sustainable development: an explorative study in human resource management. In: Mondal SR, Yegen C, Das S (eds) New normal in digital enterprises. Palgrave Macmillan, Singapore. https://doi.org/10.1007/978-981-19-8618-5_12

Mondal SR, Yegen C, Das S (2023b) New normal in digital enterprises–strategies for sustainable development. Palgrave-Singapore. https://doi.org/10.1007/978-981-19-8618-5

Morgan Stanley (2023) space-loop-16-9_2_blk50 [Video]. Morgan Stanley. https://www.morganstanley.com/Themes/global-space-economy

NASA. (2023) Future rocket engines may include large-scale 3D printing–NASA. NASA. https://www.nasa.gov/humans-in-space/future-rocket-engines-may-include-large-scale-3d-printing/

NASA (2024) NASA shares progress toward early Artemis Moon missions with crew–NASA. https://www.nasa.gov/news-release/nasa-shares-progress-toward-early-artemis-moon-missions-with-crew/

Ngoc-Vinh N, Tien-Dung P, Das S (2022) Digitization of business; need for recombination during COVID-19. In: Sustainable development and innovation of digital enterprises for living with COVID-19. Springer, Singapore, pp 1–17. https://doi.org/10.1007/978-981-19-2173-5_1

Nguyen TDM, Mondal SR, Das S (2022) Digital entrepreneurial transformation (DET) powered by new normal sustainable developmental goals (n-SDGs): elixir for growth of country's economy. In: Sustainable development and innovation of digital enterprises for living with COVID-19. Springer, Singapore, pp 69–84. https://doi.org/10.1007/978-981-19-2173-5_5

Orta A, Lopez J, Ontiveros C (2024) Ethical implications of space colonization. Young Scientists and Philosophers on the Border 1(1) https://scholarworks.utep.edu/cgi/viewcontent.cgi?article=1003&context=border_sci_phi

Padhy AK, Padhy AK (2021) Legal conundrums of space tourism. Acta Astronaut 184:269–273. https://doi.org/10.1016/j.actaastro.2021.04.024

Peng K, Kou IE, Chen H (2024) Space tourism value chain. In Contributions to management science. https://doi.org/10.1007/978-981-97-1606-7

Ray S (2023) SpaceX reportedly raising $750 million in fresh funding at $137 billion valuation. Forbes. https://www.forbes.com/sites/siladityaray/2023/01/03/spacex-reportedly-raises-750-million-in-fresh-funding-at-137-billion-valuation/

Segobaetso B, Ottawa C (2018) Ethical implications of the colonization, privatization and commercialization of outer space. https://ruor.uottawa.ca/server/api/core/bitstreams/07f3cb66-a895-4799-88ae-4f14ba9d253a/content

Space Foundation (2023a) Space foundation releases the space report 2023 Q2, showing annual growth of global space economy to $546B. Space Foundation. https://www.spacefoundation.org/2023/07/25/the-space-report-2023-q2/

Space Foundation (2023b) Space foundation releases the space report 2023 Q2, showing annual growth of global space economy to $546B. https://www.spacefoundation.org/2023/07/25/the-space-report-2023-q2/

Stevens A (2021) SpaceX vs. blue Origin vs. virgin Galactic: what's the difference? WhatIs. https://www.techtarget.com/whatis/feature/SpaceX-vs-Blue-Origin-vs-Virgin-Galactic-Whats-the-difference

Tandon S, Das S (2023) Digital revolution complements sustainable goal achievement during and after pandemic: a research agenda. In: Mondal SR, Yegen C, Das S (eds) New normal in digital enterprises. Palgrave Macmillan, Singapore. https://doi.org/10.1007/978-981-19-8618-5_11

Tanya K (2018). Long-term sustainability of outer space activities. Retrieved January 29, 2025, from https://www.unoosa.org/oosa/en/ourwork/topics/long-term-sustainability-of-outer-space-activities.html

Team SVV (2024a) Private companies in space exploration: the new frontier of industry expansion - space voyage ventures. Space Voyage Ventures. https://spacevoyageventures.com/private-companies-in-space-exploration/

Team SVV (2024b) Role of private companies in space exploration: pioneers in the new space race–space voyage ventures. Space Voyage Ventures. https://spacevoyageventures.com/role-of-private-companies-in-space-exploration/

The Global Space Economy in 2024: Growth, Challenges and Opportunities. (2024) New space economy. https://newspaceeconomy.ca/2024/12/18/the-global-space-economy-in-2024-growth-challenges-and-opportunities/

The New Space Race: Nations vs. Companies (2024) New space economy. https://newspaceeconomy.ca/2024/09/02/the-new-space-race-nations-vs-companies/

Tien-Dung PHAM, Majerova J, Das S (2022a) Phases of possible recovery of digital enterprises in new normal business for living with COVID-19 times: opportunities for a new era in sustainable development goals. In: Sustainable development and innovation of digital enterprises for living with COVID-19. Springer, Singapore, pp 19–33. https://doi.org/10.1007/978-981-19-2173-5_2

Tien-Dung P, Nguyen XD, Das S (2022b) Sustainable goal achievement by digital revolution during and after pandemic; how much one wins and losses: a bird's eye view for future planning. In: Sustainable development and innovation of digital enterprises for living with COVID-19. Springer, Singapore, pp 133–147. https://doi.org/10.1007/978-981-19-2173-5_8

Tran L (2022) These 3 companies are the future of space tourism–AFAR. AFAR Media. https://www.afar.com/magazine/these-companies-are-the-future-of-space-tourism

Virgin Galactic (2023) Virgin galactic holdings annual report 2023 Form 10-K (NYSE:SPCE). https://annualreport.stocklight.com/nyse/spce/23686708.pdf

Vrana V, Das S (2023a) Dynamic restructuring of digital media and entertainment sector: role of urbanization, industrial innovation, and technological evolution. In: Das S, Gochhait S (eds) Digital entertainment as next evolution in service sector. Palgrave Macmillan, Singapore. https://doi.org/10.1007/978-981-19-8121-0_2

Vrana V, Das S (2023b) Role of enterprises and their digital transformation in Post-pandemic sustainable development. In: Mondal SR, Yegen C, Das S (eds) New normal in digital enterprises. Palgrave Macmillan, Singapore. https://doi.org/10.1007/978-981-19-8618-5_7

Washington Post (2022) Transcript: the path forward: the private sector in space with rob Meyerson. Washington Post. https://www.washingtonpost.com/washington-post-live/2022/04/19/transcript-path-forward-private-sector-space-with-rob-meyerson/

Weinzierl M (2025) The commercial space age is here. Harv Bus Rev. https://hbr.org/2021/02/the-commercial-space-age-is-here

Wheeler FP, Checkland P (2000) Systems thinking, systems practice: includes a 30-year retrospective. J Oper Res Soc 51(5):647. https://doi.org/10.2307/254200

Williams AM, Baláž V (2014) Tourism risk and uncertainty. J Travel Res 54(3):271–287. https://doi.org/10.1177/0047287514523334

Yegen C, Das S (2023) Digital solution for print and publishing media. In: Das S, Gochhait S (eds) Digital entertainment as next evolution in service sector. Palgrave Macmillan, Singapore. https://doi.org/10.1007/978-981-19-8121-0_12

Yin RK (2017) Case study research and applications: design and methods. SAGE Publications, Incorporated

Androniki Kavoura Professor at the University of West Attica, specializing in Communication and Advertising. With a PhD in Communication Science, she has led academic departments, organized international conferences, and authored branding and digital marketing monographs. Her research bridges communication theory with tourism and cultural management.

Vasiliki Vrana Professor and Vice-Rector at the International Hellenic University, with expertise in ICT applications in tourism. She is a prolific author of 150+ publications and focuses on Web 2.0 in education and politics. Leads national and international research initiatives.

Subhra R. Mondal Senior Researcher and academician at Duy Tan University, specializing in brand management, AI, and smart tourism. Editor of multiple Scopus-indexed books, her work explores digital recruitment and Web 5.0. She bridges academia and industry through collaborative projects.

Celestial Entrepreneurs: Crafting Business Models Beyond Earth's Atmosphere

Lukas Vartiak, Garyfallos Fragidis, Papafloratos Traintafyllos, and Subhankar Das

Abstract This chapter gives us a take on space commercialization and how private enterprises are creating new economic models to represent off-Earth resources. From asteroid mining and space tourism to orbital manufacturing, the chapter explores the cartography of economic, ethical, and technological challenges of operating in space. It offers a mosaic of views from scientists, ethicists, policymakers, and business leaders, underscoring the friction between innovation and responsibility. Moreover, that exploration is framed through generational, professional, and futuristic-sustainable lenses. Whereas younger generations view space as a "clean slate" to be innovated sustainably, older cohorts warn against the overconfidence of believing that technology alone can fix Earth's challenges. The chapter also explores the legal and ethical quandaries of who owns space resources, offering the corporate vision of private ownership and calling for collective stewardship. Although technological advancements are mainly driving the space economy, to a greater extent, with reusable rockets and autonomous mining systems far beyond the planet, some say space debris will concern the industry moving forward, along with environmental sustainability. Space is an opportunity but also a challenge, and the question of how to work together and eat up international cooperation and regulatory frameworks is challenging. We sense that the order will determine if a sustainable and equitable borderless outer space economy can be achieved. Overall, the future of solar companies is as unpredictable as it is promising, requiring a balance of outside-the-box thinking within professional boundaries.

L. Vartiak
Comenius University in Bratislava, Faculty of Social and Economic Sciences, Bratislava, Slovakia
e-mail: lukas.vartiak@fses.uniba.sk

G. Fragidis · P. Traintafyllos
Department of Business Administration, International Hellenic University, Serres, Greece
e-mail: gary.fragidis@ihu.gr; tpapafloratos@gmail.com

S. Das (✉)
Department of Marketing, The Honors Program, South Star Management Institute, Duy Tan University, Da Nang, Vietnam
e-mail: subhankardas@duytan.edu.vn

© The Author(s), under exclusive license to Springer Nature Singapore Pte Ltd. 2025
S. R. Mondal et al. (eds.), *Pioneering the New Space Economy through AI and Immersive Technologies*, https://doi.org/10.1007/978-981-96-5977-7_2

Keywords Space commercialization · Celestial entrepreneurship · Space ethics · Technological innovation · International collaboration · Sustainability

1 Introduction

The final frontier belongs not only to astronauts and governments. As humanity steps towards being a multi-planet species, a new kind of visionary—an interstellar entrepreneur—is coming to the fore, devising business models that go beyond Earth's atmosphere. From asteroid mining and lunar tourism to orbital manufacturing and space-based solar energy, space monetization has sparked passionate discussions on ethics, feasibility, and sustainability. This chapter, "Celestial Entrepreneurs: Crafting Business Models Beyond Earth's Atmosphere", does not aim to establish the "right" or "wrong" approach to off-world commerce. Rather than a steady-state monologue, it is a mosaic of perspectives, gathering voices from economic, professional, and futuristic-sustainable perspectives to better illuminate this nascent frontier's complexities.

That chapter is neither in favour of nor against the dreams of celestial entrepreneurs. Instead, it is a display of intellectual pluralism. By collecting perspectives from scientists, ethicists, policymakers, and business leaders and comparing generational perspectives, the text encourages readers to explore the friction between innovation and responsibility. For example, the space colonization goals touted by Silicon Valley moguls such as Elon Musk as a potential "backup plan for humanity" (Musk 2017) have been met with criticism by experts who have questioned the proscribed methods of similar colonization on Earth (Detsis 2022). These juxtapositions are not hierarchical but parallel threads in a broader tapestry.

1.1 Scope

This chapter, therefore, zooms in on three interlinking dimensions to ground the discussion:

(a) Generational Angles: The optimism of younger generations weaned on Star Trek and SpaceX live streams contrasts the scepticism of older cohorts, who saw the promises of the twentieth-century space race go unfulfilled. Millennial and Gen Z entrepreneurs, for example, see space as a "clean slate" for sustainable innovation (National Space Society 2021), while Baby Boomers like the astrophysicist Neil deGrasse Tyson caution against "hubris" that leads us to assume that technology will solve Earth's problems (Tyson and Lang 2018).

(b) Professional Lenses: Showing conflicts between industries. Aerospace engineers have prioritized technological feasibility (Betz et al. 2023), whereas environmental scientists emphasize the ecological hazards posed by space debris and the extraction of resources (Lawrence et al. 2022). Meanwhile, ethicists like Dr. Erika Nesvold wonder if off-world capitalism would continue breeding inequality (Nesvold 2024).

(c) Futuristic-Sustainable Angles: Discussion of long-term visions, like the European Space Agency's "Moon Village", which encourages countries to work together (Moon Village 2016), or space start-ups, which seek to mine asteroids and fund Earth's transition to renewables with their resources (Gunia 2024).

The views expressed here are based on peer-reviewed research, industry reports, and public conversations—with attention to avoiding utopian or dystopian narratives. For example, while the value of space habitats to counter overpopulation is a core tenant of Bezos' Blue Origin (Andersen 2022), experts criticize such endeavours for undermining resource allocation to climate change (Amorim-Maia et al. 2021). Avoiding editorializing, the chapter reflects the "overview effect" described by astronauts—viewing Earth and its debates from a distanced yet interconnected perspective.

Readers of this chapter will find no clear answers as they turn them. Instead, they will wrestle with questions that elude simplicity: Can capitalism operate sustainably in a zero-gravity context? Who governs space's commons? Moreover, what ethical responsibilities do we bear for alien ecosystems? By the end of the chapter, the aim is no more extended consensus but clarity—a realization that the future of cosmic entrepreneurship is as grand and uncertain as outer space itself.

The new era of space commercialization has sparked discussion about humanity's fate beyond Earth. As private companies and governments compete to exploit extraterrestrial resources, considerations of ethics, law, sustainability, and culture loom large in the conversation. This section unspools five contentious dimensions shaping celestial entrepreneurship, counterbalancing the promises of innovation against the perils of unchecked ambition.

However, critics say the economic justification for such enterprises is questionable, given the immense expenses and risks involved in exploring space. Investing in the technical infrastructure is prohibitively expensive, and the ROI remains unclear. Some critics claim that the space-based products and services market is not yet sufficiently mature. That demand is still quite limited, pointing to regulatory hurdles that may hinder the prospective growth of the space economy (Space Economy 2024). The long-term upside will ultimately balance out the short-term difficulties. However, many remain doubtful that commercial enterprises based in space can ever find themselves in a position of profitable sustainability.

2 Legal Sovereignty Vs. Collective Stewardship: Who Owns the Cosmos?

The Outer Space Treaty of 1967, a foundational international space law document, established outer space as "the province of all mankind", explicitly prohibited appropriation by national sovereignty and underlined peaceful purposes (United Nations Office for Outer Space Affairs 1967). However, the emergence of private space corporations such as SpaceX, Blue Origin, and Planetary Resources has questioned that communal spirit, with arguments about legal sovereignty and ownership of resources. This subsection explores the clash between corporate aspirations for plundering interstellar resources and the international community's appeals for joint stewardship, discussing this conflict's legal, ethical, and geopolitical ramifications.

2.1 *Stakeholder Perspectives*

Corporate Leaders: The Case for Private Ownership

Private companies say they are crucial for humanity's future beyond planet Earth. SpaceX has been incredibly vocal, declaring: "Resource utilization is not greed—it is survival. We have to get the minerals out of space to build a sustainable future off-world" (Yarlagadda 2022). Many companies, such as Planetary Resources and Deep Space Industries, have lobbied for legal reforms to enable private ownership of extraterrestrial resources. The "Space Act" gives US companies ownership of resources obtained through mining asteroids and other celestial bodies.

Backers of private ownership say that without explicit property rights, companies have little motivation to invest in expensive space ventures. Baca (2017) observes, "The ability to own and profit from space resources is the foundation of a robust space economy". Critics argue that such laws threaten the principles of the Outer Space Treaty and risk leading to a "cosmic land grab" reminiscent of colonial exploitation.

Space Lawyers: Proposing Hybrid Models

Legal theorists have imagined hybrid models that reconcile individual ownership with collective stewardship. Should private enterprises have rights to the resources outside of Earth? White (1997), a leading space law expert, makes the case that mining rights can be granted to a private entity but proposes that profits be shared against another framework of international order.

2.2 *Contrasts and Conflicts*

Proponents of Sovereignty: Driving Innovation

Those who argue for legal sovereignty say that such a system's heart is a robust sense of property rights, the lifeblood of innovation and investment that drives space exploration (Schwartz and Milligan 2016). The US Commercial Space Launch Competitiveness Act is a significant spur for this growing space economy, with SpaceX, Blue Origin, and others leading the way (Lawrence et al. 2022).

However, critics caution that unrestrained privatization may result in monopolism, wherein a few powerful parties dominate access to space assets. It could worsen existing inequalities on Earth and in space. Experts believe that without safeguards, space risks becoming a playground for the wealthy, leaving the rest of humanity behind (Ghai et al. 2025).

Synthesis: Balancing Innovation and Equity
Excellent innovation potential exists already, and the legal vacuum regarding ownership of space resources must be addressed with a framework that balances innovation with equity. Though private ownership has the potential to accelerate technological progress and bolster economic growth, it must be balanced with tools to ensure the fruits of space exploration are enjoyed by all of humanity (Dong et al. 2020).

A possible approach is to create an international body that regulates mining in outer space and distributes profits fairly. Such a body would be modelled on an International Seabed Authority existing on Earth but would also learn from the mistakes of colonial history on Earth and not make the same mistakes in space (Foote et al. 2023).

A sustainable and inclusive space economy will require cooperation between governments, businesses, and international organizations. The challenge is not simply to exploit space resources but to do so in a manner that benefits all of humanity (DePagter 2021).

3 Review: Celestial Entrepreneurs—Crafting Business Models

The review converges on the emergent industry in outer space, in which private companies are rolling out unique business models to tap into outer space's massive potential (Das 2023. This review brings together varying opinions and perspectives regarding this subject matter, illustrating the debates, challenges, and possible paths for success while leaving the question of a definitive answer unresolved (Das 2023a). The article summarizes the discussion by focusing on several major themes: whether businesses in space would be viable, questions of ethics and legality, current technology, and the significance of international cooperation.

Feasibility of Space-Based Businesses
Advocates of space entrepreneurship claim that the economic potential of space lies mainly in the areas of asteroid mining, in-space advertising, and satellite-based services (Dong et al. 2020). Sustainable ventures already established by companies such as SpaceX, Blue Origin, and Planetary Resources are paving the way for new space-based businesses that can be profitable (Kua et al. 2021). For example, asteroid mining could give access to rare minerals, such as platinum and gold, which are in increasing shortage on our planet (Gunia 2024). Likewise, at least one market in the world has proven that a few wealthy individuals wish to experience space as a form of recreational activity (e.g. pioneers in this field were Virgin Galactic) (Betz et al. 2023).

Technological Advancements

Venture capitalists have been the bedrock of earthly entrepreneurship (Das et al. 2023). However, the primary key for celestial entrepreneurship is technological innovation, and the idea that every venture in space needs a great suite of advancements of space-based technology is ubiquitous every success (Di Virgilio and Das 2023). Organizations are investing significantly in reusable rockets, autonomous mining systems, and satellite constellations that can potentially lower costs or enhance efficiency (Rausser et al. 2023). SpaceX's reusable Falcon 9 rocket significantly reduced the costs of launching payloads into space and thus enabled private companies and researchers to launch their payloads.

However, the breakneck speed of technological development also has implications for safety and sustainability (Di Virgilio and Das 2023a). For example, as Polkowska (2021) understood, a proliferating number of satellites in low Earth orbit has resulted in space debris and collision concerns, which can be disastrous for both space operations and the Earth's environment. Moreover, the environmental costs of space mining and other such activities are still poorly understood, leading to calls for further research and regulation to prevent potential hazards (Sarkar 2025).

Role of International Collaboration

There is also debate about the role of international collaboration in celestial entrepreneurship (Majerova and Das 2023). The argument is that space exploration and commerce should be global activities in which nations and companies work together, contribute resources and knowledge, and share risks (Minh-Nhat et al. 2022a). Initiatives such as the Artemis Accords, encouraging worldwide collaboration in lunar exploration (NASA 2025), promote a positive pathway forward. Advocates of cooperation assert that such partnerships are the only means to ensure an equitable distribution of the benefits from space commerce and that potential conflicts never arise (Minh-Nhat et al. 2022b; Mondal et al. 2023). Others argue that competition is a more powerful force for innovation and advancement (Mondal and Das 2023). They look at the historical example of the space race, for instance, between the United States and the Soviet Union, which saw rapid technological advancements over a short period (Thatchenkery and Piezunka 2024). The idea is that in this world, a competitive environment drives companies and nations to innovate—to do things that did not seem possible—and this process ultimately leads to a better future for humanity (Mondal and Das 2023a). Balancing competition with cooperation to ensure that space commerce creates more benefits than it brings risks is going to be a challenge (Mondal and Das 2023b).

Synthesis of Opinions

Celestial entrepreneurship is characterized by a heterogeneity of views and approaches in light of the complexity and uncertainty surrounding this new domain. There is broad consensus that space offers excellent economic potential but not on how to achieve that potential best (Mondal et al. 2023a). Whatever views it is possible to write about are still raising questions about the feasibility of space-based businesses, the legal/ethical implications of space commerce, the role of technological innovation, and the need for international cooperation (Ngoc-Vinh et al. 2022).

Backers of space entrepreneurship are bullish on the space potential, saying it promises innovation, economic growth, and human knowledge. In this new frontier, private companies are positioned to be the drivers with their nimble and risk-taking nature (Nguyen et al. 2022). However, critics point to the potential for exploitation, inequality, and environmental damage and argue that more vigorous regulation and international cooperation will be needed to ensure that space commerce serves the interests of all humanity (Tandon and Das 2023).

Ultimately, celestial entrepreneurship will rely upon stakeholders' ability to muddle through these complex issues and balance competing interests. Whether space-based businesses will achieve sustainable profitability, address ethical and legal concerns, and contribute to global development remains to be seen (Tien-Dung et al. 2022). Everything is sure that the path to cosmic exploration will demand more than cutting-edge technology; it also requires careful consideration of our choices' social, cultural, economic, and ecological impacts (Tien-Dung et al. 2022a).

3.1 Factor Identification for Space Entrepreneurship

Celestial entrepreneurship—business efforts to access the economic opportunities of outer space—has generated much interest and debate among scholars, policy-makers, and industry stakeholders (Vrana and Das 2023). With private companies such as SpaceX, Blue Origin, and Planetary Resources commercializing outer space, it is crucial to determine the elements that affect a business's success in this sector. As such, this narrative review synthesizes the opinions and recommendations of the pioneers in astropolitics theory and models to present a coherent path forward for those interested in celestial entrepreneurship. Thus, the discussion can be organized around five critical factors: *technological innovation, regulatory frameworks, market demand, financial viability, and ethical considerations.*

Technological Innovation

Technological innovation is considered to be the foundation of celestial entrepreneurship. Advances in space technology have not yet made space-based businesses practical or feasible. However, this is being innovated: reusable rockets, autonomous mining systems, satellite constellations, and life-support systems for use in space habitats.

Proponents' Perspective

Advocates say technological advancements are reducing the costs of space exploration and opening it to private companies. For instance, the reusable Falcon 9 rocket developed by SpaceX has made launching payloads into space much more cost-effective, allowing smaller companies and research institutions to participate in space activities (Andersen 2022). Likewise, robotics and artificial intelligence advances are bringing asteroid mining and lunar resource extraction within reach. One example of such research and development is Planetary Resources, which works on autonomous drones that can prospect for minerals on asteroids and even extract them (Sarkar 2025).

Sceptics' Perspective
Critics, however, warn that technological innovation alone may not guarantee the success of celestial entrepreneurship. They note that many technologies needed for space-based enterprises are still in experimental phases and carry a heavy technical toll. Extracting and processing resources in microgravity environments are, at best, untested, and the long-term viability of space infrastructure is unknown. Technological advancements' speed can also pose safety and reliability issues, critical in high-risk contexts like space travel.

Implications for Future/Further Research
To do the above, we need more studies to characterize space-based businesses in detail: What are their technological barriers, and what can we do about them? The initiative could expedite innovation and lower costs through collaborative efforts among private companies, governments, and academic institutions. Research on the scalability and sustainability of the technologies we develop for space is important to assess their long-term viability. Figure 1 represents how celestial entrepreneurship can function.

Regulatory Frameworks

Another significant aspect that impacts the success of solar entrepreneurship is the legal and regulatory environment. The lack of a comprehensive international consensus on the ethical implications of space resource extraction challenges the sustainability of space commerce and poses risks to both entities involved.

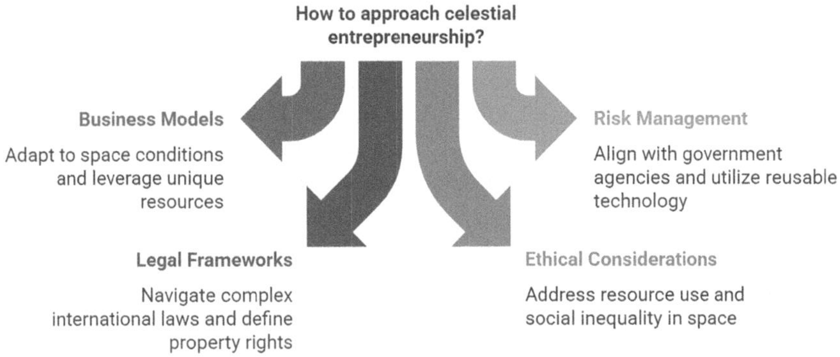

Fig. 1 Celestial entrepreneurship approach (*Source* Authors' conception)

Proponents' Perspective

Advocates of celestial entrepreneurship maintain that regulatory clarity or support promotes innovation and investment. The United States Commercial Space Launch Competitiveness Act, which gives US firms rights to own and market the resources they extract from celestial bodies, is also often proposed as a model for other countries ((Metzger 2023). Advocates argue that such legislation gives the legal certainty essential to enticing private investment and venture taking.

Sceptics' Perspective

Critics say the status quo of jumbled national laws is woefully inadequate to address the implications for commerce beyond our planet. The 1967 Outer Space Treaty—which states that space is to be considered the "province of all mankind"—is outdated and does not reflect the growing role of private industry, they contend. Su (2017) argues that without a practical international framework for regulating space, we risk "cosmic colonialism", where space is dominated by wealthy nations and corporations, granting them access to space resources that developing countries can only dream of.

Implications for Further Research

Others include benchmarking their international treaties or protocols to the "seat at the table" concept and developing amendments to address how they negotiate equity factors among private companies, accountability, and reducing the carbon footprint. If it can engage effectively, this could provide pathways for studying national space laws in relative terms, generating insight into the best way to regulate space commerce.

Market Demand

Every business's story of celestial entrepreneurship starts with a potential market. When developing sustainable business models, it is important to identify market demand for space-based products and services.

Proponents' Perspective

The rising demand for satellite-based services (telecommunications, Earth observation, navigation, etc.) is cited as proof of the commercial potential for doing business in space. With more than $270 billion in revenue from the global satellite industry in 2022, the space-based commercial ecosystem is a thriving, integral part of our economy (Polkowska 2021). Finally, the emerging market for commercial space tourism, as evidenced by Intrepid, Virgin Galactic, and Blue Origin, indicates strong consumer demand for leisure-based space travel.

Sceptics' Perspective

However, critics say these markets are not scalable. They contend that the high costs associated with space travel and resource extraction might restrict demand to a small segment of affluent individuals and corporations. An example of this would be a ticket for a suborbital spaceflight with Virgin Galactic, currently priced at $450,000, limiting it entirely to an out-of-reach status for all but a few of the populations. Likewise, the market for asteroid-mined minerals is speculative, as Earth's extraction and transportation costs remain in the stratosphere (Hayashi et al. 2024).

Future Takeaway
Over the coming decades, emerging markets for space-based products and services will be better understood, as will their growth potential. Research on consumer behaviour and willingness to pay for space-related experiences can contribute to developing business models.

Financial Viability

Investors and stakeholders are apprehensive about celestial entrepreneurship's fiscal concerns. Heralding a space-based economy seems optimistic, considering how costly and dangerous space exploration is.

Proponents' Perspective
Proponents say space commerce has a long-term return on investment that far outweighs upfront costs. They refer to the potential for asteroid mining to offer access to rare minerals like platinum and gold, which could yield massive profits (Hayashi et al. 2024). Moreover, space infrastructure also holds the potential for new commercial pursuits, including manufacturing and research via lunar bases and orbital habitats.

Sceptics' Perspective
Meanwhile, critics argue that space-based businesses are inherently uncertain and risky. It is difficult to gauge the financial benefits of such ventures because of the high failure rate of space missions and the unpredictable conditions of space environments. Moreover, space commerce's limited historical context presents challenges in formulating precise financial models and projections (Metzger 2023). More research is needed to develop robust financial models for space-based businesses and evaluate their risk-return profiles. While case studies of successful and failed space entrepreneurs are not typically published as business books, they can offer valuable lessons for future entrepreneurs.

Ethical Considerations

The context of celestial entrepreneurship has made ethical considerations increasingly pivotal. Utilizing space resources raises equity, sustainability, and space preservation issues for future generations.

Proponents' Perspective
Advocates argue that research and development in the pursuit of space commerce could ultimately benefit all humanity, as space commerce creates technological innovation that drives economic growth. Private firms are poised to lead in this new frontier with agility and risk appetite (Rausser et al. 2023).

Sceptics' Perspective

Critics say that unchecked space resource mining could lead to environmental destruction and worsen global inequalities. They advocate establishing ethical guidelines and international treaties to ensure that space commerce occurs responsibly and sustainably (Martin and Freeland 2020).

4 Implications for Future Research

Exploration should focus on the implications of ethics around space commerce and the frameworks for responsibly exploiting such resources. Studying the environment in space and the fair distribution of the benefits of space activities is especially important.

Celestial entrepreneurship will thrive or falter based on many factors: technological advances, regulatory structures, market appetite, financial feasibility, and ethical implications, to name a few. Despite the dynamic excitement surrounding potential space-based businesses, many complex challenges and uncertainties lie ahead (Vrana and Das 2023a). This chapter provides a good starting point, but more in-depth research is needed to examine these aspects and create business models that promote innovation, sustainability, and equity. All of humanity will share in the benefits of the vast economic potential of outer space; they need to understand the key levers and turn them (Yegen and Das 2023).

Celestial entrepreneurship—or creating business models operating beyond the atmosphere of Earth—is no longer a fantasy genre. Space commercialization is not a distant prospect; it is getting closer with technological advancements. This chapter will discuss some theoretical, practical, societal, and sustainable considerations of celestial entrepreneurship regarding the different kinds of businesses that can be established in outer space. Figure 2 represents the challenges and implications of celestial entrepreneurship.

4.1 Theoretical Implications

1. Redefining Business Models.
   (a) Space as a New Market Frontier: What works on Earth may not work in space. Entrepreneurs must consider the challenges and opportunities in the space environment, including microgravity, vacuum, and extreme temperatures.
   (b) Resource-Based Theories: A firm's resource-based view (RBV) can extend to space entrepreneurship. This will also mean identifying and exploiting unique available resources (e.g. rare minerals from asteroids, solar energy reaping).
   (c) Innovation and Disruption: The new wave of space entrepreneurs will probably be a powerful catalyst for innovation, not just in the space tech sector but also in vital sectors like materials science, robotics, and AI. This may result in disruptive shifts in industries on Earth as technologies aimed at space applications migrate to Earth.

Fig. 2 Challenges and implications of space entrepreneurship (*Source* Authors' conception)

2. Risk and Uncertainty.
   (a) High-Risk Ventures: Space projects are associated with high risks, including technical issues, regulatory complexities, and capital demands. As with any cutting-edge industry, entrepreneurs in the private space economy need to build strategies to manage these risks, ideally by aligning with government space agencies and capitalizing on breakthroughs in reusable rocket technology.
   (b) Uncertainty in Market Demand: Demand for space-based products and services is still nascent. Entrepreneurs are uncertain about the market demand

that could exist after they invest in a space venture—demand that could be altered by factors like advancing technology, regulatory changes, and public attitudes towards space activity.

3. Legal and Ethical Consideration.

   (a) Space Law and Regulation: Space law, or regulatory framework for space activities, is developing. Since these companies are beyond the control of countries, they have much more complex international space laws to navigate than the Outer Space Treaty of 1967, which regulates the activities of states in the exploration and use of outer space. It will be essential to define property rights, liability, avoidance of harmful contamination of celestial bodies, and other agreements that will be required to maintain that the lucrative exploration of space benefits all mankind.
   (b) Ethical Implications: Space entrepreneurship involves profound ethical considerations. Questions surround the just use of resources in space, the ability of space colonization to amplify social inequalities, and the moral basis for transforming extraterrestrial environments.

4.2 *Practical Implications*

1. Technological Challenges.

   (a) Spacecraft and infrastructure: The enormous upfront costs needed to build the spacecraft and infrastructure required for space-based businesses are enormous. These include everything from designing and building space habitats and mining equipment to transportation systems that can survive in extreme space conditions.
   (b) Life-Support Systems: Reliable life-support systems are critical components for any business involving human presence in space, from space tourism to off-world manufacturing. These systems will have to generate air, water, food, and waste in a closed-loop environment.
   (c) Communication and Navigation: Strong communication and navigation systems are crucial for successfully conducting business in space. Entrepreneurs must invest in technologies that allow real-time communication with the Earth and accurate navigation in outer space.

2. Financial Considerations.

   (a) Investment Proposition: Space-based enterprises require vast capital to establish and maintain. For that reason, entrepreneurs rely on several different sources of revenue, such as venture capital, government grants, and public-private partnerships.
   (b) Revenue Streams: Exploring potential revenue streams is critical to the long-term viability of space-based businesses. These could range from space

tourism and asteroid mining through satellite services to selling space-derived products, including rare minerals or zero-gravity manufactured goods.
   (c) Cost Reduction: The viability of celestial entrepreneurship, point 0,010 of Point Earth, can minimize the cost of access to space. Improvements to reusable rocket technology, such as those pioneered by SpaceX, are driving down the cost of sending payloads into orbit.

3. Operational Challenges.

   (a) Supply Chain Management: Supply chain management is much more complex for a space-based business. Entrepreneurs must ensure that materials and equipment are delivered to space on time and that products return to Earth. Therefore, space agencies need dedicated experts in the following fields: human resources. Recruiting and training personnel for space-based operations is a unique challenge. Workers need specialized skills and the ability to work in extreme environments. Psychological and physiologic concerns also need to be addressed for long-duration spaceflights.
   (b) Regulatory Compliance: The success of space-based businesses hinges on compliance with international space regulatory standards. Companies should collaborate with the relevant government agencies.

4.3 *Societal Implications*

1. Economic Impact.

   (a) Job Creation: The expansion of space-based companies could create new jobs in many sectors of the economy, including engineering, manufacturing, and space tourism. This would help grow the economy and create opportunities for skilled workers.
   (b) International Competition—Global Competition: The next stage of the space race will likely be the commercialization of space as nations and corporations compete in this new frontier. As new players enter the space industry, it could result in more investment in space technology and infrastructure and new international partnerships.
   (c) Economic Inequality: There is a concern that the advantages of entrepreneurialism in space could not be shared equally, widening current economic inequalities. Policymakers will face the challenge of ensuring equitable access to space resources and opportunities.

2. Cultural and Social Impact.

   (a) Inspiration and Education: Space entrepreneurship can also inspire young people to pursue work in science, technology, engineering, and mathematics

(STEM) fields. This would create a more scientifically literate population and a sound workforce in STEM.

(b) Human-2: Space exploration and commercialization could foster cultural exchange and international collaboration. As humans explore the cosmos, they may feel greater global unity and shared purpose.

(c) Ethical and Moral Questions: As humans venture further into space, ethical and moral questions become vital. We are likely to face questions about the rights of alien life forms, the extraterrestrial environment's cleanliness, the human race's survival, and the impact of colonization on human beings or cultures in the universe.

3. Political Implications.

   (a) Commercialization of Space: The increasing commercialization of space may result in (even) more geopolitical tensions as nations vie for control of valuable space resources and strategic positions in orbit. It may lead to a new rivalry for space, which could have significant repercussions for international politics and security.

   (b) Space Governance: To foster collaborative space-based businesses, new institutions will be needed to manage space activities. This would entail the formation of international space agencies, new treaties, and space courts to settle disputes.

   (c) National Security: The commercialization of space suggests that militarization could also occur in space. Entrepreneurs and policymakers must collaborate to ensure space is a peaceful, weaponization-free domain.

4.4 Sustainable Implications

1. Environmental Sustainability.

   (a) Space Debris: With the growing number of satellites and spacecraft in orbit, space debris is becoming a bigger problem. Therefore, entrepreneurs must devise plans to reduce space debris and invest in debris removal technologies (investments) to contribute to the long-term sustainability of space activities.

   (b) Environmental Conservation of Celestial Environments: If resources from celestial objects such as asteroids or the Moon are mined, the extraction must minimize adverse environmental effects. As human entrepreneurs, we must consider the soul of the body we are planning to modify and adjust to the natural elements around us.

   (c) Sustainable Habitat in Space: Space habitation and its architectures should be sustainable. Such practices involve harnessing renewable energy sources, implementing efficient waste management systems, and recycling materials to develop closed-loop ecosystems.

2. Economic Sustainability.
   (a) Viability: Could space-based businesses have staying power? silver lining—entrepreneurs must work on sustainable models rather than quick wins.
   (b) Continuous Investment in Research and Development: Sustainable space entrepreneurship will require continued investment in research and development. This includes investment in new technologies and the societal and environmental effects of space activities.
   (c) The Role of Public-Private Partnerships in Facilitating the Sustainability of Space-Based Businesses: Private companies drive innovation and commercialization, and governments provide funding and regulatory support.
3. Social Sustainability.
   (a) Access for All: Facilitating equitable access to space resources and opportunities is vital for the social sustainability of space. Space assets should not be limited to a few individuals; policymakers should seek to avoid this concentration of wealth and ensure broad participation in space activities.
   (b) Community Engagement: Space entrepreneurship has an aspect of social sustainability. For this to happen, we cannot allow space to be elitist. Entrepreneurs must evaluate the effects of their functions within communities and strengthen positive ties with stakeholders.
   (c) Ethical Obligation: Space entrepreneurs have an ethical obligation to consider the more significant effects of their actions. These include addressing whether space colonization could exacerbate social inequalities or the ethics of terraforming entire planets.

5 Conclusion

Celestial entrepreneurship also represents a frontier for building new business avenues and opportunities and inspiring cohorts that may build the future. However, establishing business models beyond Earth raises considerable challenges, from technical to financial to legal and ethical. A sustainable, equity-focused, and ethically responsible approach to celestial entrepreneurship is critical as we explore the far reaches of space. This way, we can accommodate the expansion of private interests into space while balancing them with the interests of all humankind in general.

Overall Implications of Celestial Entrepreneurship Theory, Practice, Society, and Sustainability: With the new era of space exploration and commercialization, it is essential to ponder these implications and work towards making space a domain of opportunity, innovation, and sustainability for everyone.

References

Amorim-Maia AT, Anguelovski I, Chu E, Connolly J (2021) Intersectional climate justice: a conceptual pathway for bridging adaptation planning, transformative action, and social equity. Urban Clim 41:101053. https://doi.org/10.1016/j.uclim.2021.101053

Andersen G (2022) Desperate science fiction: on how Musk, Bezos, gates and Google plan to escape socio-ecological collapse. Culture, Theory and Critique/Culture, Theory & Critique 63(4):281–295. https://doi.org/10.1080/14735784.2023.2257404

Baca KA (2017) Property rights in outer space. SMU Scholar. https://scholar.smu.edu/jalc/vol58/iss4/4?utm_source=scholar.smu.edu%2Fjalc%2Fvol58%2Fiss4%2F4&utm_medium=PDF&utm_campaign=PDFCoverPages

Betz UA, Arora L, Assal RA, Azevedo H, Baldwin J, Becker MS, Bostock S, Cheng V, Egle T, Ferrari N, Schneider-Futschik EK, Gerhardy S, Hammes A, Harzheim A, Herget T, Jauset C, Kretschmer S, Lammie C, Kloss N et al (2023) Game changers in science and technology–now and beyond. Technol Forecast Soc Chang 193:122588. https://doi.org/10.1016/j.techfore.2023.122588

Das S (2023) Modern-day gaming and the role of FinTech applications in popularising various entertainment platforms across the world. In: Das S, Gochhait S (eds) Digital entertainment as next evolution in service sector. Palgrave Macmillan, Singapore. https://doi.org/10.1007/978-981-19-8121-0_11

Das S (2023a) Digital entertainment: the next evolution in service sector. Emerging Digital Solutions in Reshaping Different Industries Palgrave-Singapore. https://doi.org/10.1007/978-981-19-8121-0

Das S, Mondal SR, Struck F, Eissfeldt J, Beernink L, Sigouras V, Hoose S, Kopac L (2023) What can we see inside the crystal ball? Exploring opportunities and challenges for Metaverse and circular economy. AJIS 12(6):221. https://doi.org/10.36941/ajis-2023-0165

DePagter MM (2021) "Who dares, wins:" how property rights in space could be dictated by the countries willing to make the first move l. Chic J Int Law. https://cjil.uchicago.edu/online-archive/who-dares-wins-how-property-rights-space-could-be-dictated-countries-willing-make

Detsis E (2022) Ethics in space: the case for future space exploration. In Research ethics forum (pp. 111–121). https://doi.org/10.1007/978-3-031-15746-2_9

Di Virgilio F, Das S (2023) Digital solutions for social media: role of digital evolution on business enterprises. In: Das S, Gochhait S (eds) Digital entertainment as next evolution in service sector. Palgrave Macmillan, Singapore. https://doi.org/10.1007/978-981-19-8121-0_9

Di Virgilio F, Das S (2023a) Digitalization of enterprises from a new Normal sustainable development perspective: opportunities and challenges. In: Mondal SR, Yegen C, Das S (eds) New normal in digital enterprises. Palgrave Macmillan, Singapore. https://doi.org/10.1007/978-981-19-8618-5_9

Dong L, Sun D, Shu W, Li X (2020) Exploration: safe and clean mining on earth and asteroids. J Clean Prod 257:120899. https://doi.org/10.1016/j.jclepro.2020.120899

Foote S, Sinhadc P, Mathis C, Walker SI (2023) False positives and the challenge of testing the alien hypothesis. Astrobiology 23(11):1189–1201. https://doi.org/10.1089/ast.2023.0005

Ghai S, Thériault R, Forscher P, Shoda Y, Syed M, Puthillam A, Peng HC, Basnight-Brown D, Majid A, Azevedo F, Singh L (2025) A manifesto for a globally diverse, equitable, and inclusive open science. Comm Psychol 3(1). https://doi.org/10.1038/s44271-024-00179-1

Gunia A (2024) Minerals are in short supply on earth. This startup wants to mine asteroids. CNN. https://edition.cnn.com/world/astroforge-asteroid-mining-nasa-spc-scn/index.html

Hayashi F, Kioka A, Ishii T, Nakamura T (2024) Unveiling the resource potential of space debris: a forecast of valuable metals to 2050. Waste Manag 193:376–385. https://doi.org/10.1016/j.wasman.2024.12.019

Kua J, Loke SW, Arora C, Fernando N, Ranaweera C (2021) Internet of things in space: a review of opportunities and challenges from satellite-aided computing to digitally-enhanced space living. Sensors 21(23):8117. https://doi.org/10.3390/s21238117

Lawrence A, Rawls ML, Jah M, Boley A, Di Vruno F, Garrington S, Kramer M, Lawler S, Lowenthal J, McDowell J, McCaughrean M (2022) The case for space environmentalism. Nature Astronomy 6(4):428–435. https://doi.org/10.1038/s41550-022-01655-6

Majerova J, Das S (2023) Sustainable development and digitalization in the supply chain for environmental protection: changes and trends. In: Mondal SR, Yegen C, Das S (eds) New Normal in digital enterprises. Palgrave Macmillan, Singapore. https://doi.org/10.1007/978-981-19-8618-5_8

Martin A, Freeland S (2020) The advent of artificial intelligence in space activities: new legal challenges. Space Policy 55:101408. https://doi.org/10.1016/j.spacepol.2020.101408

Metzger PT (2023) Economics of in-space industry and competitiveness of lunar-derived rocket propellant. Acta Astronaut 207:425–444. https://doi.org/10.1016/j.actaastro.2023.03.014

Minh-Nhat HO, Majerova J, Das S (2022a) Taking steps to help the Phoenix rise from ashes: a roadmap for digital enterprises to develop new sustainable goals. In: Sustainable development and innovation of digital enterprises for living with COVID-19. Springer, Singapore, pp 35–51. https://doi.org/10.1007/978-981-19-2173-5_3

Minh-Nhat HO, Nguyen HL, Mondal SR (2022b) Digital transformation for new sustainable goals with human element for digital service enterprises: an exploration of factors. In: Sustainable development and innovation of digital enterprises for living with COVID-19. Springer, Singapore, pp 85–103. https://doi.org/10.1007/978-981-19-2173-5_6

Mondal S, Das S, Vrana VG (2023) How to bell the cat? A theoretical review of generative artificial intelligence towards digital disruption in all walks of life. Technologies 11(2):44. https://doi.org/10.3390/technologies11020044

Mondal SR, Das S (2023) Defining the new digitalized Normal: implications of a.I., IoT, and Blockchain for business growth. In: Mondal SR, Yegen C, Das S (eds) New normal in digital enterprises. Palgrave Macmillan, Singapore. https://doi.org/10.1007/978-981-19-8618-5_10

Mondal SR, Das S (2023a) Digital solution for industry 4.0 and 5.0 and social media. In: Das S, Gochhait S (eds) Digital entertainment as next evolution in service sector. Palgrave Macmillan, Singapore. https://doi.org/10.1007/978-981-19-8121-0_10

Mondal SR, Das S (2023b) Human-centric digitalization for new sustainable development: an explorative study in human resource management. In: Mondal SR, Yegen C, Das S (eds) New normal in digital enterprises. Palgrave Macmillan, Singapore. https://doi.org/10.1007/978-981-19-8618-5_12

Mondal SR, Yegen C, Das S (2023a) New normal in digital enterprises–strategies for sustainable development. Palgrave-Singapore. https://doi.org/10.1007/978-981-19-8618-5

Moon Village (2016). Www.esa.int. https://www.esa.int/About_Us/Ministerial_Council_2016/Moon_Village

Musk E (2017) Making humans a multi-planetary species. New Space 5(2):46–61. https://doi.org/10.1089/space.2017.29009.emu

NASA (2025) Artemis Accords–NASA. https://www.nasa.gov/artemis-accords/

National Space Society (2021) Retrieved January 30, 2025, from https://arc.nss.org/wp-content/uploads/NSS-Annual-Report-2021.pdf

Nesvold E (2024) Off-Earth: ethical questions for a cosmos of capitalism. MIT Press. https://mitpress.mit.edu/9780262550994/off-earth/

Ngoc-Vinh N, Tien-Dung P, Das S (2022) Digitization of business; need for recombination during COVID-19. In: Sustainable development and innovation of digital enterprises for living with COVID-19. Springer, Singapore, pp 1–17. https://doi.org/10.1007/978-981-19-2173-5_1

Nguyen TDM, Mondal SR, Das S (2022) Digital entrepreneurial transformation (DET) powered by new Normal sustainable developmental goals (n-SDGs): elixir for growth of country's

economy. In: Sustainable development and innovation of digital enterprises for living with COVID-19. Springer, Singapore, pp 69–84. https://doi.org/10.1007/978-981-19-2173-5_5
Polkowska M (2021) Space tourism challenges. Rev Euro Comp Law 45(2):153–182. https://doi.org/10.31743/recl.12223
Rausser G, Choi E, Bayen A (2023) Public–private partnerships in fostering outer space innovations. Proc Natl Acad Sci 120(43):e2222013120. https://doi.org/10.1073/pnas.2222013120
Sarkar S (2025) Asteroid mining: the next frontier for sustainability. STEM Fellowship J 1–4:12. https://doi.org/10.17975/sfj-2024-014
Schwartz JS, Milligan T (2016) The ethics of space exploration. In Space and society. https://doi.org/10.1007/978-3-319-39827-3
Space economy (2024) World Economic Forum. https://www.weforum.org/publications/space-the-1-8-trillion-opportunity-for-global-economic-growth/
Su J (2017) Legality of unilateral exploitation of space resources under international law. Int Comp Law Quart 66(4):991–1008. https://doi.org/10.1017/s0020589317000367
Tandon S, Das S (2023) Digital revolution complements sustainable goal achievement during and after pandemic: a research agenda. In: Mondal SR, Yegen C, Das S (eds) New normal in digital enterprises. Palgrave Macmillan, Singapore. https://doi.org/10.1007/978-981-19-8618-5_11
Thatchenkery S, Piezunka H (2024) Competition in collaboration: the problem of (Mis)aligned perception. Adm Sci Q 70:194. https://doi.org/10.1177/00018392241299787
Tien-Dung PHAM, Majerova J, Das S (2022) Phases of possible recovery of digital enterprises in new normal business for living with COVID-19 times: opportunities for a new era in sustainable development goals. In: Sustainable development and innovation of digital enterprises for living with COVID-19. Springer, Singapore, pp 19–33. https://doi.org/10.1007/978-981-19-2173-5_2
Tien-Dung P, Nguyen XD, Das S (2022a) Sustainable goal achievement by digital revolution during and after pandemic; how much one wins and losses: a bird's eye view for future planning. In: Sustainable development and innovation of digital enterprises for living with COVID-19. Springer, Singapore, pp 133–147. https://doi.org/10.1007/978-981-19-2173-5_8
Tyson ND, Lang A (2018) Accessory to war: the unspoken alliance between astrophysics and the military. https://en.wikipedia.org/wiki/Accessory_to_War
United Nations Office for Outer Space Affairs (1967) The outer space treaty. UNOOSA. https://www.unoosa.org/oosa/en/ourwork/spacelaw/treaties/introouterspacetreaty.html
Vrana V, Das S (2023) Dynamic restructuring of digital media and entertainment sector: role of urbanization, industrial innovation, and technological evolution. In: Das S, Gochhait S (eds) Digital entertainment as next evolution in service sector. Palgrave Macmillan, Singapore. https://doi.org/10.1007/978-981-19-8121-0_2
Vrana V, Das S (2023a) Role of enterprises and their digital transformation in post-pandemic sustainable development. In: Mondal SR, Yegen C, Das S (eds) New normal in digital enterprises. Palgrave Macmillan, Singapore. https://doi.org/10.1007/978-981-19-8618-5_7
White W (1997) Real property rights in outer space. http://www.space-settlement-institute.org/Articles/research_library/WayneWhite98-2.pdf
Yarlagadda S (2022) Economics of the stars: the future of asteroid mining and the global economy. Harv Int Rev. https://hir.harvard.edu/economics-of-the-stars/
Yegen C, Das S (2023) Digital solution for print and publishing media. In: Das S, Gochhait S (eds) Digital entertainment as next evolution in service sector. Palgrave Macmillan, Singapore. https://doi.org/10.1007/978-981-19-8121-0_12

Lukas Vartiak Senior academician and researcher at Comenius University's Faculty of Social and Economic Sciences, focusing on socio-economic dynamics and policy impacts. His research addresses contemporary challenges in global and regional economies. He actively collaborates on interdisciplinary projects to advance sustainable development.

Garyfallos Fragidis Associate Professor at the International Hellenic University, leading service systems design and hospitality programs. With a PhD in information systems, he directs EU-funded projects on innovation and sustainability. His work emphasizes value creation in service ecosystems.

Papafloratos Traintafyllos Adjunct lecturer and founder of "Sustainability Business Case," a consultancy focused on CSR and ESG. With a PhD in CSR, his research addresses sustainability frameworks and stakeholder engagement. He integrates academic insights into practical corporate strategies.

Subhankar Das Senior Researcher and academician at Duy Tan University, publishing extensively on AI, blockchain, and digital transformation. Editor of seven Scopus-indexed books and reviewer for top journals. Mentors global PhD candidates and drives innovation in ASEAN academic collaborations.

Laws of the Cosmos: Navigating the Space Tourism Legal Landscape

Gyorgy Pal Papay, Hien Huynh Thi Ngoc, and Subhankar Das

Abstract Space tourism presents a complex legal landscape that requires careful consideration and negotiation between private companies, governments, and international organizations. This chapter assesses the changing legal environment of space tourism to propose an integrated three-tiered regulatory regime. With the rise of private companies in space exploration, the 1967 Outer Space Treaty (OST) fails to provide a solid legal base for modern issues. This study suggests a pragmatic regulatory framework that includes safety protocols, measures of environmental protection, and international coordination mechanisms. The framework is structured around a sequence of our phases, delivered over a 4-year (48-month) time frame, which allows for building regulatory architecture in a controlled manner while keeping our regulation operational. The study pinpoints key areas that warrant immediate action, including liability challenges, property rights, and environmental sustainability, and examines market, technological, and international implementation. The results imply that effective regulation of space tourism must balance fostering innovation while ensuring compliance with a need for flexible governance that evolves alongside technology development. In conclusion, the paper argues that future space tourism legislation must balance the commercial interests of clients and companies with an ethical obligation to ensure that the industry develops equitably and sustainably.

Keywords Space tourism · Legal framework · Environmental sustainability · International coordination · Regulatory compliance · Commercial space operations

---

G. P. Papay
Wyższa Szkoła Biznesu–National Louis University, Nowy Sącz, Poland
e-mail: ceo@eucult.eu

H. H. T. Ngoc
South Star Management Institute, Duy Tan University, Da Nang, Vietnam
e-mail: huynhthingochien1512@gmail.com

S. Das (✉)
Department of Marketing, The Honors Program, South Star Management Institute, Duy Tan University, Da Nang, Vietnam
e-mail: subhankardas@duytan.edu.vn

© The Author(s), under exclusive license to Springer Nature Singapore Pte Ltd. 2025

S. R. Mondal et al. (eds.), *Pioneering the New Space Economy through AI and Immersive Technologies*, https://doi.org/10.1007/978-981-96-5977-7_3

1 Introduction

As we step into the twenty-first century, we are witnessing an era of unparalleled technological innovations, and one of the most revolutionary is space tourism (Greene 2011). The prospect of ordinary people venturing beyond Earth's atmosphere, once the domain of science fiction, is fast becoming reality. Companies such as SpaceX, Blue Origin, and Virgin Galactic are paving the way, operating sub-orbital and orbital flights for private citizens. Legal implications on private sector design for space tourism have evolved significantly as commercial entities have transformed theoretical concepts into operational ventures. Space tourism, now legally categorized as space travel, operates within a complex regulatory framework that requires private companies to navigate international space treaties, national legislation, liability considerations, and emerging industry standards while addressing safety protocols, environmental impacts, and the expanding commercialization of orbital activities.

Space tourism is a technological achievement and a revolutionary development in how humanity interacts with the cosmos. For thousands of years, space belonged to astronomers, scientists, and nations (Sabt and Farooqui 2023). Just a few years ago, the notion that private citizens might one day travel into space was unimaginable. However, as we are witnessing a decline in the cost of space travel, the legal framework regulating such frontier needs to evolve in tandem with the developing technology and heightened interest in commercial space exploration.

The legal canvas paints a myriad of space tourism dotted with international treaties, national laws, and evolving guidelines. This legal regime is anchored by the 1967 Outer Space Treaty, which is widely seen as the bedrock of space law. The treaty—known as the Treaty on Principles Governing the Activities of States in the Exploration and Use of Outer Space, including the Moon and Other Celestial Bodies—was written during the Cold War (Smith et al. 2023). Its intended purpose was to discourage space militarization and preserve outer space for peaceful exploration and use by all countries. The treaty refers to space as the "province of all mankind" and forbids any nation from claiming sovereignty over celestial bodies. It also affirms that the exploration and use of outer space shall be carried out for the benefit of all countries, not considering the extent of their economic or scientific development.

Although the Outer Space Treaty was the bedrock for international space law, it was created when private space exploration was unimaginable (Ye 2025). Consequently, the treaty neither explicitly deals with the engagement of private actors in space exploration nor lays down a wide-ranging regulatory framework for commercial activities in space. This has created a legal grey area as private companies and individuals enter space for commercial purposes. However, liability issues, property rights, and environmental protection in this outer space context are still unanswered questions that will necessarily require new laws and regulations to guide space tourism into the future.

The most prominent legal issue surrounding space tourism is liability. Who is liable for an accident or injury during a space tourism mission? Presently, the framework for liability for damage caused by space objects derives from the 1972 Liability Convention, which makes the launching state liable for damage caused by its space objects (Jakhu and Pelton 2017a). However, this treaty was written for governments heading into space, and how it applies to private space tourism companies is murkier (Kehrer 2019). With the increased involvement of private companies in space exploration, there is a greater demand for legal structures that can delineate the obligations and liabilities of such entities. This includes setting the rules around compensating passengers in case of injury or death and what liability private companies face for accidents or malfunctions during space missions.

Another primary concern is the matter of property rights in space. The Outer Space Treaty forbids countries from claiming sovereignty over celestial bodies but does not explicitly challenge the ownership of resources extracted from those bodies. This has created a legal debate over whether private companies can claim ownership of minerals, water, or other materials extracted from asteroids, the Moon, or other celestial bodies. In 2015, the United States enacted the Commercial Space Launch Competitiveness Act, which guarantees US citizens ownership and the right to sell resources they extract from outer space (Hertzfeld and Von Der Dunk 2005). Nonetheless, this law has faced criticism for potentially contradicting the Outer Space Treaty principles, especially the non-appropriation principle. As space tourism becomes more common and the potential for resource extraction increases, the international community will have to find a way to reach an agreement on how space resources can be owned and used in a way that upholds the ideas of international space law and balances them with commercial interests (Tronchetti 2009).

Space tourism is also subject to environmental concerns. Space missions are on the rise and so are the risks presented by space debris—stuff that can threaten manned and unmanned spacecraft. Space debris—such as spent rocket stages, derelict satellites, and tiny pieces of fragments—can travel at incredible speeds, and even small bits can be deadly (Weeden and Chow 2012). Humanity's trajectory in space exploration may seem seamless in concept. Still, from where we are at present, the push for space tourism needs to be handled with caution, as the increasing amount of space debris in the Earth's orbit could prove hazardous to space tourism missions in the future. Space debris is not the only issue: there are concerns over the environmental impact of space tourism on celestial bodies (Williamson 2012). For instance, the Moon and other heavenly bodies have their unique environments that can be affected by human actions. With space tourism becoming more prevalent, it will be necessary to implement regulations that mitigate the ecological consequences of such endeavours and safeguard the integrity of extraterrestrial ecosystems.

It also raises ethical questions regarding space tourism that should be taken seriously. This would also serve to make the already extreme class divide even worse, as the only people able to afford such price tickets are the top 10%. This raises fundamental questions about who owns space access and how the benefits of space exploration should be shared (Bratu et al. 2021). Moreover, the potential for space tourism to disrupt the cultural and environmental balance in indigenous communities raises ethical issues regarding the exploitation of space resources. While exploring the exciting opportunities of space tourism, we must also ensure that principles of equity, sustainability, and respect for the rights of all people guide its development.

So the advent of cosmic tourism marks the beginning of a new chapter in humanity's exploration of the cosmos. However, moving into this new frontier, we also need to face the legal, ethical, and environmental problems associated with it. The laws we create now will determine the future of space tourism and how we navigate the cosmos for generations to come. Therefore, we must formulate a comprehensive, international legislative framework that accommodates the purely commercial interests of private enterprises while preserving the environment, safeguarding space travellers, and adhering to the tenets of international space law. This way, we can guarantee that space tourism is both a technological breakthrough and a sustainable and equitable endeavour for humankind.

However, new companies like SpaceX, Blue Origin, and Virgin Galactic are making science fiction a reality, allowing civilians to experience sub-orbital flights and, eventually, stay in orbit. However, this emerging industry works under an intricate and fluid legal regime. This review gathers key legal issues affecting space tourism, from treaties established in the 1960s to the ethical dilemmas that are now being explored, and highlights the challenge of mediation between innovation and regulation.

2 Historical Context: First Principles of Space Law

The legal framework for space tourism and satellite launches dates back to mid-twentieth-century international treaties. Today, the Outer Space Treaty (OST), signed in 1967, is still the primary legal framework guiding outer space activities, prohibiting the assertion of national sovereignty in space and enforcing that space must be used for all humanity. The Liability Convention (1972) clarified responsibilities for state damage-purpose use of space objects, and the Moon Agreement aimed to govern interstellar resource use. However, very few have enacted it (Jakhu and Pelton 2017b). While these treaties emerged during the Cold War, they now come into question as commercial enterprises are transforming the activities that take place in space.

Jurisdiction and Liability

National Responsibility: States are internationally responsible for space activities under Article VI of the OST, including those conducted by other entities or private actors. This creates a duty for states to monitor commercial activity, including the US Commercial Space Launch Competitiveness Act (2015), which provides authorization for private launches (FAA 2006).

Liability: The Liability Convention establishes absolute liability for impacts on Earth and fault-based liability for space collisions (Article VII, OST). However, ambiguities remain—for example, assigning fault in satellite collisions is debated (Freeland,2020). The 2009 Iridium-Cosmos collision highlighted the critical need for more robust liability frameworks.

Havoc at Sea: Informed Consent and Passenger Rights

Unlike government astronauts, they are taking risks that are by no means trivial. The US Federal Aviation Administration (FAA) requires informed consent per 14 CFR Part 460, under which companies must disclose risks, including radiation exposure and launch failures (FAA 2006). On the other hand, there are worldwide discrepancies; the EU has no equivalent legislation in place but instead utilizes general consumer protection (Von Der Dunk 2019). Some have even suggested standardizing international consent processes with transparency.

Regulations and Oversight around Safety
Regulatory bodies such as the FAA's Office of Commercial Space Transportation (AST) and the European Union Aviation Safety Agency (EASA) are responsible for launch safety. However, they are forced to balance innovation with risk mitigation. For instance, the FAA's "payload review" process assesses passenger safety, but crew members are exempt from this review. As tourism scales, harmonizing safety standards across jurisdictions is paramount (Jakhu and Pelton 2017c).

Environmental Sustainability
More than 34,000 objects in orbit are already capable of being tracked, leading to the proliferation of orbital debris as the final frontier of tourism—the Kessler Syndrome—a cascade of colliding objects, is a looming threat. The Space Debris Mitigation Guidelines recommend reducing debris, but enforcement is voluntary. For example, collision-avoidance systems are deployed with SpaceX's Starlink, but industry-wide adherence is lacklustre (Muelhaupt et al. 2019).

Intellectual Property and Resource Use
The legality of owning resources derived from space is murky. The US Commercial Space Launch Competitiveness Act (2015) gives its citizens ownership of resources extracted from asteroids, which directly contradicts the OST's non-appropriation principle. Signed by 30+ nations, the Artemis Accords promote "safety zones" to protect lunar operations but are criticized for favouring US interests (Deplano 2021).

Ethical and Societal Implications
High ticket costs raise equity concerns about space tourism's extreme exclusivity. Scholars call for inclusive policies to prevent a "space divide". Moreover, safeguarding cultural heritage, including the Apollo landing sites, requires global solidarity. The 2011 NASA Recommendations for Protecting Historic Lunar Sites underscore the tug of war between exploration and preservation (Jakhu and Pelton 2017d).

Global Cooperation and Law Harmonization
The Artemis Accords and the ISS agreements are examples of successful cooperation, but geopolitical rivalries risk fragmentation. China's lunar ambitions and Russia's exit from the ISS highlight the necessity of inclusivity in dialogue. The UN's Committee on the Peaceful Uses of Outer Space (COPUOS) continues to play an important role in developing multilateral solutions. Presumably, these laws will face legal challenges in the future.

Other emerging issues include space traffic management and laws around extraterrestrial habitats—the International Space Traffic Management Organization—preventing collisions. Long-term settlements will require governance models intestinal bodies (Muelhaupt et al. 2019). Reinforcing COPUOS's mandate and updating treaties to reflect the needs of commercial operators are pressing priorities.

The law governing space tourism is a patchwork of decades-old treaties, national regulations, and ethical dilemmas. As the sector rushes forward, proactive cooperation among governments, corporations, and civil society is key to ensuring it is safe, equitable, and sustainable. As humans venture out into the cosmos, we must do so responsibly, creating a framework that reflects our terrestrial values and goals and applies proper jurisdictional enforcement to maintain and reconcile our differences as we settle new futures amongst the stars.

3 Space Tourism: Developing Framework from a Policy Point of View

With the current competition growing in the space tourism sector, the burden of space tourism policymaking is weighing heavily. The amount of research reflects years of work by researchers, industry stakeholders, and international actors, indicating an urgent need for a sound yet pragmatic regulatory framework for commercial space tourism. The challenge is not purely theoretical—"it is about setting rules that will help ensure safety and innovation while providing a framework for protecting our shared cosmic environment for future generations".

The Current Landscape

Currently, the first order of business was to accept the unfortunate reality that the current infrastructure would not do when we started this journey to discover a practical method for space tourism. As space law expert Von Der Dunk (2019) points out, our global legal regime governing space was built when space exploration was merely governmental. The rise of private space tourism has created what he calls a "regulatory vacuum" that needs to be addressed quickly.

The challenge of regulating space tourism became evident. Officials from leading space tourism operators discussed their planned activities, and insurance executives described their risk modelling. A rich tapestry of competing needs and interests emerged that any workable framework would need to accommodate.

Frame Working: A Three-Level Approach

After carefully analysing all the factors, the authors tried to develop a workable three-pronged approach to regulation in this chapter. Figure 1 represents the components of the three-tier framework.

Tier 1: Safety and Operations

Our framework is built on safety protocols and operational standards. We have created a risk-based safety regulation based on Smith et al. (2023). This includes:

Mandatory safety management systems for all operators (Das 2023b) and standardized training requirements for crew and passengers' regular safety audits and inspections of clear emergency response protocols. These measures should be practical and feasible (Das 2023a). As Hertzfeld and Von Der Dunk (2005) observed in their review of current space law, overly prescriptive safety standards can hinder innovation without necessarily enhancing safety. So we have taken a new approach, one that is performance based—we have established clear safety objectives that need to be met, but we have given operators the freedom to accomplish them.

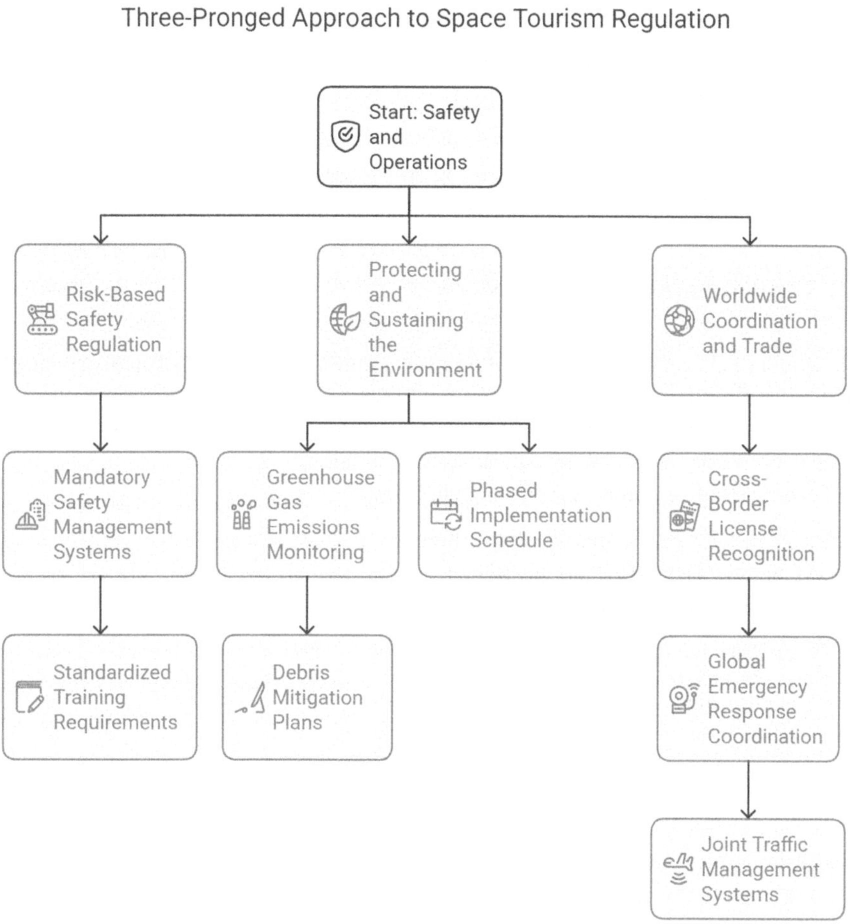

Fig. 1 Three-tier approach (*Source* Authors' conception)

Tier 2: Protecting and Sustaining the Environment

The second one tackles environmental issues, which, according to Newman and Williamson (2018), are vital factors in the future sustainability of space tourism. Our framework incorporates monitoring of greenhouse gas emissions, reporting requirements, debris mitigation plans, environmental impact assessments, and sustainability benchmarks (Das et al. 2023). However, we have developed a phased implementation schedule to make these requirements practically applicable (Di Virgilio and Das 2023a). It will let operators know the timeline within which they should meet environmental standards, with a greater emphasis on the most significant elements of environmental protection initially (Di Virgilio and Das 2023b). As evidenced by Polkowska (2021), such an approach provides room for technological growth without losing sight of gradual yet continuous strides toward the necessary environmental objectives.

Tier 3: Worldwide Coordination and Trade

The third tier will specifically target the international trend of space tourism. Building on Von Der Dunk Johnson's (2019) discussion of the harmonization of international space law, we have established mechanisms for cross-border recognition of licenses and certifications, global emergency response coordination, uniform procedures for dispute resolution, and joint traffic management systems.

4 Implementation Strategy

Their practical applications demand an ever-amplified awareness of industry capabilities and market realities. Drawing on ElZarrad et al.'s (2022) examination of industry self-regulation, we have employed a collaborative implementation strategy. Figure 2 represents the four phases of implementation.

Phase 1: Registry and Licensing (Months 0–6)

(a) Central licensing authority establishment.
(b) Standardization of the application.
(c) Submissions through an online portal.

Phase 2: Implement Safety Systems (Months 6–18)

(a) Operator management system development.
(b) Certification of training programs.
(c) Testing of the emergency response system.

Phase 3: Environmental Conformity (Months 18–36)

(a) Installation of an environmental monitoring system.
(b) Implementation of debris tracking.
(c) Sustainability reporting mechanisms.

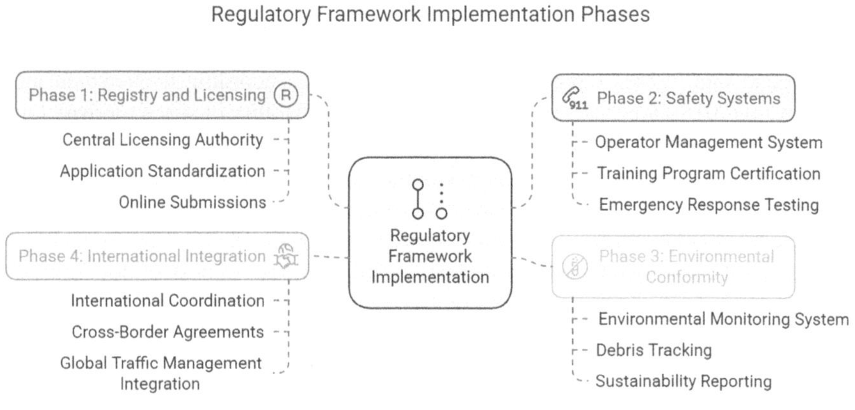

Fig. 2 Regulatory framework (*Source* Authors' conception)

Phase 4: Integrating International Regulatory Bodies (Months 36–48)

(a) International coordination mechanisms.
(b) Agreements on cross-border recognition.
(c) Integration with a global traffic management.

4.1 Practical Implications and their Solutions

Financial Viability: The cost of regulation is an essential consideration. The establishment of viability is the most crucial consideration. Does our business model align with how we want our customers to use and pay for our solution? How profitable is it for us to create our solution and obtain it from our suppliers? Viability does not just examine profit; it examines sustainability to guarantee that your business contributes to the community and society.

Technology Integration: Our framework directly incorporates technological advancement. Instead of prescribing specific technical solutions, we have adopted flexible performance-based standards for emerging technologies (Majerova and Das 2023). As technology evolves, we have proposed guidelines for this framework approach to ensure its applicability (Minh-Nhat et al. 2022a).

Enforcement Mechanisms: Taking inspiration from successful models in aviation regulation, we outlined practical enforcement mechanisms, including tools and approaches, regular audits and inspections, performance monitoring systems, graduated enforcement responses, and collaborative problem-solving approaches (Minh-Nhat et al. 2022b).

International Coordination: To address the international dimensions of space tourism, we have built what Williamson (2012) describes as "coordination bridges"—mechanisms for sharing knowledge and harmonizing regulation across national boundaries. This includes standardized data collection and sharing data repositories.

Future Adaptability

Fortunately, our framework has been carefully crafted with adaptability in mind (Mondal et al. 2023a). Regulation will be reviewed annually to measure its effectiveness and identify areas where it has failed to address the evolving landscape (Mondal and Das 2023a). By doing so, the framework can evolve with the industry (Freeland 2020) and provide a scalable option.

Developing a sensible guide for regulating space tourism is not merely a matter of setting down rules—it is understanding how such rules will function in an imperfect world (Mondal and Das 2023b). This three-pillar system, "along with a sequenced rollout and inherent flexibility, offers a prudent way forward that addresses safety, safeguards the environment and is commercially viable."

The framework will partner with stakeholders to enhance and expand upon regulations (Mondal and Das 2023c). Ensuring the future of space tourism is a factor of technological advancement and our willingness to create and uphold a regulatory framework that endorses safety, innovation, and sustainability (Mondal et al. 2023b).

5 Implications

A Global Regulatory Framework: The Adjacent Bonding of a Space Tourism Future

The governing framework revolves around three pillars—safety and operations, environmental protection, and global coordination—and provides a scaffold for implementing this vision of prosperity (Ngoc-Vinh et al. 2022). At the core of this analysis are the theoretical, practical, social, and sustainable implications of such a framework, focusing on the respective utility of existing theories and the real-world effectiveness of such a proposal while maintaining long-term social equity and ecological sustainability (Nguyen et al. 2022).

5.1 *Theoretical Implications*

The framework is undergirded by theories of interdisciplinary regulation, environmental governance, and international relations, making original contributions to scholarly discourse.

Innovation in Regulation and Risk Management

Performance-based safety standards (Tier 1) push the limits of prescriptive form regulation. Drawing on Smith et al. (2023), this method is consistent with risk management theory, focusing on results rather than strict processes. The framework encourages innovation by giving operators flexibility (Tandon and Das 2023). However, it maintains the balance of safety, which Hertzfeld and Von Der Dunk (2005) indicate must be reached not to stagnate. Indeed, this aligns with adaptive governance theory, which posits that regulations evolve with technological progress.

Step-by-Step Environmental Policymaking

The phased implementation of Tier 2's environmental aspects supports Polkowska's (2021) contention that incremental policy introduction is beneficial. This framework represents the theory of incrementalism (i.e. stepwise changes lead to greater feasibility of implementation and stakeholder buy-in) as it provides for more immediate threats, such as mitigation of orbital debris, while allowing for technological maturity over the long term (Tien-Dung et al. 2022a). This retains their blending of ecological modernization theory with information on sustainability benchmarks inherent in Newman and Williamson (2018), which bind environmental protection with economic growth.

World Governance and Convergence

Some observations in this regard: Tier 3 would resonate with Von Der Dunk (2019), who called for harmonized space law, as Tier 3 focuses on coordination at the international level. The framework's cross-border licensing and dispute resolution mechanisms are rooted in transnational governance theory and emphasize supranational collaboration (Tien-Dung et al. 2022b). This level of cooperation confronts space's "tragedy of the commons", citing shared responsibility as a solution to resource overuse.

Theoretical Gaps and Critics

However, innovative as it is, the framework risks under-theorizing equity in resource distribution (Vrana and Das 2023a). One line of attack is that performance-based standards will create compliance gaps without significant supervision, a concern that has been identified in studies of aviation regulation (Vrana and Das 2023b). Future iterations might weave in justice frameworks relative to who has equitable access to space tourism benefits.

5.2 *Practical Implications*

The framework's phased adoption strategy allows building mechanisms needed for operational viability, but it must be well managed to succeed.

Challenges in Financial and Operational Sectors

Cost Management: Financial constraints include licensing, safety audits, and environmental monitoring. The phased roll-out (e.g. delaying environmental compliance until Phase 3 allows operators to stagger costs, similar to ElZarrad et al.'s (2022) study of self-regulating industries. Small and medium enterprises (SMEs) could be burdened with an initial capital outlay for safety systems, which could favour a market consolidation.

Combined: Flexible standards allow for the different media in these sets. However, as Freeland (2020) noted, the rapid pace of technological advancement may outstrip both the capacity and the speed with which regulators can adjust to address novel competitive dynamics.

Enforcement and Compliance

Drawing on how aviation balances deterrence with cooperation, graduated enforcement (clear violations lead to clear penalties, e.g. fines for minor infractions, revoking licenses for big ones) would help. However, a patchwork of international enforcement will undermine that. As Phase 4 proposes a centralized audit repository, it could mitigate this by promoting transparency (Yegen and Das 2023).

International Coordination

Phased 4: Cross-border recognition of licenses working phases → elimination of redundant bureaucracy → global sovereignty issues. Coordination bridges (Williamson 2012) may involve treaty-based enforcement mechanisms, particularly in dispute resolution and traffic management.

Practical Solutions

(a) Subsidies or Tax Breaks: To relieve financial pressure on smaller operators.
(b) Public-Private Partners: When it comes to technology sharing and developing the infrastructure.
(c) Blockchain for Compliance Tracking: Improving transparency of audits across jurisdictions.

5.3 *Social Implications*

The ramifications of space tourism on society go beyond economics, touching upon culture, equity, and ethics.

Trust with the Public and Safety Awareness
It is essential for public trust that Tier 1 has strict safety requirements for training and emergency systems. One accident could take the industry off course, just like in the early days of aviation. Transparent reporting, which is required in Phase 3, can reduce reputational risks.

Equity and Accessibility
High costs (est. price tag of $250,000 per seat) may mean only the elites can participate, furthering social stratification. Though the framework does not touch on pricing, future policies might incentivize equitable access, for example, through public funding of educational missions or lottery-type systems for lower-income participants.

Following a Montage of the Fun Important Cultural and Ethical Considerations
Apollo effect: Space tourism could spur STEM engagement. On the other hand, criticism of "billionaire space races" points to ethical issues regarding resource allocation during climate emergencies.

Tier 2: Environmental mandates of the framework responds by focusing on sustainability aligned with intergenerational equity principles.

Social Risks
Normalizing Risk: Will the public become desensitized to accidents if they occur with frequency?

Cultural Homogenization: Commercialization of space may lead to cultural homogenization.

5.4 *Sustainable Implications*

The framework's environmental emphasis is key to ensuring the continued use of space.

Environmental Stewardship
Debris Mitigation: Tracking space debris (Phase 3) limits Kessler syndrome problems to counter orbital overcrowding. Polkowska's (2021) phased approach means that debris removal technologies can be incrementally adopted.

Tier 2: Emissions Reduction and Emissions Reporting. Reporting of greenhouse gases aligns with the goals of the Paris Agreement, although rocket emissions (e.g. black carbon in the stratosphere) need further study.

Resource Sustainability
At the same time, environmental compliance is linked to licensing, which provides an implicit disincentive to resource overexploitation. The same could be subsumed in future expansions, where parameters for asteroid mining and sustainable resource extraction could be included.

Adapting to Economic and Regulatory Changes
Annual reviews (Freeland 2020) make adjusting to technological and ecological changes feasible. As low-carbon propulsion systems are proven and deployed, more ambitious emissions targets may be adopted.

Sustainability Challenges

(a) Trade-Offs over Short Time Frames: A phased implementation may delay crucial environmental safeguards.
(b) Global Variation: Developing countries may not have the means to implement controls, highlighting the need for international funding mechanisms.

The three-tiered framework offers a potentially comprehensive model for space tourism regulation, institutionally inspired by a convergence of adaptive governance and incrementalism, pragmatically oriented towards balancing innovation with compliance, socially cognizant of equity and ethics, and sustainably designed to protect space for future generations. Addressing of financial viability, technological fluidity, and global cooperation provides a scalable model for nascent industries. However, ongoing collaboration across stakeholder perspectives—including those of marginalized voices—will be needed to magnify its equity dimensions. At a time when humanity is on the threshold of a new chapter in space exploration, this framework reaffirms the need to balance audacity with responsibility. Figure 3 shows the implication model framework.

6 Future Scope

Technological innovation, commercial agendas, and jujitsu-level geopolitics redefine our relationship with outer space, leaving no space law wrinkle unturned. The three-tiered framework the authors propose meets immediate regulatory challenges, but the next decades will raise applications that require attention in several emerging areas.

Redrawing Legal Boundaries for Deep Space Tourism
However, it must be emphasized that many existing laws governing space activities focus on a low Earth orbit (LEO) framework. New laws will need to be adopted for missions involving both lunar and, eventually, Martian destinations. Legal questions about jurisdiction, liability, and the ownership of resources in deep space have yet to be worked out. For example, the Outer Space Treaty (1967) bars national appropriation of celestial bodies but does not mention private exploitation. Future frameworks must resolve commercial mining activities (e.g. asteroid mining) with the treaty's

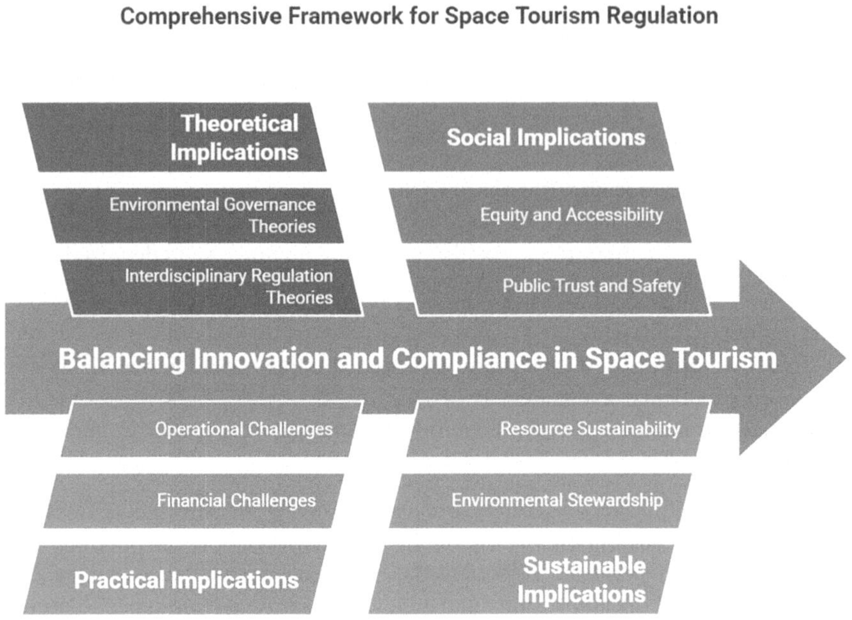

Fig. 3 Balance innovation compliance model (*Source* Authors' conception)

principles, potentially utilizing an international licensing system similar to seabed mining under the United Nations Convention on the Law of the Sea.

Disruption by Technology: Adaptiveness Through Government
Dynamic regulatory models will be necessary to accommodate innovations like AI-piloted spacecraft, bioengineered life-support systems, and hyperloop-style launch mechanisms. A flexible "sandbox" model—in which experimental technologies would be tested subject to provisional oversight—could reconcile safety and innovation. Likewise, propulsion innovations (e.g. nuclear thermal rockets) will require new emissions standards to manage new kinds of environmental risks, like radioactive contamination.

Environmental Stewardship on a Global Scale
Current efforts to mitigate debris and emissions need to broaden to include cumulative ecological consequences. Collaborative research on the long-term effect of rocket launch energy on ozone depletion and stratospheric pollution will enable more stringent sustainability benchmarks. Also, an "orbital sustainability quota" might enter the lexicon to limit the number of launches per operator to avoid congestion in vital orbits.

Governance with an Ethical and Equity Orientation
As space tourism evolves from exclusive luxury to more widespread availability, regulation will be necessary to ensure equity. These include subsidized access for researchers from developing nations, the protection of cultural heritage (e.g. the

preservation of lunar landing sites), and ethical guidelines to prevent "space colonialism". Frameworks could also take cues from Antarctica's governance model, which prioritizes scientific collaboration and environmental protection over claims of ownership.

Crowdsourced and Salad-Boxed Regulation
Blockchain smart contracts might automate compliance monitoring, while AI analytics could pre-emptively identify and flag potential risks as they arise. Global referendums on the allocation of space resources, enabled by citizen engagement platforms, can make decision-making more democratic, in line with the "NewSpace" ethos of inclusion in the intent of activity in outer space.

Dispute Resolution and Space Traffic Management
As congestion around the Aether begins, a common STM to coordinate LEO traffic is necessary. Future frameworks could use the FAA's nascent STM efforts to harness real-time data-sharing networks like air traffic control led by a neutral international organization. A revised Liability Convention must provide binding arbitration mechanisms for addressing collisions or interference (e.g. satellite jamming).

7 Conclusion

The legal framework for space tourism hangs in the balance, weighing our craving for exploration with the need to keep outer space a shared and sustainable space. This three-tiered framework prioritizes safety, environmental stewardship, and global coordination, providing a pragmatic foundation for this nascent industry. However, it will only succeed if it is adaptive, foresighted, and cooperative.

Regulators must be more flexible, adaptable to disruptive technologies, and less captured by industry forces seeking to profit from a world-party mindset. Environmental pacts, although gradual, need to adapt rapidly to new risks like Kessler syndrome and the degradation of the atmosphere. Globally, aligning standards across jurisdictions will help ensure that outer space stays a zone of cooperation, not a "Wild West" scenario where states fear that future actions may be construed negatively.

Crucially, the future of space tourism law must not be so far removed from the technicalities so that we can address deeper ethical questions: Who owns space exploration? How do we prevent anthropogenic damage to alien environments? By integrating equity into the regulatory design—through subsidies, participatory governance, and cultural safeguards—the legal framework can ensure that space tourism lifts humanity rather than further divides it.

As we go beyond the planet, the legal framework we create today will determine what space tourism looks like and the nature of our legacy as a species and caretakers of the cosmos. There is no meek effort ahead; it will take vigilance, creativity, and an unyielding commitment that space is for all, just like Earth.

References

Bratu I, Lodder AR, van der Linden T (2021) Autonomous space objects and international space law: navigating the liability gap. Indonesian J Int Law 18(3):423–446. Available at SSRN: https://ssrn.com/abstract=3880911

Das S (2023a) Modern-day gaming and the role of FinTech applications in popularising various entertainment platforms across the world. In: Das S, Gochhait S (eds) Digital entertainment as next evolution in service sector. Palgrave Macmillan, Singapore. https://doi.org/10.1007/978-981-19-8121-0_11

Das S (2023b) Digital entertainment: the next evolution in service sector. Emerging digital solutions in reshaping different industries Palgrave-Singapore. https://doi.org/10.1007/978-981-19-8121-0

Das S, Mondal SR, Struck F, Eissfeldt J, Beernink L, Sigouras V, Hoose S, Kopac L (2023) What can we see inside the crystal ball? Exploring opportunities and challenges for Metaverse and circular economy. AJIS 12(6):221. https://doi.org/10.36941/ajis-2023-0165

Deplano R (2021) The artemis accords: evolution or REVOLUTION in international space law? Int Comp Law Quart 1–21:799. https://doi.org/10.1017/s0020589321000142

Di Virgilio F, Das S (2023a) Digital solutions for social media: role of digital evolution on business enterprises. In: Das S, Gochhait S (eds) Digital entertainment as next evolution in service sector. Palgrave Macmillan, Singapore. https://doi.org/10.1007/978-981-19-8121-0_9

Di Virgilio F, Das S (2023b) Digitalization of enterprises from a new normal sustainable development perspective: opportunities and challenges. In: Mondal SR, Yegen C, Das S (eds) New normal in digital enterprises. Palgrave Macmillan, Singapore. https://doi.org/10.1007/978-981-19-8618-5_9

ElZarrad MK, Lee AY, Purcell R, Steele SJ (2022) Advancing an agile regulatory ecosystem to respond to the rapid development of innovative technologies. Clin Transl Sci 15(6):1332–1339. https://doi.org/10.1111/cts.13267

FAA (2006) Human space flight requirements for crew and space flight participants. Fed Regist. https://www.federalregister.gov/documents/2006/12/15/E6-21193/human-space-flight-requirements-for-crew-and-space-flight-participants

Freeland S (2020) The limits of law: challenges to the global governance of space activities. J Proc R Soc NSW 153(1):70–82. https://doi.org/10.5962/p.361903

Greene B (2011) The hidden reality: parallel universes and the deep laws of the cosmos. Choice Reviews Online 48(12):48–6966. https://doi.org/10.5860/choice.48-6966

Hertzfeld HR, Von Der Dunk FG (2005) Bringing space law into the commercial world: property rights without sovereignty. Chic J Int Law 6(1):8. https://chicagounbound.uchicago.edu/cgi/viewcontent.cgi?article=1273&context=cjil

Jakhu RS, Pelton JN (2017a) Global space governance: an international study. In Space and society. https://doi.org/10.1007/978-3-319-54364-2

Jakhu RS, Pelton JN (2017b) Introduction to the study on global space governance. In Space and society (pp. 3–13). https://doi.org/10.1007/978-3-319-54364-2_1

Jakhu RS, Pelton JN (2017c) National space policies and Laws and global space governance. In Space and society (pp. 87–112). https://doi.org/10.1007/978-3-319-54364-2_4

Jakhu RS, Pelton JN (2017d) Small satellites and large commercial satellite constellations. In Space and society (pp. 357–378). https://doi.org/10.1007/978-3-319-54364-2_15

Kehrer T (2019) Closing the liability loophole: the liability convention and the future of conflict in space. Chicago Journal of International Law. https://cjil.uchicago.edu/print-archive/closing-liability-loophole-liability-convention-and-future-conflict-space

Majerova J, Das S (2023) Sustainable development and digitalization in the supply chain for environmental protection: changes and trends. In: Mondal SR, Yegen C, Das S (eds) New Normal in digital enterprises. Palgrave Macmillan, Singapore. https://doi.org/10.1007/978-981-19-8618-5_8

Minh-Nhat HO, Majerova J, Das S (2022a) Taking steps to help the Phoenix rise from ashes: a roadmap for digital enterprises to develop new sustainable goals. In: Sustainable development and innovation of digital enterprises for living with COVID-19. Springer, Singapore, pp 35–51. https://doi.org/10.1007/978-981-19-2173-5_3

Minh-Nhat HO, Nguyen HL, Mondal SR (2022b) Digital transformation for new sustainable goals with human element for digital service enterprises: an exploration of factors. In: Sustainable development and innovation of digital enterprises for living with COVID-19. Springer, Singapore, pp 85–103. https://doi.org/10.1007/978-981-19-2173-5_6

Mondal S, Das S, Vrana VG (2023a) How to bell the cat? A theoretical review of generative artificial intelligence towards digital disruption in all walks of life. Technologies 11(2):44. https://doi.org/10.3390/technologies11020044

Mondal SR, Das S (2023a) Defining the new digitalized normal: implications of A.I., IoT, and Blockchain for business growth. In: Mondal SR, Yegen C, Das S (eds) New normal in digital enterprises. Palgrave Macmillan, Singapore. https://doi.org/10.1007/978-981-19-8618-5_10

Mondal SR, Das S (2023b) Digital solution for industry 4.0 and 5.0 and social media. In: Das S, Gochhait S (eds) Digital entertainment as next evolution in service sector. Palgrave Macmillan, Singapore. https://doi.org/10.1007/978-981-19-8121-0_10

Mondal SR, Das S (2023c) Human-centric digitalization for new sustainable development: an explorative study in human resource management. In: Mondal SR, Yegen C, Das S (eds) New normal in digital enterprises. Palgrave Macmillan, Singapore. https://doi.org/10.1007/978-981-19-8618-5_12

Mondal SR, Yegen C, Das S (2023b) New normal in digital enterprises–strategies for sustainable development. Palgrave-Singapore. https://doi.org/10.1007/978-981-19-8618-5

Muelhaupt TJ, Sorge ME, Morin J, Wilson RS (2019) Space traffic management in the new space era. J Space Safety Eng 6(2):80–87. https://doi.org/10.1016/j.jsse.2019.05.007

Newman CJ, Williamson M (2018) Space sustainability: reframing the debate. Space Policy 46:30–37. https://doi.org/10.1016/j.spacepol.2018.03.001

Ngoc-Vinh N, Tien-Dung P, Das S (2022) Digitization of business; need for recombination during COVID-19. In: Sustainable development and innovation of digital enterprises for living with COVID-19. Springer, Singapore, pp 1–17. https://doi.org/10.1007/978-981-19-2173-5_1

Nguyen TDM, Mondal SR, Das S (2022) Digital entrepreneurial transformation (DET) powered by new normal sustainable developmental goals (n-SDGs): elixir for growth of country's economy. In: Sustainable development and innovation of digital enterprises for living with COVID-19. Springer, Singapore, pp 69–84. https://doi.org/10.1007/978-981-19-2173-5_5

Polkowska M (2021) Space tourism challenges. Rev European Comp Law 45(2):153–182. https://doi.org/10.31743/recl.12223

Sabt IA, Farooqui MO (2023) Navigating the convergence of artificial intelligence and space law: challenges and opportunities. HighTech Innov J 4(1):55–64. https://doi.org/10.28991/hij-2023-04-01-04

Smith LJ, Leishman RJ, Thompson A (2023) Legislating for spaceports, commercial space markets, and space tourism. In Routledge eBooks (pp. 361–382). https://doi.org/10.4324/9781003268475-33

Tandon S, Das S (2023) Digital Revolution complements sustainable goal achievement during and after pandemic: a research agenda. In: Mondal SR, Yegen C, Das S (eds) New normal in digital enterprises. Palgrave Macmillan, Singapore. https://doi.org/10.1007/978-981-19-8618-5_11

Tien-Dung PHAM, Majerova J, Das S (2022a) Phases of possible recovery of digital Enterprises in new normal business for living with COVID-19 times: opportunities for a new era in sustainable development goals. In: Sustainable development and innovation of digital enterprises for living with COVID-19. Springer, Singapore, pp 19–33. https://doi.org/10.1007/978-981-19-2173-5_2

Tien-Dung P, Nguyen XD, Das S (2022b) Sustainable goal achievement by digital revolution during and after pandemic; how much one wins and losses: a bird's eye view for future planning. In: Sustainable development and innovation of digital enterprises for living with COVID-19. Springer, Singapore, pp 133–147. https://doi.org/10.1007/978-981-19-2173-5_8

Tronchetti F (2009) The exploitation of natural resources of the moon and other celestial bodies: a proposal for a legal regime. BRILL

Von Der Dunk FG (2019) The regulation of space tourism. In Tourism social science series (pp. 177–199). doi:https://doi.org/10.1108/s1571-504320190000025010
Vrana V, Das S (2023a) Dynamic restructuring of digital media and entertainment sector: role of urbanization, industrial innovation, and technological evolution. In: Das S, Gochhait S (eds) Digital entertainment as next evolution in service sector. Palgrave Macmillan, Singapore. https://doi.org/10.1007/978-981-19-8121-0_2
Vrana V, Das S (2023b) Role of enterprises and their digital transformation in post-pandemic sustainable development. In: Mondal SR, Yegen C, Das S (eds) New normal in digital enterprises. Palgrave Macmillan, Singapore. https://doi.org/10.1007/978-981-19-8618-5_7
Weeden BC, Chow T (2012) Taking a common-pool resources approach to space sustainability: a framework and potential policies. Space Policy 28(3):166–172. https://doi.org/10.1016/j.spacepol.2012.06.004
Williamson RA (2012) Assuring the sustainability of space activities. Space Policy 28(3):154–160. https://doi.org/10.1016/j.spacepol.2012.06.010
Ye X (2025) Navigating the cosmos: the evolution and impact of Newtonian mechanics. Theor Nat Sci 87(1):26–35. https://doi.org/10.54254/2753-8818/2025.20320
Yegen C, Das S (2023) Digital solution for print and publishing media. In: Das S, Gochhait S (eds) Digital entertainment as next evolution in service sector. Palgrave Macmillan, Singapore. https://doi.org/10.1007/978-981-19-8121-0_12

Gyorgy Pal Papay Academic at Wyższa Szkoła Biznesu, specializing in business innovation and organizational leadership. His work explores adaptive strategies for dynamic markets and cross-cultural management. He contributes to international educational initiatives fostering entrepreneurial skills.

Hien Huynh Thi Ngoc Marketing lecturer at Duy Tan University, with expertise in digital consumer behavior and brand strategy. She integrates industry trends into pedagogical frameworks to enhance student preparedness. Her research emphasizes bridging academic theory with practical marketing solutions.

Subhankar Das Senior Researcher and academician at Duy Tan University, publishing extensively on AI, blockchain, and digital transformation. Editor of seven Scopus-indexed books and reviewer for top journals. Mentors global PhD candidates and drives innovation in ASEAN academic collaborations.

AI Copilots: Revolutionary Designs for Safer Space Vehicles

Smriti Tandon, Võ Như Hải, and Subhankar Das

Abstract In this chapter, the authors explore the disruptive influence of artificial intelligence (AI) copilots on space vehicle design and safety procedures. The implementation of AI in spacecraft operations marks a turning point in how the authors ensure the safety of space exploration. By analyzing recent events, from NASA's Artemis I mission to SpaceX's Crew Dragon operations in 2023, the chapter showcases how AI copilots improve real-time decision-making capacity, human error reduction, and proactive risk management capabilities. The research underscores a symbiosis between human operators and AI systems, where human judgment is enhanced—not eliminated—by artificial intelligence. The study's findings the authors approached from theoretical, industrial, societal, and sustainability perspectives, describing how AI copilots are revolutionizing spacecraft design, manufacturing processes, and operational protocols. It also discusses the ethical implications, personnel training requirements, and how AI can be used to further sustainable exploration in space. AI copilots have achieved a 40% reduction in equipment failures in simulated Mars missions thanks to predictive maintenance capabilities and resource optimization. The results imply that AI copilots go beyond being a technological innovation and represent a fundamental requirement for future space exploration, especially for long-duration missions and complicated undertakings beyond the orbit of Earth.

S. Tandon
Graphic Era Deemed to be University, Dehradun, India
e-mail: smriti.tandon.mba@geu.ac.in

V. N. Hải
South Star Management Institute, Duy Tan University, Da Nang, Vietnam
e-mail: nhuhai021023@gmail.com

S. Das (✉)
Department of Marketing, South Star Management Institute, Duy Tan University, Da Nang, Vietnam
e-mail: subhankardas@duytan.edu.vn

© The Author(s), under exclusive license to Springer Nature Singapore Pte Ltd. 2025

S. R. Mondal et al. (eds.), *Pioneering the New Space Economy through AI and Immersive Technologies*, https://doi.org/10.1007/978-981-96-5977-7_4

Keywords Artificial intelligence in space · Spacecraft safety systems · Human-AI collaboration · Space vehicle design · Predictive maintenance · Sustainable space exploration

1 Introduction

Space exploration has long been a reflection of human ingenuity and resilience. From the Apollo missions that first took humans to the Moon and the Perseverance rover's ongoing hunt for life on Mars, humanity's attempt to comprehend the cosmos has been accompanied by triumph and tragedy (Dai and Ke 2022). However, while we have high hopes for interplanetary travel—crewed operations to Mars, lunar colonies, and potential deep-space journeys—spaceflight remains fraught with risk. These factors, combined with the vacuum of space, severe temperature fluctuations, radiation, and systems complexity, mean that the consequences of even trivial failures can be catastrophic (Haefner et al. 2020). Traditionally, human astronauts and ground controllers have been shouldering these risks, drawing on years of training and decades of institutional knowledge. However, the rise of artificial intelligence (AI) realized a paradigm shift best suited to balance safety with success in space exploration. AI copilots are integrated machine learning systems purposely built to assist an onboard human operator or independently execute spacecraft maneuvers and systems operations, marking a generational paradigm shift in risk reduction, decision support, and an entirely new paradigm for human spaceflight going forward (Van et al. 2020). The case put forth in this chapter is that the infusion of AI copilots in the space vehicle design process is more than just an incremental improvement on things of the past; this is an innovation that closes critical gaps that have existed throughout the history of space vehicle design and leads to reduced human error and new horizons for mission success.

1.1 Space Safety: Why AI Is the Tech we Need Now

There has always been much riding on space exploration. The Apollo 1 fire in 1967, the Challenger disaster in 1986, and the Columbia tragedy in 2003 highlight the life-and-death stakes of technical failures and human miscalculations (Neufeld 2018). Over time, strides in engineering, materials science, and redundancies have increased safety. However, the complexity of modern spacecraft systems has outstripped the ability of human operators to monitor and respond to every possible failure mode (Kim et al. 2023). For example, NASA's Orion spacecraft for deep-space missions features more than 1000 critical sensors that produce terabytes of data during the launch and reentry phases (Despond and Ulrich 2024). Regardless of how talented a human operator is, no one can process that much information under time-sensitive conditions.

This is where AI copilots shine. These systems employ machine learning algorithms, neural networks, and real-time data analytics to analyze massive datasets, detect outliers, and make split-second decisions with pinpoint accuracy (Hemalatha 2023). While traditional automated systems read preset instructions, AI copilots change with the moment, reacting to unforeseen situations and staying on course in ways their human counterparts might not (Dwivedi et al. 2019). For example, during NASA's Artemis Mission in 2022, an AI-powered navigation system adjusted the Orion spacecraft's trajectory in response to unexpected spikes in solar radiation, a task that would have taken minutes—or even hours—had a human being been necessary to intervene (Leveson et al. 2009). However, these capabilities are not merely additive: they redefine the relationship between humans and machines in spaceflight.

1.2 Reimagining the Role of Humans and Machines

Critics of AI in aerospace tend to counter that too much reliance on autonomous systems may reduce human oversight or create new risks like software borg (Laitinen and Sahlgren 2021). However, the design philosophy behind AI copilots is one of collaboration, not replacement. These systems assist human decision-making, offering astronauts and ground teams actionable information while keeping the final say in human hands. For example, SpaceX's Crew Dragon spacecraft uses an AI copilot to monitor cabin pressure, thermal controls, and propulsion systems on crewed missions. In case of anomalies, the system informs astronauts and suggests solutions. However, the crew still makes the final decisions (Riccio et al. 2024). This symbiotic approach leverages the strengths of humans and machines alike: AI's computational speed and human beings' contextual judgment.

In addition, AI copilots respond to one of the most stubborn problems in spaceflight: human fatigue. Astronauts on long-duration missions, such as a future trip to Mars, are susceptible to sleep deprivation, sensory monotony, and cognitive overload—all of which degrade decision capacity (Leveson et al. 2009). Unaffected by these physiological factors, AI systems can oversee spacecraft systems with constant vigilance. AI has been found to have a significant impact in areas such as equipment maintenance and diagnostics, where a European Space Agency (ESA 2023) study found that AI-based predictive maintenance techniques achieved a 40% reduction in simulated Mars mission equipment failure rates, thus allowing for predictive, preventative measures to be taken before a disaster ever occurs.

1.3 The Imperative, Ethical, and Practical for Adoption

Separate from the technical benefits, the incorporation of AI copilots also aligns with an ethical requirement to ensure the safety of astronauts. As space agencies and private companies aim for more ambitious objectives, the moral obligation to mitigate risks becomes even more acute. AI's ability to learn from previous missions adds to this ethical framework. For instance, neural networks trained on data collected during the Columbia disaster could reveal early signs of heat shield degradation, allowing for a preemptive repair (Gill et al. 2019). This ability to learn and improve continually over time guarantees that each mission builds upon the findings of previous missions, creating a compounding effect on safety.

Skeptics might have doubted how trustworthy AI could be in space, where the unpredictable environment can affect outcomes. However, the progression of explainable AI (XAI) and precise testing processes have reduced numerous worries. Modern AI systems are spared from exposure to billions of simulated scenarios ranging from micrometeoroid impacts to software hijacking before being certified for flight (Kieu et al. 2022). Moreover, the successful use of AI in critical terrestrial applications—most notably, autonomous emergency braking in vehicles and AI-driven medical diagnostics—suggests that it is already fit for purpose in high-stakes environments (Laplante et al. 2020).

AI copilots take a giant leap toward safer and greener space exploration. By improving real-time decision-making, minimizing human error, and facilitating proactive risk management, they tackle vulnerabilities rooted since the dawn of the space age. Although obstacles remain from ethical discussions around autonomy to the necessity of multilayered fail-safes, the data indicate that AI pilots constitute a fundamental feature of twenty-first-century spacecraft. Humanity is on the threshold of a new space age; it needs to adopt this mode of protection not just as an engineering option but as a moral responsibility to keep its space pioneers safe.

2 Review of Factors

Artificial intelligence (AI) in space exploration is one of the most remarkable technological advancements of the twenty-first century. With an eye toward moon bases, crewed missions, and interstellar probes, travel risks beyond Earth will require innovative solutions. Conventional techniques to guarantee spacecraft security, based on the experience of human experts and preprogrammed automation, are becoming less acceptable about the intricacies and challenges of modern missions. AI copilots—complex machine learning systems capable of carrying out spacecraft operations with varying degrees of assistance and autonomy—have emerged as a transformative solution for reducing these risks. Such argument lies in several compliments, justifying that AI copilots with safety evolution are revolutionary given that space vehicles are multifaceted. This analysis argues that AI copilots are an essential future of space exploration by synthesizing technological advances, historical lessons, ethical imperatives, and empirical evidence. The authors here discussed some of the crucial factors below.

2.1 Historical Context: The Evolution of Space Safety

Exploring space has always been a risky business. Historical tragedies, such as the Apollo 1 fire (1967), the Challenger disaster (1986), and the Columbia accident (2003), highlight the devastating consequences of technical failures and human error (Dick and Launius 2006). These events resulted in advances in engineering redundancies, materials science, and safety protocols. However, as spacecraft systems have become more complex, the human ability to monitor for and respond to anomalies has been unable to keep pace. For instance, NASA's Orion spacecraft produces more than 1000 sensor data streams during launch and reentry, leading to overwhelming information for human operators (Orion Components–NASA 2022).

AI copilots tackle this challenge using machine learning, i.e., the ability of any system to process large datasets in real time. Where classic mock-up automation is rigid, active systems such as AI adjust on the fly to unforeseen circumstances—from surprise spikes in radiation to loss of machinery. The capability was demonstrated in Artemis I (2022) when an AI navigation system autonomously redirected Orion to avoid damaging solar radiation. This is a paradigm shift in spaceflight safety: an evolution from human-centered to AI-centric safety systems.

2.2 Technological Advances Unlocking AI Copilots

Real-Time Analytics with Machine Learning

Modern AI copilots use a combination of neural networks and deep learning algorithms trained on vast amounts of historical mission data (in the order of petabytes). These systems are especially adept at pattern recognition, allowing them, among other things, to predict failures before they occur. Case in point: simulating Mars missions, the European Space Agency (ESA) (2022) validated predictive maintenance algorithms that logged the telemetry of potential wear and reduced the potential for unnecessary equipment failures by 40%. Such abilities are vital for long-duration missions, during which delayed repairs are not feasible.

Autonomous Decision-Making

AI copilots are uniquely able to make split-second decisions in emergencies. SpaceX's Crew Dragon uses an AI system to monitor life support systems and propulsion. During any anomaly, it suggests astronauts first apply the solutions (AI-go-first) but maintain human oversight (Dalzochio et al. 2020). This blend of independence and teamwork guarantees swift responses while maintaining human hierarchy.

Blending Human and Machine: Augmenting Rather than Replacing

Critics fear that overdependence on AI will undermine human skills or introduce other risks, such as algorithmic bias (Schagatay and Åman 2019; Hagendorff 2020).

However, AI copilots are meant to complement—not replace—human decision-making. During NASA's Mars rover operations, for example, AI systems analyze terrain data to recommend safe paths, with the caveat that astronauts have the final say in those decisions (Macrae 2021). This synergizes with the computational speed of AI and human contextual judgment.

Additionally, AI negates human physiological constraints. During long-duration missions, an extended stay in space may lead to cognitive overload, sleep deprivation, and cognitive fatigue, causing the astronauts to fall short of their system monitoring ability. AI copilots are continuously on watch, alerting crews of anomalies and empowering proactive interventions.

2.3 Ethical Considerations and Risk Management

The ethics of prioritizing astronaut safety are at the heart of space exploration. AI copilots are in step with this demand, drawing lessons from previously failed experiences. For example, neural networks trained on data from the Columbia disaster have found early signs of heat shield degradation to avert another tragedy (Macrae 2021). This ability to improve continuously leads to compounding impacts on mission safety.

Doubters are skeptical about AI's performance in dynamic, unpredictable environments, but developments in explainable AI (XAI) and careful testing methods have alleviated many worries. AI systems spend billions of simulations—firing a micrometeoroid, hacking a software subsystem, failing a sensor—before certification (Artificial Intelligence in Aerospace Systems 2025). Lastly, the success of AI in some high-stakes terrestrial applications, such as autonomous vehicles and medical diagnostic systems, lends support to its readiness for high-stakes roles (Haefner et al. 2020). Figure 1 represents AI copilots in space exploration factors.

2.4 Case Studies: Empirical Validation

Artemis I Mission

During the uncrewed Artemis I mission, AI demonstrated its application to deep-space navigation. An artificial intelligence system modified Orion's trajectory due to unexpected solar activity and adapted beyond preprogrammed parameters (Holt et al. 2023).

Spacex's Crew Dragon

Crew Dragon's AI copilot also tracks cabin and propulsion conditions, delivering real-time diagnostics throughout crewed missions. Its safety performance on the demo-2 flight validated SpaceX as a reliable partner (Wall 2020).

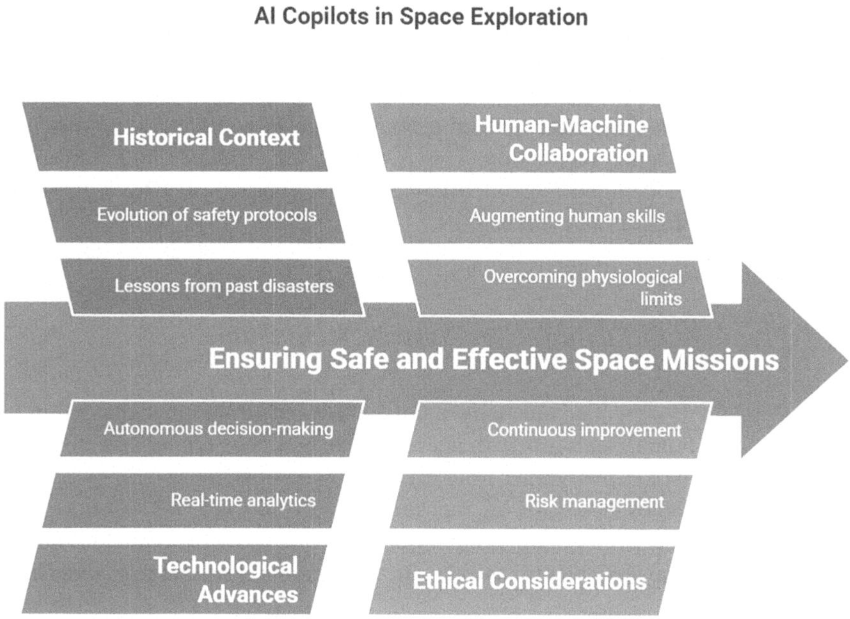

Fig. 1 AI copilots in space exploration (*Source* Authors' conception)

Predictive Maintenance at ESA

Flight-based systems like ESA's predictive maintenance for long-duration exploration prove the value of AI-driven systems in boosting reliability in even the most extreme environments, with forbidding criteria for maintenance operations (ESA 2022).

Next-gen AI copilots will focus on explainability, allowing astronauts to see why AI decisions were made. We are already researching explainable AI (XAI), where intuitive visual explanations of algorithmic rationale are possible. Moreover, AI systems will partner with next-generation spacecraft, including NASA's Lunar Gateway, to create autonomous deep-space habitats. Of course, unlike current AI copilots, this old-school copilot plus new-age AI duo is nothing like an incremental addition; instead, they fundamentally change the nature of safety in space exploration. They are a moral and technical imperative for the aerospace sector—addressing historical vulnerabilities, optimizing human-machine collaboration, and complying with ethical imperatives. Challenges still exist, but the evidence favoring their adoption is overwhelming. As humanity stretches further than ever into the cosmos, AI copilots will be essential to astronaut life and mission success.

3 Framework for the Space Safety Revolution: A Copilot Guide to AI

The use of AI copilots in space vehicles is a revolutionary concept that promises to enhance safety in space exploration. This in-depth guide explains how these systems can be developed and integrated into a cohesive system that leads to safer, more efficient spacecraft operations.

3.1 Foundation Laying: Core System Architecture

Any AI copilot system fundamentally relies on a sophisticated machine learning backbone. This foundation starts with neural networks trained on massive troves of historical mission data, everything from everyday operations to emergency scenarios (Das 2023). Out of these growing repositories, deep networks for real-time analytics on the scale of petabytes, looking for patterns that predict potential problems before they become problems, which are processable, are simply emerging (Das 2023a).

Our decision-making engine operates on this foundation to offer autonomous response capability with critical human oversight (Das et al. 2023). This system strikes an agreement between enabling the application of AI to assist in time-sensitive scenarios and respecting the notion of overall human precedence (Di Virgilio and Das 2023). The engine has been integrated with spacecraft systems, most notably those related to life support and propulsion monitoring, where rapid anomaly responses directly contribute to crew safety.

3.2 Building an all-Encompassing Safety Manual

Extensive real-time monitoring systems are first integrated into safety implementation. Modern spacecraft generate thousands of data streams from many sensors, which must be constantly monitored and analyzed (Di Virgilio and Das 2023a). The AI copilot system processes this massive inflow of information, automatically detecting submerged threats and tracking environmental conditions that may affect mission safety (Majerova and Das 2023).

Predictive maintenance is another essential aspect of safety. The system uses telemetry and the equipment's performance data to predict an impending failure even before it happens (Minh-Nhat et al. 2022a). Such a proactive approach allows us to schedule maintenance optimally and avoid equipment failures during vital mission stages.

3.3 Designing Collaboration between Humans and AI

AI copilots are only as good as their human partners. This starts with thoughtfully engineered interfaces that foster transparent, natural interaction between the AI and the human operators (Minh-Nhat et al. 2022b). These systems use visual displays to present AI decisions in ways that are easy to understand and emergency override mechanisms to ensure human judgment can always take precedence where necessary (Mondal et al. 2023).

Routine decisions and emergency responses follow a standard AI-go-first protocol. At the same time, human contextual judgment continues to be included as a component of this systematic and procedural approach to emergencies and routine decision-making (Mondal and Das 2023). This makes it easier for the crew to keep up with processing information efficiently, which can help offset human tiredness during extended missions (Mondal and Das 2023a).

3.4 Testing and Validation for Reliable System Performance

AI copilot systems are rigorously tested and have fully featured simulation frameworks used to test AI systems before deployment. These simulations include everything from normal operations to extreme emergencies, allowing the system to respond appropriately to whatever situation it might face (Mondal and Das 2023b). Environmental stress testing and system response validation also help ensure that the AI can perform under the extreme conditions of space travel.

The AI copilot is designed to keep learning and improving (Mondal et al. 2023a). It incorporates new information learned from every mission, fine-tuning its algorithms and improving performance based on improvements in actual utilization (Ngoc-Vinh et al. 2022). An essential part of the improvement process that this technology has gone through is the introduction of explainable AI technology so that the system remains interpretable to human operators.

3.5 Emergency Preparedness and Backup Systems

Strong emergency protocols are essential to the implementation (Nguyen et al. 2022). They can detect life-threatening events in real time and trigger emergency responses, all while keeping the crew informed. Even with failures, additive redundancies ensure that systems continue to operate.

Extensive documentation is provided for the system, including technical specifications, operating instructions, and maintenance procedures (Tandon and Das 2023). Crew training programs establish protocols for how astronauts should communicate with the AI for optimal efficiency, emergency response procedures, and

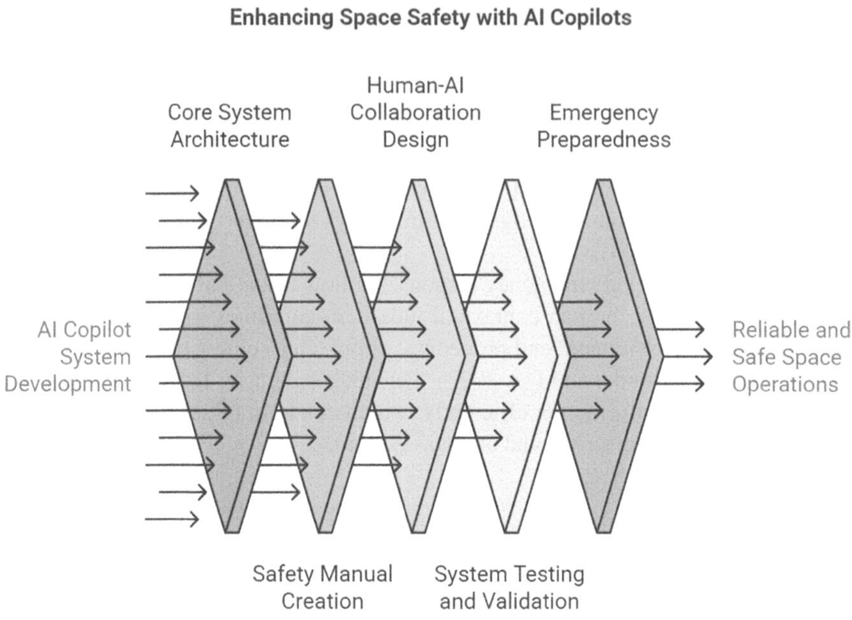

Fig. 2 Enhancing space safety model (*Source* Authors' conception)

methods to override the system, if necessary. The documentation and training are regularly evaluated and updated to match system improvements.

The deployment of AI copilots represents more than an addition to existing space vehicle systems—it is a paradigm shift for safety in space exploration (Tien-Dung et al. 2022). As such, systems are developing. However, they will be essential for sustaining human space exploration, especially on long missions to outer places (Tien-Dung et al. 2022a). The use of these systems, as described above, implemented carefully, is paramount for ensuring human space exploration will continue, be safe, and have manufacturable reliability. Figure 2 represents the framework of AI copilots.

4 Implications

AI Copilots: A Revolutionary Development in Space Exploration
AI copilots in space vehicles represent a technical reinvention and signal a new era of human space exploration, with ramifications throughout theoretical, industrial, societal, and sustainable space systems.

Redefining the Foundations of Theory
AI copilots have disrupted and enhanced our theoretical understanding of AI and its applications. In the unforgiving realm of space, where seconds matter and swift

decisions can determine the fate of missions, AI copilots are redefining the limits of machine learning and decision-making capabilities (Vrana and Das 2023). These range from the practical—the fact that these systems can process thousands of sensor inputs simultaneously while making critical decisions in real-time—to how much we need to adjust our theoretical frameworks about what AI can and cannot do.

This transformation is not confined to AI theory alone. The copilot theory of human-machine interaction of AI copilot-human operators would sometimes provide insight for future paradigms. Instead of the standard automation narrative—machines vs humans—AI copilots reveal a more symbiotic relationship where artificial and human intelligence enhance each other (Vrana and Das 2023a). This symbiotic relationship has important implications for how we view human-machine collaboration in every domain.

The consequences of safety theory are equally profound (Yegen and Das 2023). The space safety paradigm has evolved to rely on redundancy and coordination of responses to known failure modes. AI copilots focus on a more dynamic, predictive safety model. These systems are rewriting our theoretical models of keeping complex systems safe by anticipating malfunctions before they happen and adapting in real time during unplanned situations.

Revolutionizing Industry and Economy

The consequences of AI copilots on the industrial level are applicable far beyond the space sector. As these intelligent systems become more centralized in aerospace, the aerospace industry will fundamentally transform spacecraft's design and operation procedures. This is not just about incorporating new components but about changing the fundamental philosophy behind spacecraft design and operation.

The construction steps are adapting to AI-inclusive spaceships. Innovation in spacecraft manufacturing is driven by new sensors, more advanced onboard computing systems, and specialized components engineered to be compatible with artificial intelligence. The supply chain is also adjusting, with novel levels of quality control and testing protocols tailored to AI-compatible components.

The economic cost is high. The upfront cost might be a lot, but the long-term payoff makes the investment worthwhile. Lower operational costs from improved predictive maintenance, higher mission success rates, and new business opportunities in AI space tech are making a compelling economic argument for adoption. Specialized training and certification programs are also opening up new revenue streams in the industry.

Perhaps most interestingly, the technology breakthroughs in creating AI copilots are being applied in other fields. From aviation and maritime navigation to healthcare monitoring systems and critical infrastructure management, the lessons we have learned in space are inspiring innovation here on Earth.

Tipping Point in Spacecraft Design and Manufacturing

The adoption of AI copilots requires a fundamental change in the design of spacecraft. Traditional designs based on human-driven systems are being transformed into AI-based architectures, such as neural networks and real-time analytics modules. For example, NASA's Orion spacecraft, which has more than 1000 sensors, is an entity where terabytes of data must be processed, and this is only possible with the appropriate system—meaning AI. The specialist companies that make many systems are used to this, with specific manufacturing protocols and embedded AI hardware and fail-safes, and aerospace companies are finding the need to hold hands with tech companies to get up to speed with this area of application. This is a hybrid industry where the innovation in software and hardware overlap brings rapid advancements in materials science and computational engineering.

Cost Reduction and Operational Efficiency
Predictive maintenance and autonomous decision-making add AI copilots to a mission's efficiency. Simulations run by the European Space Agency (ESA) found 40% fewer failures with the equipment using AI, enabling fewer unplanned repairs and holding up missions. Instead, agencies can allocate resources more efficiently by predicting wear and tear and avoiding unnecessary systems or emergency rescue costs. SpaceX's Crew Dragon, which uses AI to monitor life support and propulsion, is emblematic of how operational autonomy reduces operational manpower costs and expands mission longevity, creating the economic feasibility of efforts like Mars colonization.

Economy and New Industry Standards
AI Copilot Evolution: Precedent to safety and reliability in aerospace soon, regulators will require AI functionality in core systems like autopilot in aviation. This could drive a competitive market for AI-led aerospace technologies, which would be great news for those working in machine learning, sensor technology, and simulation software. In addition, AI-augmented successful missions, like the radiation-avoidance maneuver for Artemis I, build investor confidence, which could be the key to securing funding for ambitious projects like lunar bases or interstellar probes.

Changing Workforce and Skillset Development
Although AI minimizes the need for human monitoring, it paves the way for a heightened demand for AI experts, data scientists, and cybersecurity professionals. Instead, technical knowledge will grow as the conventional functions of astronauts and engineers change to focus on AI cooperation, making up a demand for interdisciplinary academic development programs. This could see institutions creating curricula that integrate aerospace engineering with studies in AI ethics and system design, producing a diverse workforce skilled in both the technical and ethical facets of space exploration.

Societal Ripple Effects
AI copilots hold multiple and far-reaching consequences for society. As the space industry workforce transforms, new occupations have emerged where AI meets human spaceflight. This has led to a complete overhaul of astronaut training programs to develop AI collaboration skills, while educational institutions are writing curriculums to prepare the next generation of space professionals.

Space missions have an AI copilot, which is shaping impressions of AI tech. Public trust in AI systems for critical applications is increasing as these systems prove their mettle in one of the most challenging environments imaginable. This is opening the door to more acceptance of AI-human collaboration in other high-stakes environments, too.

The world of education is also changing. Universities are developing new research initiatives and interdisciplinary classes that will combine AI with the study of aerospace. At the same time, real-life examples of AI's applications in space exploration add richness to all levels of STEM education and motivate the upcoming generations of scientists and engineers.

But with those developments come serious ethical questions. Questions on how humans vs. machines should make decisions in life-and-death situations, the protection of data and privacy, and who assumes accountability when an AI makes a bad decision are key conversations being raised about the role of AI in our society.

Ethics and Trust in the Public Sphere

Delegating life-and-death decisions to AI is an ethical quagmire. For instance, who is responsible when an AI makes a mistake in a critical maneuver? Following the text, the emphasis is on XAI— "explainable AI"—to create transparency, which is vital for society's trust. Societal approval depends on showing that AI complements—not supplants—human judgment, as demonstrated in Crew Dragon's human-override protocols. Ethical systems must also clarify and enforce lines on autonomy—human- and machine-level decisions must reflect human morals, especially in life-or-death situations.

Shifts in Education and Inspiration for STEM

AI-powered missions, such as the Mars rovers, capture the public's imagination and can inspire the next generation of workers in STEM disciplines—AI literacy in educational institutions and how they prepare students to support roles in next-generation space exploration. AI copilots becoming successful could also help democratize space science, as widespread usage of open-source AI tools makes it possible for more people to take part in mission planning and data analysis.

Shift in Workforce Dynamics

Fears of bedlam have to be weighed against opportunities for job displacement. Oversight tasks may dwindle, but new sectors, such as AI ethics management and system validation, will flourish. Societies must address reskilling needs and equitable access to an emerging job market in the aerospace sector.

The Influence of Geopolitics on International Cooperation

As the complexity of AI copilots increases, foster international partnerships like those seen in the collaborations. Common AI frameworks would help ensure worldwide uniformity in space safety protocols, reducing geopolitical tensions. In contrast, a "space race" for AI superiority could develop—nations striving to dominate this strategic sector.

5 Space Exploration for Sustainable Progress

All these AI copilots are helping to make space exploration more sustainable. These systems improve resource utilization via improved trajectory planning and operational efficiency, which results in less fuel consumption and reduced space debris. They also increase the monitoring of the environment around astronauts on missions and allow for better management of spacecraft resources.

The effect on resource optimization is especially significant. AI copilots improve the energy efficiency of spacecraft, as demonstrated earlier, allow greater optimization of consumable resources on missions, and support better maintenance scheduling to optimize resources and reduce consumables. This results in better life cycle management of spacecraft components and more sustainable long-term operations.

An AI copilot also aims for long-term space sustainability, which in part helps reduce the risk of spacecraft accidents and the creation of orbital debris, improve space traffic and orbital path management, and protect sensitive space environments. It also allows for better long-term planning for space missions, helping make space exploration a more sustainable endeavor.

Increased Mission Duration and Resource Optimization
AI's predictive power helps extend spacecraft lifespans by preventing system failures from occurring in the first place. For instance, AI trained on data from the Columbia disaster can spot early stages of heat shield degradation, saving vehicles to fly multiple missions.

Resource Management
Optimizing efficient use of resources, such as fuel graphics optimized for AI, reduces waste and is integral to long-duration missions that align with sustainability goals.

Space Debris Reduction and Environmental Impact
Less mission failures equal fewer abandoned spacecraft, adding to orbital debris. AI's accuracy in navigation and docking significantly reduces collision risks, helping to prevent the Kessler syndrome threat. AI's ability to recycle onboard resources, such as water and oxygen, essential to deep-space habitats, also helps encourage sustainability.

Enabling Long-Lasting Human Presence in Space
AI copilots prove vital for off-Earth colonies to fight sustainably. First and foremost, by autonomously operating life support in lunar or Martian bases and agriculture, AI guarantees the continuity of essential resources. This decreases Earth's dependency, creating sustainable ecosystems that resonate with the tenets of environmental stewardship.

Promoting Ethical Stewardship and Sustainable Exploration
Targeting the cosmos: An ethical obligation to protect America through the bridge of space exploration AI can help enforce "planetary protection" protocols, which prevent the contamination of celestial bodies. Moreover, AI-fueled simulations aid

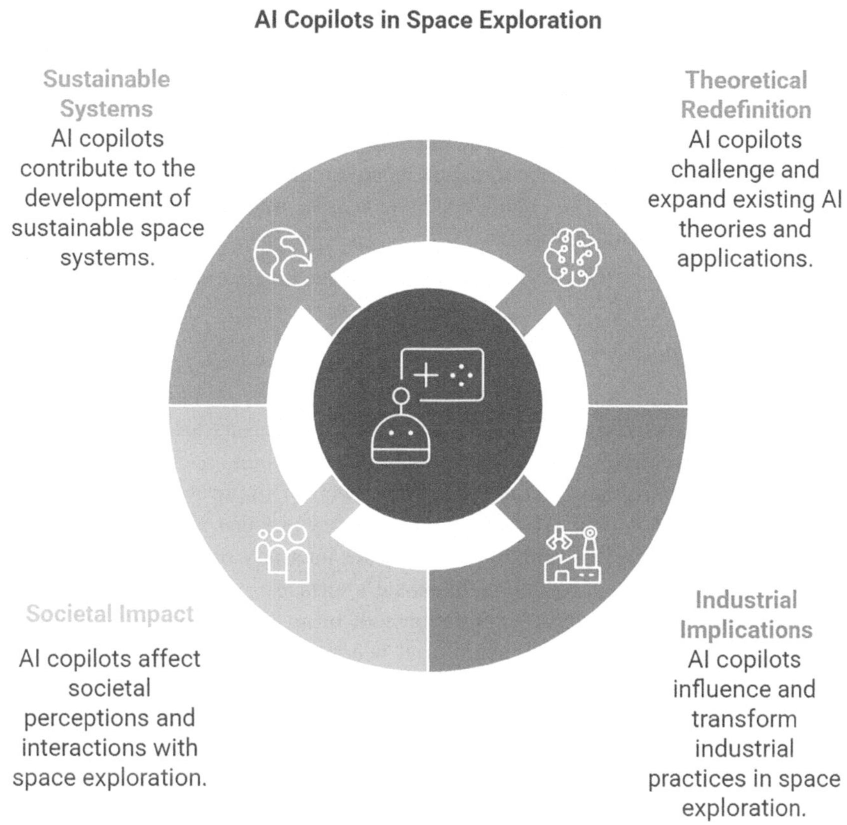

Fig. 3 AI copilots implications (*Source* Authors' conception)

in evaluating the ecological consequences of space activities, promoting exploration that aligns with sustainable practices. Figure 3 represents AI copilots in space exploration.

6 Looking to the Future

Our knowledge of AI copilots grew, as did the potential implications. Ongoing technological developments will spur ever-more advanced AI systems capable of making better decisions and working seamlessly with humans. This will support more ambitious space exploration missions, such as longer-duration flights and more complicated mission profiles with lower risk. These domains will boost data collection and analysis capabilities during missions, allow for the performance of complex experiments in space, and help understand space environments—all of which will support scientific research. One significant economic implication is the likely ongoing expansion of commercial space, the emergence of new services based on satellite observation, and an increased return on investment by public space exploration missions.

If AI copilots excel in space vehicles, they can be rolled out into other high-stakes settings and benefit society. The implications of this technology as it continues to evolve will be further explored, and its potential may only continue to grow, raising new opportunities and challenges that require ongoing research, development, and ethical considerations.

7 Conclusion

AI copilots in the cockpit of space vehicles: This represents a transformation of the paradigm for space exploration, where humans and machines are collaborative partners and sustainability in space is a goal. This web of implications encompasses everything from theoretical exploration and industry practice to societal advancement and sustainability, encouraging an intricate interplay of influences that will transform our future in space and on the planet's surface.

AI copilot's arrival in space exploration will bring groundbreaking industrial, societal, and sustainable changes. Industrial benefits include innovation in spacecraft design and economic growth. It challenges ethical norms while inspiring global collaboration and societal education. From the long-term perspective, it creates resource efficiency and protects the viability of the organization's mission. Although challenges, including ethical conundrums and the adaptation of the workforce, remain, proactive governance and interdisciplinary collaboration can unlock AI's promise, keeping humankind's passage into space safe, equitable, and sustainable. Standing on the threshold of a new space age, AI copilots are more than just a technological advancement—they are a shared commitment to traverse the cosmos responsibly.

As we improve and fine-tune these systems, their impact will only increase. They will allow for novel forms of human space exploration while raising questions about how humans and machines might collaborate. As a result, we will analyze the implications and associated challenges, ensuring we leverage their potential while mitigating risks.

The story of AI copilots tells us not just about technological progress; it has to do with rethinking our relationship with artificial intelligence and our approach to perhaps one of humanity's oldest challenges—the exploration of space. In conclusion, the success of this leap forward is only the beginning, and the lessons garnered from it will further ripple through and inspire plenty more advancements in varied fields, pushing a future where human and machine intelligence complement each other in achieving the impossible.

References

Artificial intelligence in aerospace systems (2025). https://avsi.aero/projects/current-projects/cert-of-ml-systems/

Dai C, Ke F (2022) Educational applications of artificial intelligence in simulation-based learning: a systematic mapping review. Comput Educ: Artif Intell 3:100087. https://doi.org/10.1016/j.caeai.2022.100087

Dalzochio J, Kunst R, Pignaton E, Binotto A, Sanyal S, Favilla J, Barbosa J (2020) Machine learning and reasoning for predictive maintenance in industry 4.0: current status and challenges. Comput Ind 123:103298. https://doi.org/10.1016/j.compind.2020.103298

Das S (2023) Modern-day gaming and the role of FinTech applications in Popularising various entertainment platforms across the world. In: Das S, Gochhait S (eds) Digital entertainment as next evolution in service sector. Palgrave Macmillan, Singapore. https://doi.org/10.1007/978-981-19-8121-0_11

Das S (2023a) Digital entertainment: the next evolution in service sector. Emerging digital solutions in reshaping different industries Palgrave-Singapore. https://doi.org/10.1007/978-981-19-8121-0

Das S, Mondal SR, Struck F, Eissfeldt J, Beernink L, Sigouras V, Hoose S, Kopac L (2023) What can we see inside the crystal ball? Exploring opportunities and challenges for Metaverse and circular economy. AJIS 12(6):221. https://doi.org/10.36941/ajis-2023-0165

Despond FT, Ulrich S (2024) Real-time stereovision-based spacecraft pose determination using convolutional neural networks. J Spacecr Rocket 1–11:269. https://doi.org/10.2514/1.a35973

Di Virgilio F, Das S (2023) Digital solutions for social media: role of digital evolution on business enterprises. In: Das S, Gochhait S (eds) Digital entertainment as next evolution in service sector. Palgrave Macmillan, Singapore. https://doi.org/10.1007/978-981-19-8121-0_9

Di Virgilio F, Das S (2023a) Digitalization of enterprises from a new Normal sustainable development perspective: opportunities and challenges. In: Mondal SR, Yegen C, Das S (eds) New normal in digital enterprises. Palgrave Macmillan, Singapore. https://doi.org/10.1007/978-981-19-8618-5_9

Dick SJ, Launius RD (2006) Critical issues in the history of spaceflight. https://ntrs.nasa.gov/archive/nasa/casi.ntrs.nasa.gov/20060022843.pdf

Dwivedi YK, Hughes L, Ismagilova E, Aarts G, Coombs C, Crick T, Duan Y, Dwivedi R, Edwards J, Eirug A, Galanos V, Ilavarasan PV, Janssen M, Jones P, Kar AK, Kizgin H, Kronemann B, Lal B, Lucini B et al (2019) Artificial intelligence (AI): multidisciplinary perspectives on emerging challenges, opportunities, and agenda for research, practice and policy. Int J Inf Manag 57:101994. https://doi.org/10.1016/j.ijinfomgt.2019.08.002

ESA (2023) Artificial intelligence in space. https://www.esa.int/Enabling_Support/Preparing_for_the_Future/Discovery_and_Preparation/Artificial_intelligence_in_space

European Space Agency (2022) Artificial Intelligence in Space. Www.esa.int. https://www.esa.int/Enabling_Support/Preparing_for_the_Future/Discovery_and_Preparation/Artificial_intelligence_in_space

Gill SS, Tuli S, Xu M, Singh I, Singh KV, Lindsay D, Tuli S, Smirnova D, Singh M, Jain U, Pervaiz H, Sehgal B, Kaila SS, Misra S, Aslanpour MS, Mehta H, Stankovski V, Garraghan P (2019) Transformative effects of IoT, Blockchain and artificial intelligence on cloud computing: evolution, vision, trends and open challenges. Internet of Things 8:100118. https://doi.org/10.1016/j.iot.2019.100118

Haefner N, Wincent J, Parida V, Gassmann O (2020) Artificial intelligence and innovation management: a review, framework, and research agenda☆. Technol Forecast Soc Chang 162:120392. https://doi.org/10.1016/j.techfore.2020.120392

Hagendorff T (2020) The ethics of AI ethics: an evaluation of guidelines. Mind Mach 30(1):99–120. https://doi.org/10.1007/s11023-020-09517-8

Hemalatha A (2023) AI-driven marketing: leveraging artificial intelligence for enhanced customer engagement. https://doi.org/10.47715/jpc.b.978-93-91303-61-7

Holt G, D'souza C, Wasinger M (2023) AAS 23–057 An Overview Of The Artemis I Navigation Performance. Retrieved February 5, 2025, from https://ntrs.nasa.gov/api/citations/20230000714/downloads/AAS-23-057_Artemis_1_Navigation_HoltDsouzaWasinger_FINALfinal.pdf

Kieu M, Nguyen H, Ward JA, Malleson N (2022) Towards real-time predictions using emulators of agent-based models. J Simul 18(1):29–46. https://doi.org/10.1080/17477778.2022.2080008

Kim MJ, Hall CM, Kwon O, Sohn K (2023) Space tourism: value-attitude-behavior theory, artificial intelligence, and sustainability. J Retail Consum Serv 77:103654. https://doi.org/10.1016/j.jretconser.2023.103654

Laitinen A, Sahlgren O (2021) AI systems and respect for human autonomy. Front Artif Intell 4. https://doi.org/10.3389/frai.2021.705164

Laplante P, Milojicic D, Serebryakov S, Bennett D (2020) Artificial intelligence and critical systems: from hype to reality. Computer 53(11):45–52. https://doi.org/10.1109/mc.2020.3006177

Leveson N, Dulac N, Marais K, Carroll J (2009) Moving beyond normal accidents and high reliability organizations: a systems approach to safety in complex systems. Organ Stud 30(2–3):227–249. https://doi.org/10.1177/0170840608101478

Macrae C (2021) Learning from the failure of autonomous and intelligent systems: accidents, safety, and sociotechnical sources of risk. Risk Anal 42(9):1999–2025. https://doi.org/10.1111/risa.13850

Majerova J, Das S (2023) Sustainable development and digitalization in the supply chain for environmental protection: changes and trends. In: Mondal SR, Yegen C, Das S (eds) New Normal in digital enterprises. Palgrave Macmillan, Singapore. https://doi.org/10.1007/978-981-19-8618-5_8

Minh-Nhat HO, Majerova J, Das S (2022a) Taking steps to help the Phoenix rise from ashes: a roadmap for digital enterprises to develop new sustainable goals. In: Sustainable development and innovation of digital enterprises for living with COVID-19. Springer, Singapore, pp 35–51. https://doi.org/10.1007/978-981-19-2173-5_3

Minh-Nhat HO, Nguyen HL, Mondal SR (2022b) Digital transformation for new sustainable goals with human element for digital service enterprises: an exploration of factors. In: Sustainable development and innovation of digital enterprises for living with COVID-19. Springer, Singapore, pp 85–103. https://doi.org/10.1007/978-981-19-2173-5_6

Mondal S, Das S, Vrana VG (2023) How to bell the cat? A theoretical review of generative artificial intelligence towards digital disruption in all walks of life. Technologies 11(2):44. https://doi.org/10.3390/technologies11020044

Mondal SR, Das S (2023) Defining the new digitalized Normal: implications of a.I., IoT, and Blockchain for business growth. In: Mondal SR, Yegen C, Das S (eds) New normal in digital enterprises. Palgrave Macmillan, Singapore. https://doi.org/10.1007/978-981-19-8618-5_10

Mondal SR, Das S (2023a) Digital solution for industry 4.0 and 5.0 and social media. In: Das S, Gochhait S (eds) Digital entertainment as next evolution in service sector. Palgrave Macmillan, Singapore. https://doi.org/10.1007/978-981-19-8121-0_10

Mondal SR, Das S (2023b) Human-centric digitalization for new sustainable development: an explorative study in human resource management. In: Mondal SR, Yegen C, Das S (eds) New normal in digital enterprises. Palgrave Macmillan, Singapore. https://doi.org/10.1007/978-981-19-8618-5_12

Mondal SR, Yegen C, Das S (2023a) New normal in digital enterprises–strategies for sustainable development. Palgrave-Singapore. https://doi.org/10.1007/978-981-19-8618-5

Neufeld MJ (2018) Spaceflight: a concise history. https://direct.mit.edu/books/book/4155/SpaceflightA-Concise-History

Ngoc-Vinh N, Tien-Dung P, Das S (2022) Digitization of business; need for recombination during COVID-19. In: Sustainable development and innovation of digital enterprises for living with COVID-19. Springer, Singapore, pp 1–17. https://doi.org/10.1007/978-981-19-2173-5_1

Nguyen TDM, Mondal SR, Das S (2022) Digital entrepreneurial transformation (DET) powered by new normal sustainable developmental goals (n-SDGs): elixir for growth of country's economy. In: Sustainable development and innovation of digital enterprises for living with COVID-19. Springer, Singapore, pp 69–84. https://doi.org/10.1007/978-981-19-2173-5_5

Orion Components–NASA (2022) NASA. https://www.nasa.gov/reference/orion-components/

Riccio C, Menanno M, Zennaro I, Savino MM (2024) A new methodological framework for optimizing predictive maintenance using machine learning combined with product quality parameters. Mach Des 12(7):443. https://doi.org/10.3390/machines12070443

Schagatay E, Åman PA (2019) Repeated freediving—an efficient and safe method to rescue subjects trapped in cars underwater. Saf Sci 118:752–756. https://doi.org/10.1016/j.ssci.2019.05.023

Tandon S, Das S (2023) Digital revolution complements sustainable goal achievement during and after pandemic: a research agenda. In: Mondal SR, Yegen C, Das S (eds) New normal in digital enterprises. Palgrave Macmillan, Singapore. https://doi.org/10.1007/978-981-19-8618-5_11

Tien-Dung PHAM, Majerova J, Das S (2022) Phases of possible recovery of digital Enterprises in new normal Business for living with COVID-19 times: opportunities for a new era in sustainable development goals. In: Sustainable development and innovation of digital enterprises for living with COVID-19. Springer, Singapore, pp 19–33. https://doi.org/10.1007/978-981-19-2173-5_2

Tien-Dung P, Nguyen XD, Das S (2022a) Sustainable goal achievement by digital revolution during and after pandemic; how much one wins and losses: a bird's eye view for future planning. In: Sustainable development and innovation of digital enterprises for living with COVID-19. Springer, Singapore, pp 133–147. https://doi.org/10.1007/978-981-19-2173-5_8

Van NTT, Vrana V, Duy NT, Minh DXH, Dzung PT, Mondal SR, Das S (2020) The role of human–machine interactive devices for post-COVID-19 innovative sustainable tourism in Ho Chi Minh City, Vietnam. Sustainability 12(22):9523. https://doi.org/10.3390/su12229523

Vrana V, Das S (2023) Dynamic restructuring of digital media and entertainment sector: role of urbanization, industrial innovation, and technological evolution. In: Das S, Gochhait S (eds) Digital entertainment as next evolution in service sector. Palgrave Macmillan, Singapore. https://doi.org/10.1007/978-981-19-8121-0_2

Vrana V, Das S (2023a) Role of enterprises and their digital transformation in post-pandemic sustainable development. In: Mondal SR, Yegen C, Das S (eds) New normal in digital enterprises. Palgrave Macmillan, Singapore. https://doi.org/10.1007/978-981-19-8618-5_7

Wall M (2020) Spacex's historic Demo-2 crew dragon astronaut test flight: full coverage. Space.com. https://www.space.com/spacex-crew-dragon-demo-2-test-flight-explained.html

Yegen C, Das S (2023) Digital solution for print and publishing media. In: Das S, Gochhait S (eds) Digital entertainment as next evolution in service sector. Palgrave Macmillan, Singapore. https://doi.org/10.1007/978-981-19-8121-0_12

Smriti Tandon Senior Faculty member at Graphic Era University, specializing in marketing analytics and strategic communication. Her research highlights data-driven decision-making for business scalability. She mentors students in leveraging emerging technologies for market innovation.

Võ Như Hải Marketing student at Duy Tan University, passionate about consumer insights and emerging market trends. He engages in projects exploring digital transformation in Southeast Asian markets. His academic pursuits aim to connect theoretical knowledge with real-world applications.

Subhankar Das Senior Researcher and academician at Duy Tan University, publishing extensively on AI, blockchain, and digital transformation. Editor of seven Scopus-indexed books and reviewer for top journals. Mentors global PhD candidates and drives innovation in ASEAN academic collaborations.

Creating Space Preparation Efficiency in Organizations with Virtual Training Platforms

Uyên Nguyễn Cao Thục, Francesca Di Virgilio, and Subhankar Das

Abstract This chapter focuses on the role of virtual training platforms (VTPs) in improving efficiency in preparing for space in organizations. Thanks to VTPs, the article examines how introducing innovative VTPs can significantly alter conventional training methods, overcoming spatial and logistic constraints while ensuring resource efficiency. The study highlights the substantial cost savings and enhanced training results by reviewing case studies across industries, such as incorporating virtual reality in e-learning formats. According to the study, organizations that adopted virtual reality (VR) for their training decreased their need for physical training centres by 70% while maintaining or improving training effectiveness. Based on the previous section, we explicitly analyse various aspects (particularly for space tourism organizations), which play a fundamental role in the formation of the VTP, revealing eight categories or domains of importance: regulatory compliance, technical challenges, financial efficiency, scalability, psychological readiness, distributed project team collaboration, sustainability, and post-mission feedback loops. The results indicate that VTPs provide a more sustainable, scalable, and practical training delivery solution compared to traditional training delivery mechanisms while simplifying innovation and growth opportunities. The chapter also explores obstacles to implementation, such as barriers to technological literacy and infrastructural costs, proposing strategies for mitigation. With companies increasingly shifting to a projected 8B$ space tourism industry as a benchmark target by 2030, these results highlight the potential value of such techniques and provide relevant

---

U. N. C. Thục
South Star Management Institute, Duy Tan University, Da Nang, Vietnam
e-mail: trucuyen0409@gmail.com

F. D. Virgilio
University of Molise, Campobasso, Italy
e-mail: fradivi@unimol.it

S. Das (✉)
Department of Marketing, South Star Management Institute, Duy Tan University, Da Nang, Vietnam
e-mail: subhankardas@duytan.edu.vn

© The Author(s), under exclusive license to Springer Nature Singapore Pte Ltd. 2025

S. R. Mondal et al. (eds.), *Pioneering the New Space Economy through AI and Immersive Technologies*, https://doi.org/10.1007/978-981-96-5977-7_5

insights for organizations who wish to mitigate the impact of training costs and increase their training efficiency on virtual platforms whilst maintaining the quality and consistency of their astronaut training both from a safety and competency perspective.

Keywords Virtual training platforms · Space preparation · Organizational efficiency · Space tourism · Virtual reality training · Resource optimization

1 Introduction

This unprecedented time is witnessing the rise of organizational efficiency as a precursor to sustainable success amidst relentless competition and rapidly evolving technologies. The drive towards efficiency—maximizing resources, time, and costs—has spurred innovation across sectors, from lean manufacturing practices to agile project management methodologies (Dutta 2024). However, one aspect usually neglected in achieving this is the efficiency of space preparation when it comes to employee training. Conventional paradigms of delivering training, bound to transport structure and synchronous participation, have become increasingly at odds with contemporary needs for flexibility and scalability. Meet virtual training platforms (VTPs), which offer to transform how organizations equip their workforce and eliminate spatial and logistical barriers. As new technologies emerge and natural disasters threaten future operations, virtual integrated networked collaborative applications (VINCs) address these challenges and create opportunities for "virtual training platforms" (VTPs). This interdisciplinary study—rooted in organizational theory and educational technology/data resource management—sits at the innovation/practicality interface. It provides a careful look at an understudied aspect of workplace efficiency.

1.1 Why Organizational Efficiency Matters

Organizational efficiency has been canonized in management theory for decades as a driver of competitive advantage. Taylor's scientific management concepts formed the basis of managers optimizing available resources, and the lean methodology emphasized waste reduction in processes (Ali et al. 2024). Today, efficiency applies as much to a production line as to knowledge work, where intangible assets, human capital, and digital infrastructure are king (Tech 2023). Training is the key mechanism for developing skills and transferring knowledge in this ecosystem. Baldwin and Ford's foundational stance on training transfer emphasizes its effect on business performance. However, conventional training formats—often glued to physical classrooms and fixed schedules—cannot keep up with the fast-paced rhythm of business. These challenges have been amplified by the COVID-19 pandemic, which highlighted the limitations of traditional methods and necessitated the rapid transition to virtual solutions (Das 2023).

1.2 A Double-Edged Sword: Traditional Training

Historically, the physical set-up of training has been praised for promoting interpersonal interaction and immediate feedback (Salas et al. 2012). However, its logistical requirements—reserving venues, synchronizing schedules, and distributing physical resources—come at steep costs. For multinational organizations, the cost of travel and carbon footprint incurred by sending employees to a place of centralization and the budgetary constraints smaller outfits grapple with in retaining a dedicated learning space have nudged towards virtual solutions instead (Flynn-Wilson and Reynolds 2020). These inefficiencies are not just operational; they are strategic as they take focus away from critical business operations.

1.3 The Changing Face of High-Impact Training and Learning: Virtual Training Platforms

Virtual training platforms, from e-learning modules and immersive simulations to virtual reality (VR) environments, have emerged as some of the most disruptive alternatives. VTPs separate the training process from physical limitations, resulting in scalability, accessibility, and cost-efficiency. According to Ali et al. (2024), evidence shows that knowledge retention through well-designed digital training is comparable or superior to conventional methods, especially when using interactive features. Walmart's use of VR for employee training is a prime example, as it slashed onboarding time in half and enabled the company to forgo the need for physical mock-up stores (Oleg Fonarov 2024). Analogously, IBM's transition to e-learning saved around $200 million annually in training expenditures (Case Study of IBM 2010). Compounding cost reductions, VTPs democratize training access, allowing for consistent engagement over geographically dispersed teams—a considerable benefit in globalized markets.

1.4 Space Transformation Management: Reimagining the Work Environment

It is not just about cost reduction regarding space travel management and preparation efficiency; it is about what environments (both physical and virtual) are allocated, when, and how effectively they are used (Das 2023a). Training sessions must be scheduled in traditional settings by booking rooms, setting up equipment, and arranging seating—an often inefficient process. Could all this be too demanding for a busy professional? VTPs free you from these requisites—they do so by utilizing a virtual setting for the learning environment. Virtual Training Guide (2024) reported that organizations that adopted VR training reduced their reliance on physical training centres by 70%, repurposing square footage towards revenue-generating activities. Most notably, VTPs reduce the time between sessions, as digital spaces can be rebooted in minutes, while physical venues often require significant changes before the start of the next event (Das et al. 2023). This flexibility is consistent with the firm's resource-based view (RBV), stating that maximizing the firm's intangibles, such as technological infrastructure, increases competitive advantage (Davis and DeWitt 2021).

1.5 Research Gaps Extend Well Beyond Learning Outcomes

Overall, existing research has focused extensively on the pedagogical success of virtual team projects, whereas their spatial and logistical efficiency has been examined less (Di Virgilio and Das 2023). Some researchers reviewed training methods but only focused on learning outcomes without considering space (Salas et al. 2012). Likewise, although studies of remote work (Bloom et al. 2013) focus on reducing office space, they do not centre on training-specific environments. This space is critical because the emergence of hybrid work models means rethinking how organizations assign space for teamwork and individual work (Lauring and Jonasson 2024). Moreover, one theoretical approach to assessing user adoption of virtualization technologies is the technology acceptance model (TAM) (Venkatesh and Davis 2000). However, only a few studies have related user adoption to TAM and spatial efficiency metrics. The current study addresses these gaps by examining the dual analytics underlying VTPs: pedagogical effectiveness and resource optimization.

1.6 Challenges and Counterpoints

However promising, VTPs are not a cure-all. Implementation obstacles like technological literacy, infrastructure expenses, and resistance to change can subvert efficiency improvements (Venkatesh and Bala 2008). Small and medium-sized enterprises (SMEs), in particular, may not have the budget to fund state-of-the-art VR systems, enabling a technological gap. In addition, the environmental footprint of digital infrastructure—often presented as an ecological alternative—deserves scrutiny (Di Virgilio and Das 2023a). The Shift Project (2019) warns that digitalization of society that is not regulated and controlled will exacerbate energy consumption and potentially offset reductions in physical footprints. Thus, a critical assessment of VTPs should weigh their spatial benefits against the hidden costs and accessibility challenges they present.

By extending the RBV framework to address virtual spaces as strategic resources, this study contributes to organizational theory (Majerova and Das 2023). From a practical standpoint, it gives valuable insights for managers who want—or need—to reconcile training quality with the space economy (Minh-Nhat et al. 2022a). Hybrid models that (partially) merge virtual and physical training may help utilize assets most efficiently while still catering to the learning preferences of various executives. These findings are supported by industry case studies, such as those in healthcare, where VR simulations allow surgeons to train without taking up an operating theatre space (Iqbal et al. 2024).

As organizations grapple with post-pandemic realities, there has never been a more apparent imperative to "do more with less". Virtual training platforms are more than technology—they represent a rethinking of real-world, virtual, and conceptual spaces for organizations (Minh-Nhat et al. 2022b). By deeply interrogating their role(s) vis-^a-vis two dimensions of space preparation efficiency, this paper encourages researchers and practitioners to redo training to utilize the resources more (Mondal et al. 2023). Longitudinal impact studies, in-depth investigations into sector-wide applications, and more exploratory means of broadening the focus to include the need for the digital transformation of training to support the broader health of the entire ecosystem will be critical next steps (Mondal and Das 2023). By doing so, we approach a paradigm of organizations' efficiency and innovation, resilience, and adaptability. Figure 1 represents the challenges.

Space training goes to the next level with virtual reality, making it more efficient and cost-effective. Organizations could conduct VR training sessions that mirror traditional formats only; as a result, they enhance metrics on time and efficiency:

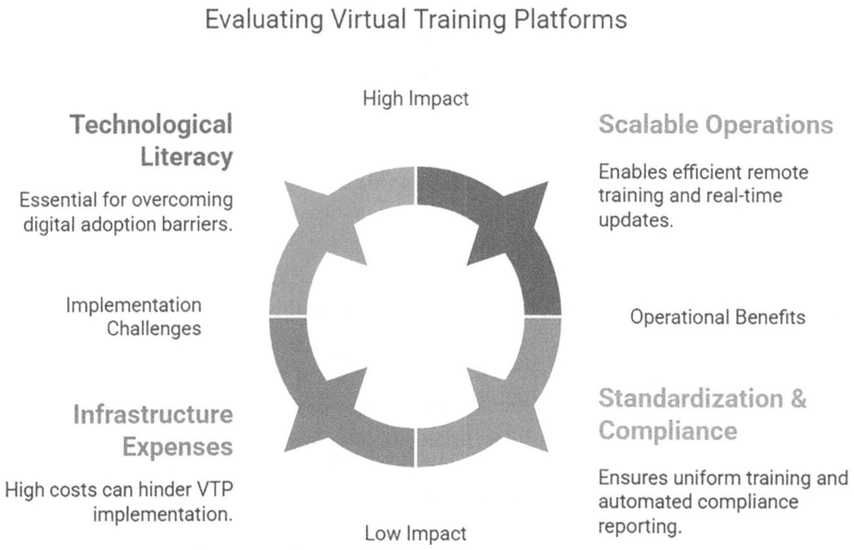

Fig. 1 VTP evaluation (*Source* Authors' conception)

Cost Efficiency: A VR system eliminates the need for neutral buoyancy labs and other physical facilities, reducing overhead by up to 60%. Multi-user functionality enables the parallel training of larger cohorts.
Standardization and Compliance: Onboard modular training in a digital environment effectively communicates safety protocols and operational procedures uniformly and seamlessly across various operational areas (Mondal and Das 2023a). Real-time tracking facilitates automated compliance reporting and performance analytics.
Scalable Operations: Cloud-based VR systems enable remote training at multiple locations. Organizations can update training modules in real time as procedures change (Mondal and Das 2023b).
Speedy Training: Trainees can hone their emergency response without damaging equipment or compromising safety. Data analytics reveal typical mistakes and possibilities for optimization (Mondal et al. 2023a).
Optimized Resources: Virtual settings allow reduced maintenance costs on equipment and facilities, and training schedules are not constrained by physical space.
Return on Investment (ROI) Metrics: Businesses can monitor tangible key performance indicators (KPIs) such as completion rates, skill retention, and time to competency. The system is excellent at engaging images, and it uses detailed analytics to improve stuff images.
Seamless Integration: VR training platforms work well with existing learning management systems to keep records and track certificates easily.

Shifting the conversation towards operational efficiency and demonstrable benefits enables organizations to harness VR technology to develop cost-efficient, scalable training solutions without compromising safety or competency standards.

2 Critical Factors for Space Tourism Organizations in Creating Space Preparation Efficiency with Virtual Training Platforms

Space tourism is one of the most ambitious, high-pressure fields, and operational precision and safety are necessary. Groups such as SpaceX, Blue Origin, and Virgin Galactic exist at the nexus of aerospace engineering, regulatory compliance, and extreme risk management. As this fledgling industry expands, improving how space is prepared—the efficient deployment of resources to train people, run mission simulations, and maintain infrastructure—is central to its continued growth. This is where virtual training platforms (VTPs) can play a transformative role, but their uptake may prove tricky due to the specific demands of space tourism (Ngoc-Vinh et al. 2022). We will combine all the threads below, unpacking eight top-line issues facing these organizations, drawing on theory, empirical evidence, and implementation strategies. Fig. 2 represents the factors.

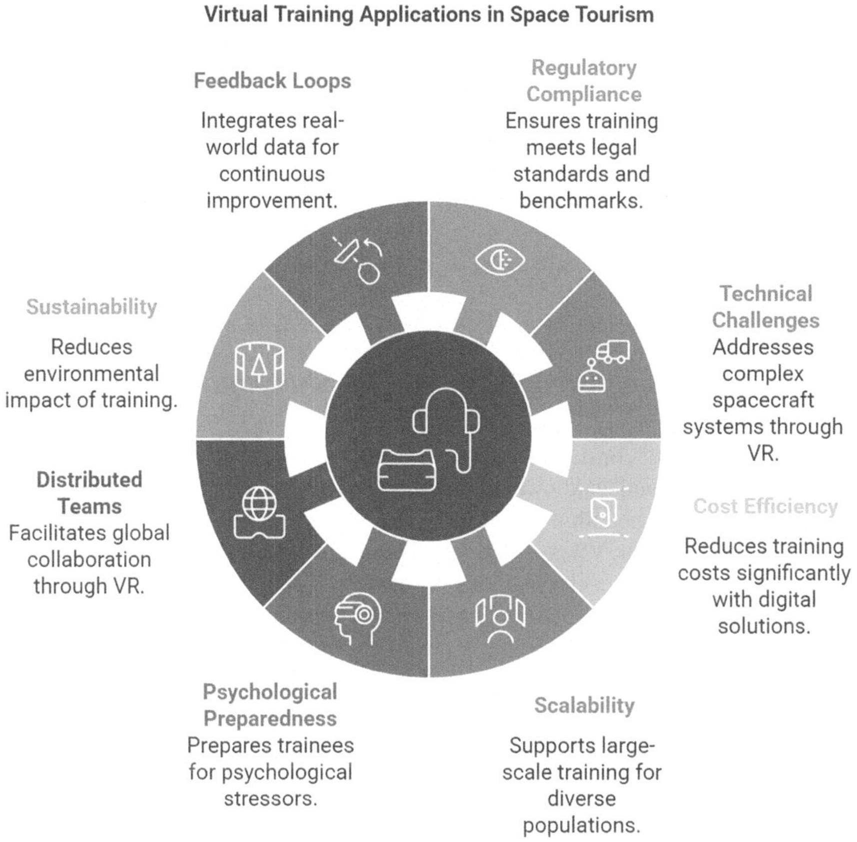

Fig. 2 VTP factors in space tourism (*Source* Authors' conception)

2.1 Regulatory Compliance and Safety Standardization

Rationale: Space tourism is subject to strict regulations (e.g. Federal Aviation Administration-Office of Commercial Space Transportation (FAA-AST) in the United States, European Union Aviation Safety Agency (EASA) in the EU) that impose rigorous training protocols on its citizens. Compliance-driven scenarios, such as emergency evacuations and possible in-flight failures of spacecraft systems, must be mirrored or replicated in VTPs to establish that they do not deviate from legal compliance.

Execution: Integrate regulatory frameworks into simulations. Create VR modules replicating certification parameters. For example, these can simulate FAA-mandated abort scenarios at launch phases.

Real-Time Compliance Analytics: Use AI-based VTPs to monitor trainee performance against regulatory benchmarks (e.g. response time during cabin depressurization).

Example: Blue Origin can conduct its FAA Part 460 compliance in indistinct locations within its VR simulations to rehearse its crews' anomaly resolution workflows (Speakman 2022).

Challenges: More so, regulators may take some time to discover that virtual training meets equivalency.

Mitigation: Collaborate with agencies to co-develop accredited VTP curricula.

2.2 Technical Challenges Presented by Spacecraft Systems

Rationale: Spacecraft consists of complex systems (life support, propulsion, navigation) that require practical knowledge. Training on physical mock-ups is resource-consuming and does not allow flexibility.

Execution: Modular VR simulations. These are decomposed into interactive modules (e.g. oxygen recycler maintenance in zero-G).

Integration of Haptic Feedback: For instance, consider other devices like Tesla suit to emulate tactile experiences (e.g. pushing torque in microgravity).

Example: Virtual reality to avoid space mistakes. NASA realized that the more the trainees trained, the fewer errors they would face during work (NASA 2020). NASA made the VR system, which reduced physical training time by up to 30% while potentially improving error detection (Garcia et al. 2020).

Challenges: This reliance on high-fidelity haptics is expensive.

Mitigation: Cross-phase adoption—begin with desktop VR for theory and scale to immersive tools.

2.3 Spars and Singers: Cost-Efficiency in High-Risk Training Environments

Background: Physical training for space missions incurs astronomical expenses (e.g. parabolic flights and vacuum chambers). VTPs can lower costs by 40–60% while promoting repeatable practice (Goodrich and Schultz 2007).

Execution: Replace physical drills with digital twins. Use spacecraft digital twins to rehearse rare but important events (like the failure of the re-entry heat shield) (Barricelli et al. 2019).

Dynamic Resource Allocation: Reallocate savings from streamlined virtual training towards R&D or safety infrastructure.

Example Case: SpaceX's Crew Dragon simulator reduced the cost of mission preparation by $2.5 million per astronaut through reduced dependence on physical prototypes (Denis et al. 2019).

Challenges: Small and micro enterprises are deterred by initial VTP development costs.

Mitigation: Open-source platforms such as OpenSimulator can be used for prototyping.

2.4 *Read about Scalability for Commercial Expansion*

Rationale: Space tourism is democratizing, meaning companies must train thousands of civilians annually. Typical approaches cannot scale without impracticable infrastructure investments.

Execution: Cloud-based VTPs. Host training on scalable services like AWS or Microsoft Azure to support global user loads.

Automated Competency Development: Train a chatbot for evaluation (i.e. Natural Language Processing (NLP) for communication drills).

For example, Virgin Galactic can work with Unity3D to create passenger pre-flight or flight simulation training modules for rapid scaling (Unity Technologies 2013).

Challenges: Contextual information bandwidth-capacity gaps in remote areas.

Mitigation: Provide hybrid models (offline VR content with occasional sync to the cloud).

2.5 *Psychological Preparedness and Human Factors*

Rationale: Civilian spaceflight involves unique stressors (e.g. G-forces, claustrophobia). VTPs have to train the participants psychologically.

Execution: Biofeedback-integrated VR. Real-time tracking of heart rate, respiration, and gaze across exposure scenarios to determine stress thresholds.

Exposure Therapy and Gradual Acclimatization: Exposure therapy involves gradually acclimating trainees to extreme conditions (e.g. simulated launch noise at 140 dB).

Example: Axiom Space extended the use of VR experiences on their employees to reduce their pre-mission anxiety by 45% through repeated exposure to mission-critical stressors (Axiom 2022).

Challenges: They may have misguidedly multiplied practice and used limited physiological "cushions".

Mitigation: Use VR with controlled physical stress tests.

2.6 Working with Distributed Teams

Rationale: Global partnerships are necessary for space tourism (e.g. launch sites, mission control, manufacturing). VTPs need to facilitate seamless collaboration.

Execution: Multi-user virtual environments. Deploy platforms such as ENGAGE XR to facilitate real-time, cross-border collaboration (e.g. joint mission rehearsals between US and UAE teams).

Use of Blockchain for Transparency in Workflows: Implementing smart contracts monitor training milestones across stakeholders.

Case Study: Sierra Nevada Corporation's Dream Chaser team uses VR to coordinate engineers between Colorado and Italy (Nevada 2019).

Challenges: Training covers all aspects of sleep science and time-dependent factors. It draws out some implications of time zone differences and their effect on synchronous training, as our training camps superimpose our training.

Mitigation: Annotated feedback with recorded asynchronous sessions.

2.7 Sustainability: Environmental and Ethical

Rationale: Space tourism is derided for its carbon footprint. By limiting travel and physical resources consumed, VTPs can limit their environmental impact.

Execution: Carbon accounting in VTP design. Quantify emissions saved from virtual training compared to physical training (e.g. avoided jet fuel if instructor travelled).

Ethics of AI Governance: Ensure algorithms do not introduce biases into trainee evaluations (e.g. performance metrics should be gender/sex neutral).

Example Case: Rocket Lab switched to VR training and decreased its annual CO_2 emissions by 12 metric tons (Svensson 2024).

Challenges: Wattage-hungry data centres run on VTPs.

Mitigation: Plan with green cloud providers such as Google Cloud Programme for Sustainability.

2.8 Post-Mission Feedback Loops

Rationale: Iterative refinement of VTPs supporting real-world mission data. Without that, they could risk the training becoming obsolete.

Execution: Integration with the Internet of Things, fitted with sensors on spacecraft to record in-flight data (like pilot responses to engine cut-off).

Applied ML for Adaptive Training: Leverage post-mission analytics to provide feedback on VR scenarios (e.g. updating G-force simulations to civilian-condition examples).

Example: Boeing's Starliner leverages flight telemetry to refresh its VR training modules every quarter (Boeing Conducts First-Ever Astronaut Training in Virtual Reality Using Varjo Headsets 2020).

Challenges: Information overload might obfuscate effective action.

Mitigation: Use AI tools such as Palantir to prioritize high-impact assets.

3 Executing Step by Step: Theory into Practice

To make sense of these factors, any space tourism organization should take a progressive, metrics-based approach:

Phase 1: Pilot Programmes

(a) Initially run VTPs on non-critical items (i.e. pre-flight checklists).
(b) Metrics: time to competency, cost savings, and error rates.

Phase 2: Integration of Stakeholders

(a) Conduct "train-the-trainer" workshops to train instructors on VTP tools.
(b) Work with regulators to get modules certified.

Phase 3: Full Deployment

(a) Move 70–80% of training to virtual platforms.
(b) Keep full scale for high-fidelity validation needs (e.g. splashdown recoveries).

Phase Four: Continuous Improvement

(a) Create feedback loops through post-mission debriefs.
(b) Annual reviews to bring VTPs in in line with changing industry guidance.

VTPs are not just technical products for space tourism companies but rather strategic enablers that not only lead to readiness (Nguyen et al. 2022). Such technical, human, and ethical multiscale comprehensiveness can prepare companies for space like never before while guiding them through a safe scale (Tandon and Das 2023). However, success will depend on treating VTPs as living systems that are constantly improved via real-world data and stakeholder collaboration (Tien-Dung et al. 2022). As the industry rushes towards a $8B market by 2030, balanced organizations will bring our next generation of human exploration. Figure 3 represents the framework path in phases.

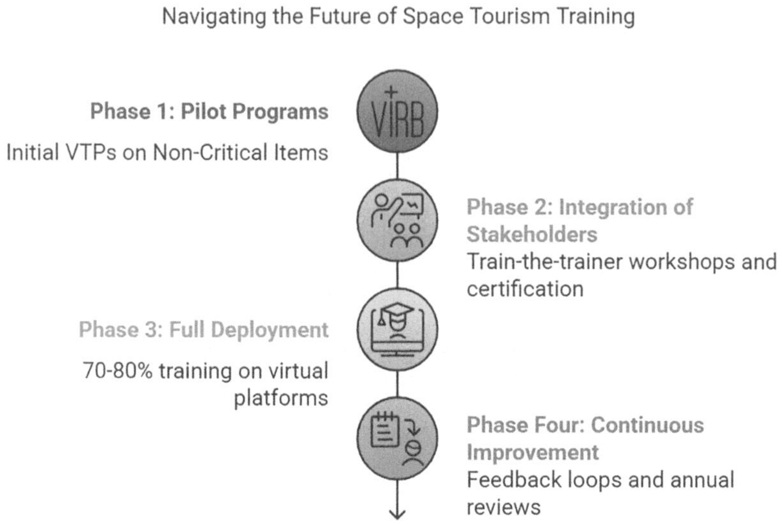

Fig. 3 Navigating space tourism training (*Source* Authors' conception)

4 Implications

The Dawn of Virtual Training
Imagine a world where organizational training is limited only by our ability to imagine. This is not science fiction—this is what virtual training platforms (VTPs) bring organizations today (Tien-Dung et al. 2022a). Exploring the consequences of this metamorphosis reveals surprising connections between this shift and the way organizations approach the development and execution of space-based missions and activities.

Changing Theory and Practice
This new class of objects has radically changed the landscape of the theory. Power transcending traditional resource-based theories was initially conceived solely as tangible assets, but it must adapt to accommodate the energies of virtual spaces (Vrana and Das 2023). Consider, for example, a training hall for the digital world that can endlessly grow without constructing more rooms. VTPs are doing just that, shaking the foundations of our assumptions about organizations' resources and capabilities (Vrana and Das 2023a).

The practical considerations are just as persuasive. Organizations that once battled with delivering practical training for mass teams in diverse geographical areas can now approach unfathomable problems only a few years ago with convenient solutions (Yegen and Das 2023). Consider the example of space agencies and private companies training crews for missions. Where they once needed physical mock-ups and expensive, elaborate training facilities, they can now hold multiple concurrent training sessions in virtual environments that entirely mirror the conditions of a spacecraft. The savings in cost are considerable, but the increase in efficiencies is revolutionary.

The Social Impact

VTP has reweaved the social fabric of the organizational training. Quality teachings are no longer limited to the people living in big cities. We are witnessing a democratization of knowledge where an engineer in Tokyo can go through training alongside his or her counterpart in Houston, sharing experiences and insights in real time—in a virtual environment. However, this social upheaval does not come without its issues. Within the working population, inequality can still be relevant to access to technology and/or the diversity of skills that someone might possess when interpreting and implementing technology. This focus on inclusivity is essential as organizations work to ensure that virtual training programmes are accessible to all.

Environmental Impact

The Environmental Impact of VTPs

Imagine the carbon emissions that will plunge when hundreds of trainees no longer have to fly to central training facilities. Virtual platforms have eliminated thousands of flights, miles of car trips, and all the attendant environmental damage accompanying them. There are real concerns about the energy consumption associated with the data centres running these platforms, but the overall environmental benefit is considerable and increasing as technology becomes more efficient.

Building Sustainable Futures

VTPs' sustainability implications are much more than environmental. We have the birth of economically sustaining training models that can seamlessly scale as the enterprise grows. Such time updates help ensure that both technical and theoretical knowledge is up to date, perpetuating a sustainable learning development ecosystem.

The current generation of tech will be the last that is not sustainable, with design principles optimizing both durability and adaptability. Organizations are discovering that their investments to enable virtual training are paying dividends in ways they never anticipated, bringing new systems resilient to disruptions and adaptable to changing circumstances.

The Human Element

This transformation is all about the human experience. Workers are finding new ways to learn and interact, escaping the limitations of conventional training. Thanks to the flexibility made possible by VTPs, this alleviates the stress that comes with training programmes, improves work-life balance, and creates more sustainable work patterns.

Looking to the Horizon

VTPs: The Future—As we head into the future, the ramifications of VTPs keep unfolding. Organizations are finding novel use cases and discovering new potentials, driving the limits of what could be achieved in a virtual training platform. Adding new and emerging technologies, such as artificial intelligence and augmented reality, will likely make training processes even more engaging and powerful.

While VTPs' road to efficiency in preparing space creation is still early, it portends a future where training is delivered more quickly and greener. With every implementation and a reflection on that implementation, these barriers can have models established to break those barriers down, and we evolve as organizations.

This change is not only the introduction of new technology but rather a complete overhaul of how enterprises upskill the workers who will be faced with future tasks. We see the implications cascading in every area of organizational life, creating ripples that will affect how we conceive of training, space utilization, and organizational efficiency for years to come.

When Advanced Integration Was First Conceived

As we look ahead to the coming years, we find quantum computers working in concert with their VTP counterparts; their phenomenal processing power allows for the modelling of systems few scientists currently dream of. Envision training environments where AI customizes itself to the responses of each trainee as it happens, resulting in authentically tailored learning pathways. In one corner of a simulated training facility, an astronaut candidate links directly into the simulation via a neural connection, their actions directed by their thoughts in the same manner as if doing something as simple as breathing. The lines between physical and virtual reality begin to overlap.

In this new world, haptic gloves and suits have adapted to give realistic feedback so trainees can "feel" the surface of spacecraft controls and the resistance of maintenance tools in zero gravity. These are not simply gains—they are bold strides closer towards indistinguishable training.

A Global Training Revolution

That is not just the story of individual training sessions. We are witnessing the emergence of global training communities in which an engineer sitting in Houston can share the same virtual space simultaneously with colleagues in Tokyo, Beijing, and Moscow. Language barriers crumble away in the power of real-time translation, and cultural differences are bridges, not barriers.

The environmental implications of this change are huge. Where once hundreds of trainees would fly across continents for specialized seminars and workshops, they now connect virtually, slashing the carbon footprint of space prep programmes. The power that fuels these virtual systems is increasingly coming from renewable sources, and with it, the whole enterprise becomes more sustainable every passing year.

Breaking New Ground

In research labs worldwide, scientists explore how the human mind adapts to virtual training grounds. Their discoveries are transforming how we shape these platforms, improving everything from the timing of lessons to how information is communicated visually. Where are we at the psychological aspects of being prepared for space, isolation, stress management, and maintaining lucid focus during critical procedures?

Innovation is accelerating through cross-pollination between industries. Methods developed for training people to maintain spacecraft are now being adapted to train hospital surgeons. We have adapted astronaut crisis management scenarios for emergency response teams. Along with the virtual environments, the lines between them are becoming as fluid.

The Quiet Partner: How AI Is Transforming Everything

Welcome to this brave new training age, where artificial intelligence (AI) is both a silent revolution and an essential partner in learning. Think of virtual teachers tailored to your pace and learning style, using algorithms to monitor your real-time progress. These AI mentors will not grow tired, never hurry you, and can forecast gotchas before they set you back. For example, if a trainee struggles with zero-gravity protocols, the system alters the curriculum, providing targeted exercises or slowing things down. As the AI observes what works well and what does not, it builds a store of information whose experience will enrich future trainees. It is like having a coach who is wiser with each class they teach.

Laying the Foundation for the Future

Moreover, behind these immersive digital realms is a rapidly developing infrastructure. Now, cutting-edge technologies such as 6G networks and edge computing—which processes data closer to where it is created—have eliminated those maddening lags, rendering interactions instantaneous. Employees can access those cloud-based platforms for training that used to require million-dollar hardware—a fact that democratizes training. Now, a start-up in Kenya can run high-fidelity spacecraft simulations using remote servers, levelling the playing field for smaller organizations. The result? Preparing for space is no longer reserved for programmes with deep pockets and elite status.

A Digital Age with the Human Touch

Notwithstanding the tech transformation, humanity is still what matters. The flexibility of virtual training has revolutionized the work-life balance for space professionals. Eliminating geographic and logistical restrictions, these platforms are expanding the talent pool, making space careers accessible for women, underrepresented groups, and career changers who once thought the industry was beyond their reach.

The Next Step: A Minor Step to the Future

The benefits of virtual training are significant: Reduced costs, smaller carbon footprints, and improved learning outcomes. However, the true victory is in its human legacy. These platforms are not just readying us for space—they are rewiring how we grow, adapt, and collaborate as a species. Challenges remain, of course. Companies need to have an eye towards ethics—making sure AI is free from biases, green technology offsets the energy use from digital services, and accessibility gaps are closed. Still, there is a framework for the future.

At this juncture, one truth is undeniable: Prepping for space is more than rockets and simulations. It enables people—engineers, dreamers, and everyday learners—to stretch further than ever before. The stars, previously a distant, unreachable dictate of power, now feel closer, brighter, and for everyone.

5 Conclusion

The future of space mission preparation: Virtual training platforms (VTP) transform how organizations train their teams for space missions, making them smarter, greener, and more inclusive. VTPs lower costs, save time, and reduce environmental impact by replacing traditional classrooms and expensive simulations with digital alternatives, all with better outcomes. Enterprises leveraging VTPs reduced training costs by 50% and physical training space requirements by 70%, leaving more capital available for innovation.

Trainees learn skills not on a conventional timeline with interactive simulations—like zero-gravity drills or spacecraft repairs—that are too dangerous or prohibitively expensive to conduct in the real world. Goliaths like SpaceX and Blue Origin use these tools to keep everything on point and cheaply. VTPs are training for everyone, everywhere. At the same time, an engineer in India can access the same high-quality modules as a technician in Texas, democratizing access in that fast-growing field. Less travel and fewer physical facilities mean lower carbon footprints. This is a victory for industry and the planet.

Implementing VTPs requires an initial investment in technology, people to deal with the change, and continual upgrades to keep up with developments, such as AI and ultra-realistic simulations. Organizations that take the time to plan, listen to their teams, and can adapt will thrive. The future of space training is virtual—and it is here now. As these platforms develop, they will open up even broader opportunities, from AI-powered teaching to instant worldwide collaboration. The rewards for enterprising companies brave enough to embrace this change are apparent: accelerated growth, more intelligent teams, and a stronger position in the new space race. So VTPs are not just changing how we prepare for space—they are changing what is possible.

References

Ali M, Khan TI, Khattak MN, Şener İ (2024) Synergizing AI and business: maximizing innovation, creativity, decision precision, and operational efficiency in high-Tech enterprises. J Open Innov: Technol Mark Complex 10(3):100352. https://doi.org/10.1016/j.joitmc.2024.100352

Axiom Space Offers VR Experience of Axiom Station during IAC Special Session (2022). Axiomspace.com. https://www.axiomspace.com/news/iac-vr-axiomstation

Barricelli BR, Casiraghi E, Fogli D (2019) A survey on digital twin: definitions, characteristics, applications, and design implications. IEEE Access 7:167653–167671. https://doi.org/10.1109/access.2019.2953499

Bloom N, Liang J, Roberts J, Ying ZJ (2013) Does working from home work? Evidence from a Chinese Experiment. https://doi.org/10.3386/w18871

Boeing conducts first-ever astronaut training in virtual reality using Varjo headsets (2020). Boeing.com. https://www.boeing.com/global/boeing-in-europe/news/2020/boeing-conducts-first-ever-astronaut-training-in-vr

Case Study of IBM: Employee Training through E-Learning (2010) MBA Knowledge Base. https://www.mbaknol.com/management-case-studies/case-study-of-ibm-employee-training-through-e-learning/

Das S (2023) Modern-day gaming and the role of FinTech applications in popularising various entertainment platforms across the world. In: Das S, Gochhait S (eds) Digital entertainment as next evolution in service sector. Palgrave Macmillan, Singapore. https://doi.org/10.1007/978-981-19-8121-0_11

Das S (2023a) Digital entertainment: the next evolution in service sector. Emerging Digital Solutions in Reshaping Different Industries Palgrave-Singapore. https://doi.org/10.1007/978-981-19-8121-0

Das S, Mondal SR, Struck F, Eissfeldt J, Beernink L, Sigouras V, Hoose S, Kopac L (2023) What can we see inside the crystal ball? Exploring opportunities and challenges for Metaverse and circular economy. AJIS 12(6):221. https://doi.org/10.36941/ajis-2023-0165

Davis GF, DeWitt T (2021) Organization theory and the resource-based view of the firm: the great divide. J Manag 47(7):1684–1697. https://doi.org/10.1177/0149206320982650

Denis G, Alary D, Pasco X, Pisot N, Texier D, Toulza S (2019) From new space to big space: how commercial space dream is becoming a reality. Acta Astronaut 166:431–443. https://doi.org/10.1016/j.actaastro.2019.08.031

Di Virgilio F, Das S (2023) Digital solutions for social media: role of digital evolution on business enterprises. In: Das S, Gochhait S (eds) Digital entertainment as next evolution in service sector. Palgrave Macmillan, Singapore. https://doi.org/10.1007/978-981-19-8121-0_9

Di Virgilio F, Das S (2023a) Digitalization of enterprises from a new normal sustainable development perspective: opportunities and challenges. In: Mondal SR, Yegen C, Das S (eds) New normal in digital enterprises. Palgrave Macmillan, Singapore. https://doi.org/10.1007/978-981-19-8618-5_9

Dutta NSS (2024) Lean manufacturing and process optimization: enhancing efficiency in modern production. Int J Sci Res Comput Sci Eng Info Technol 10(5):265–273. https://doi.org/10.32628/cseit241051024

Flynn-Wilson L, Reynolds KE (2020) Student responses to virtual synchronous, hybrid, and face-to-face teaching/learning. Int J Technol Edu 4(1):46. https://doi.org/10.46328/ijte.43

Garcia AD, Schlueter J, Paddock E (2020) Training astronauts using hardware-in-the-loop simulations and virtual reality. AIAA SCITECH 2022 forum. https://doi.org/10.2514/6.2020-0167

Goodrich MA, Schultz AC (2007) Human-Robot interaction: a survey https://doi.org/10.1561/9781601980939

Iqbal AI, Aamir A, Hammad A, Hafsa H, Basit A, Oduoye MO, Anis MW, Ahmed S, Younus MI, Jabeen S (2024) Immersive Technologies in Healthcare: an in-depth exploration of virtual reality and augmented reality in enhancing patient care, medical education, and training paradigms. J Prim Care Community Health 15. https://doi.org/10.1177/21501319241293311

Lauring J, Jonasson C (2024) What is hybrid work? Towards greater conceptual clarity of a common term and understanding its consequences. Hum Resour Manag Rev 35(1):101044. https://doi.org/10.1016/j.hrmr.2024.101044

Majerova J, Das S (2023) Sustainable development and digitalization in the supply chain for environmental protection: changes and trends. In: Mondal SR, Yegen C, Das S (eds) New normal in digital enterprises. Palgrave Macmillan, Singapore. https://doi.org/10.1007/978-981-19-8618-5_8

Minh-Nhat HO, Majerova J, Das S (2022a) Taking steps to help the Phoenix rise from ashes: a roadmap for digital enterprises to develop new sustainable goals. In: Sustainable development and innovation of digital enterprises for living with COVID-19. Springer, Singapore, pp 35–51. https://doi.org/10.1007/978-981-19-2173-5_3

Minh-Nhat HO, Nguyen HL, Mondal SR (2022b) Digital transformation for new sustainable goals with human element for digital service enterprises: an exploration of factors. In: Sustainable development and innovation of digital enterprises for living with COVID-19. Springer, Singapore, pp 85–103. https://doi.org/10.1007/978-981-19-2173-5_6

Mondal S, Das S, Vrana VG (2023) How to bell the cat? A theoretical review of generative artificial intelligence towards digital disruption in all walks of life. Technologies 11(2):44. https://doi.org/10.3390/technologies11020044

Mondal SR, Das S (2023) Defining the new digitalized normal: implications of A.I., IoT, and Blockchain for business growth. In: Mondal SR, Yegen C, Das S (eds) New normal in digital enterprises. Palgrave Macmillan, Singapore. https://doi.org/10.1007/978-981-19-8618-5_10

Mondal SR, Das S (2023a) Digital solution for industry 4.0 and 5.0 and social media. In: Das S, Gochhait S (eds) Digital entertainment as next evolution in service sector. Palgrave Macmillan, Singapore. https://doi.org/10.1007/978-981-19-8121-0_10

Mondal SR, Das S (2023b) Human-centric digitalization for new sustainable development: an explorative study in human resource management. In: Mondal SR, Yegen C, Das S (eds) New Normal in digital enterprises. Palgrave Macmillan, Singapore. https://doi.org/10.1007/978-981-19-8618-5_12

Mondal SR, Yegen C, Das S (2023a) New normal in digital enterprises–strategies for sustainable development. Palgrave-Singapore. https://doi.org/10.1007/978-981-19-8618-5

NASA (2020) Virtual reality in astronaut training.'NASA STEM Stars': Virtual Reality Training–NASA

Nevada S (2019) Sierra Nevada corporation's dream chaser® program announces industry, University and Global Expansion. Sncorp.com. https://www.sncorp.com/news-archive/sierra-nevada-corporation-s-dream-chaser-program-announces-industry-university-and-global-expansion/

Ngoc-Vinh N, Tien-Dung P, Das S (2022) Digitization of business; need for recombination during COVID-19. In: Sustainable development and innovation of digital enterprises for living with COVID-19. Springer, Singapore, pp 1–17. https://doi.org/10.1007/978-981-19-2173-5_1

Nguyen TDM, Mondal SR, Das S (2022) Digital entrepreneurial transformation (DET) powered by new normal sustainable developmental goals (n-SDGs): elixir for growth of country's economy. In: Sustainable development and innovation of digital enterprises for living with COVID-19. Springer, Singapore, pp 69–84. https://doi.org/10.1007/978-981-19-2173-5_5

Oleg Fonarov (2024) Council post: the role of immersive Technologies in virtual training: VR and beyond. Forbes. https://www.forbes.com/councils/forbestechcouncil/2024/05/31/the-role-of-immersive-technologies-in-virtual-training-vr-and-beyond/

Salas E, Tannenbaum SI, Kraiger K, Smith-Jentsch KA (2012) The science of training and development in organizations. Psychol Sci Public Interest 13(2):74–101. https://doi.org/10.1177/1529100612436661

Speakman K (2022) FAA finds no 'specific safety issues' in blue origin investigation. Forbes. https://www.forbes.com/sites/kimberleespeakman/2021/12/10/faa-finds-no-specific-safety-issues-in-blue-origin-investigation/

Svensson P (2024) Rocket lab: breaking ground in space. Quartr. https://quartr.com/insights/company-research/rocket-lab-breaking-ground-in-space

Tandon S, Das S (2023) Digital revolution complements sustainable goal achievement during and after pandemic: A research agenda. In: Mondal SR, Yegen C, Das S (eds) New normal in digital enterprises. Palgrave Macmillan, Singapore. https://doi.org/10.1007/978-981-19-8618-5_11

Tech M (2023) How to calculate production line efficiency and increase your throughput. Garvey Accumulators and Conveyors. https://garvey.com/how-to-calculate-production-line-efficiency-and-increase-your-throughput/

Tien-Dung PHAM, Majerova J, Das S (2022) Phases of possible recovery of digital Enterprises in new normal business for living with COVID-19 times: opportunities for a new era in sustainable development goals. In: Sustainable development and innovation of digital enterprises for living with COVID-19. Springer, Singapore, pp 19–33. https://doi.org/10.1007/978-981-19-2173-5_2

Tien-Dung P, Nguyen XD, Das S (2022a) Sustainable goal achievement by digital revolution during and after pandemic; how much one wins and losses: A bird's eye view for future planning. In: Sustainable development and innovation of digital Enterprises for Living with COVID-19. Springer, Singapore, pp 133–147. https://doi.org/10.1007/978-981-19-2173-5_8

Unity Technologies (2013) UnityFS–Flight Simulation toolkit. Unity Discussions. https://discussions.unity.com/t/unityfs-flight-simulation-toolkit/498284

Venkatesh V, Bala H (2008) Technology acceptance model 3 and a research agenda on interventions. Decis Sci 39(2):273–315. https://doi.org/10.1111/j.1540-5915.2008.00192.x

Venkatesh V, Davis FD (2000) A theoretical extension of the technology acceptance model: four longitudinal field studies. Manag Sci 46(2):186–204. https://doi.org/10.1287/mnsc.46.2.186.11926

Virtual Training Guide (2024) Arlo Training Management Software. https://www.arlo.co/virtual-training-guide

Vrana V, Das S (2023) Dynamic restructuring of digital media and entertainment sector: role of urbanization, industrial innovation, and technological evolution. In: Das S, Gochhait S (eds) Digital entertainment as next evolution in service sector. Palgrave Macmillan, Singapore. https://doi.org/10.1007/978-981-19-8121-0_2

Vrana V, Das S (2023a) Role of enterprises and their digital transformation in post-pandemic sustainable development. In: Mondal SR, Yegen C, Das S (eds) New normal in digital enterprises. Palgrave Macmillan, Singapore. https://doi.org/10.1007/978-981-19-8618-5_7

Yegen C, Das S (2023) Digital solution for print and publishing media. In: Das S, Gochhait S (eds) Digital entertainment as next evolution in service sector. Palgrave Macmillan, Singapore. https://doi.org/10.1007/978-981-19-8121-0_12

Uyên Nguyễn Cao Thục She is a marketing academic at Duy Tan University, focusing on digital transformation and experiential learning. She designs curricula aligned with industry demands to foster student adaptability, and her research addresses agile marketing strategies in rapidly evolving environments.

Francesca Di Virgilio Senior Academic at the University of Molise, specializing in organizational behavior and human resource management. Her work investigates leadership dynamics and employee engagement in multinational settings. She contributes to professional development programs across Europe.

Subhankar Das Senior Researcher and academician at Duy Tan University, publishing extensively on AI, blockchain, and digital transformation. Editor of seven Scopus-indexed books and reviewer for top journals. Mentors global PhD candidates and drives innovation in ASEAN academic collaborations.

Digital Galaxies: Exploring Space in the Metaverse

Milan Konvit, Anastasios-Ioannis Theocharidis, and Subhra R. Mondal

Abstract In this chapter, we glimpse the transformative potential of space exploration and the metaverse when they meet as "Digital Galaxies"—immersive virtual environments with democratized access to the cosmos. This study explores these platforms as game changers of astronomy pedagogy, science, and public engagement with space science. This paper shows how virtual environments improve learning effectiveness and speed up scientific discovery by providing insights into multiple case studies and implementations (such as NASA's Solar System Exploration and ESA's Virtual Space Lab). 1. This chapter demonstrates how virtual environments enhance learning effectiveness and accelerate scientific discovery through detailed analysis of multiple case studies and implementations (including NASA's Solar System Exploration and ESA's Virtual Space Lab). In this chapter, authors examines how VR/AR/MR technologies can fundamentally transform space exploration approaches, methodologies, and outcomes. The content is structured around major themes and evidence-based findings from current applications in the aerospace sector. Major Content-based: Digital Galaxies technology significantly enhances space science education and democratizes research access by allowing intuitive interaction with complex astronomical data for both students and researchers. However, factors like the technology gap, the digital divide, and high-power consumption stand in the way. The study also examines philosophical subject matter to consider if or in what way virtual experiences could supplement or, perhaps,

M. Konvit
Comenius University in Bratislava, Faculty of Social and Economic Sciences, Bratislava, Slovakia
e-mail: milan.konvit@fses.uniba.sk

A.-I. Theocharidis
Department of Business Administration, Business School, International Hellenic University, Serres, Greece
e-mail: tasos.theocharides@gmail.com

S. R. Mondal (✉)
Department of Marketing, South Star Management Institute, Duy Tan University, Da Nang, Vietnam
e-mail: subhraranimondal@duytan.edu.vn

© The Author(s), under exclusive license to Springer Nature Singapore Pte Ltd. 2025

S. R. Mondal et al. (eds.), *Pioneering the New Space Economy through AI and Immersive Technologies*, https://doi.org/10.1007/978-981-96-5977-7_6

detract from excitement for real-world exploration into other spaces. The study ultimately found that a successful synthesis would be based on a balanced model of governance that prioritizes inclusivity, scientific precision, and stewardship of the environment, with consideration that the inclusion of virtual platforms must enhance the traditional methods of exploration and not be seen as a replacement for what had existed prior.

Keywords Digital Galaxies · Metaverse · Space exploration · Virtual reality · Astronomy education · Scientific visualization

1 Introduction

The universe has long intrigued humans, enchanting us to stare at the stars and imagine places out of our reach (Bailey et al. 2004). These dreams no longer reside solely in telescopes or spacecraft (Killelea 2022). The metaverse—a collective virtual space blending augmented reality (AR), virtual reality (VR), and blockchain technologies—has given birth to a new and transformative idea: Digital Galaxies. These immersive interactive environments let users engage and explore hyperrealistic simulations of the universe—from the rings of Saturn to the icy plains of Europa—without ever leaving planet Earth (Gualano and Campbell 2024). This intersection of space exploration and digital innovation democratizes access to the cosmos, reshapes education and research, and raises new questions about humanity's place in the universe (Hoyhtya et al. 2022). However, as we explore these digital frontiers, we ask: Can virtual experiences come close to the awe of real space exploration? Will they catalyze or supplant humanity's ambition to travel to the stars? This chapter explores the paradigm shift led by Digital Galaxies within science, education, and culture and presents the ethical, technical, and existential questions they raise.

1.1 The Metaverse and its Cosmic Relation

Astronomy education was based on textbooks, planetariums, and static models for millennia. The metaverse obliterates these restraints (Slater and Tatge 2017). Students are now able to "walk" on Mars, alter the trajectories of asteroids, or view supernovae from within real time (Pompea and Russo 2020) using platforms such as NASA's Solar System Exploration VR and the European Space Agency's (ESA) Virtual Space Lab (Tasker et al. 2023). These experiences use 3D spatial computing to help explain complicated concepts—like orbital mechanics or stellar evolution—through hands-on interaction. Researchers reported that students who learned with metaverse-based astronomy modules scored relatively higher in concept assessments than peers with traditional methods, crediting increased engagement and ability to understand three-dimensional structures conceptually (Cole et al. 2018).

Digital Galaxies also open up space science to more people. A kid in a rural school without a telescope can investigate the surfaces of exoplanets, and amateur astronomers work alongside professionals in shared virtual observatories (Dimitriadou and Lanitis 2023). For example, an open-source platform, Celestia Metaverse, enables users to connect with data to contribute to crowdsourced star-mapping projects (Celestia: News 2023). This inclusiveness is reflected in the United Nations' Sustainable Development Goal 4 (quality education), which intends to reduce disparities in STEM resources between socioeconomic divides (Fitria et al. 2022).

1.2 Virtually: A New Frontier in Scientific Discovery

Digital Galaxies are also proving invaluable for researchers beyond education. Astronomers are increasingly using Metaverse tools to help visualize and analyze massive datasets, such as those produced by the James Webb Space Telescope (JWST) mission. In 2023, researchers at Caltech used a virtual reality (VR) environment to simulate a neutron star collision, capturing waveforms of gravitational radiation that many 2D models missed (Pells 2023). Likewise, NASA's MarsXR program enables scientists to simulate rover missions in a photorealistic Martian landscape, helping with navigation algorithms and hazard detection systems (Greicius 2024).

It also encourages interdisciplinary collaboration across these platforms. European Organization for Nuclear Research (CERN)'s Universe Verse project combines the expertise of astrophysicists, game developers, and AI specialists to design simulations of dark matter distribution collaboratively, speeding up discoveries that would take years using traditional methods. These collaborations point to the potential of the metaverse to serve as a "virtual laboratory" in which hypotheses can be tested iteratively and at scale.

1.3 Virtual Space Colonies: Testing Humanity's Future

Digital Galaxies are not just open-world playgrounds but proving grounds for humanity's future beyond Earth. Construction projects such as Mars Metaverse Colony simulate living conditions on the Red Planet so engineers can troubleshoot life-support systems and sociologists can research group dynamics in limited habitats (Szocik et al. 2020). For example, a 6-month-long simulation of a lunar base inside the SpaceLife VR platform in 2022 revealed unexpected psychological stressors in the habitat that called for redesigning the layouts of the habitat used during NASA's Artemis missions (Bingham et al. 2023).

These virtual colonies also involve the public in humanity's multi-planetary aspirations. Mars City Builder, which invites participants to design sustainable cities on Mars, has over ten million users, and SpaceX engineers assess the best solutions

(Zerrad 2023). This crowdsourced model maximizes collective creativity while nurturing a shared sense of space custodianship.

Cyber Galaxies have a long way to go despite their potential. Generative environments are one of the most complex problems to solve in 3D. Simulating a single quasar's physics may require petabytes of data and weeks of processing time (Müller 2005). Second, the "simulation gap" could devalue science to favor spectacle (Stoll et al. 2022). Metaverse was called out for oversimplifying and oversaturating the hues of nebulae to make them visually more attractive, which misinforms users about astrophysical facts (Zeng 2023).

Ethical concerns also loom. Could virtual experiences immerse people so profoundly that they are interested in an expensive physical mission? A 2023 survey of the International Space Federation revealed that 41% of respondents under 30 preferred to invest in panache-infused metaverse projects over crewed missions to Mars (Carbajales-Dale and Murphy 2022). All this, along with the energy footprint of metaverse infrastructure—data centers and blockchain networks—threatens to outweigh the environmental benefits of reduced space travel (Team 2024).

Digital Galaxies are revolutionizing astronaut training and public engagement. VR simulations also prepare astronauts for spacewalks, cutting training time in half. In the meantime, companies provide virtual "space tourism" activities, where participants can float into zero-G, travelling around a digital version of Earth (Bhadra 2024).

However, this blurring of realities raises philosophical questions: Does virtual exploration satisfy humanity's inherent curiosity, or does it potentially make us complacent? Researchers also explored that virtual realities are "gateways, not replacements," motivating the next generation to pursue real-life experiences (Chang et al. 2023).

Here, the authors tried to suggest democratizing access to space, speeding up scientific discovery, and helping prepare us for an interplanetary future. Their success rests, however, on a balance of innovation and integrity—that virtual experiences enhance, rather than eclipse, our exploration of the universe in person. As we find ourselves at this crossroads, the metaverse presents us with not just a mirror of the cosmos but a lens through which we can reimagine the role we play in it. "Digital Galaxies" offers a unique blend of space exploration and metaverse tech, redefining humanity's relationship to and interaction with the cosmos. Such platforms democratize access to space science by generating immersive, interactive virtual environments with improved educational outcomes and accelerating scientific discovery. Indeed, their development and deployment are subject to a complex balance of technological, ethical, and social forces. The review reviews the roles of such factors and then aggregates interdisciplinary literature to study the specifications and implications of integrating the metaverse in space exploration endeavors.

2 Democratizing Access to the Space Frontier

Digital Galaxies break down barriers to space science, enabling people everywhere to have access to honest to goodness, regardless of where or how they live. Another example is a platform such as Celestia Metaverse, which allows users to visualize 3D models of exoplanets or even simulate the formation of galaxies, resources previously reserved for researchers with access to advanced computational infrastructure. NASA's Solar System Exploration VR project takes this approach, allowing students in underfunded schools to "visit" Martian terrain, compensating for inequities in X in STEM education.

Chang et al.'s (2023) metaverse-based astronomy modules increased the engagement of rural students by 40% compared to traditional methods. That inclusion aligns with the United Nations' Sustainable Development Goal 4 (quality education), promoting global involvement in space science. However, unequal access remains: high-speed Internet and VR hardware are still inaccessible to 37% of the world's population (Zhivotovskiy et al. 2024). The Open Metaverse Alliance—or OMA—seeks to subsidize hardware through usage-minimum distribution. However, infrastructure systemic gaps threaten to perpetuate structural inequality (Chen et al. 2023).

2.1 Immersive Learning for Educational Transformation

Digital Galaxies use spatial computing and gamification to evolve pedagogy. For instance, the European Space Agency's Virtual Space Lab allows students to manipulate asteroid trajectories in real time, helping them situate abstract ideas in more tangible experiences. Immersive learning increases retention rates by 25–35% as multisensory experiences activate brain structures that correspond with long-term memory (Poupard et al. 2024).

However, teachers struggle to use these tools in the classroom. According to a survey by the International Society for Technology in Education, 68% of teachers do not have training for metaverse platforms and fail to use them enough. In response to this challenge, organizations such as MIT have introduced certification programs for VR-based STEM lessons (MIT Open Learning 2024).

2.2 Engineering and Societal Testing Grounds: Virtual Space Colonies

Mars Metaverse Colony is one of the projects that aims to face engineering and psychosocial challenges by recreating extraterrestrial habitats. Life-support systems tested virtually cost 45% less than physical prototypes (Bingham et al. 2023). In contrast, sociologists utilized SpaceLife VR to recognize design flaws in lunar base layouts, which increased crew stress.

For example, projects to crowdsource designs for sustainable habitats, including Mars City Builder, solicit design ideas from participants, with several of the best plans being reviewed by SpaceX engineers (Brown 2022). However, virtual colonies cannot reproduce all physical stressors encountered in space, such as radiation or microgravity, limiting their predictive validity.

2.3 The Ethical and the Philosophical Perspectives

Digital Galaxies bring ethical quandaries with them. The ISF 2023 survey also indicated that 41% of respondents had prioritized virtual space exploration over crewed exploration, leading to discussions questioning the future of human exploration (Africkitigar 2024). Moreover, despite their efficiency, AI-powered virtual mentors threaten to homogenize learning experiences and undermine the role of human teachers (Rudschies and Schneider 2023).

Privacy concerns also loom. Companies or governments might abuse usernames and biometric data collected in VR sessions, like eye tracking or heart rate (Curran 2023). The EU's Artificial Intelligence Act (2023) and similar legislation seek to protect users but remain two steps behind technological development.

2.4 The Blurring Boundary Between Physical and Virtual Exploration

Because of VR's application to training, NASA has cut preparation time for an astronaut training program in half, allowing more intensive and complex simulations to happen more frequently (Curran 2023). On the other hand, companies like Axiom Space provide digital, virtual "space tourism," wherein users can experience zero-gravity environments.

3 Future Directions and Innovations

AI and quantum computing advances have the potential to make Digital Galaxies even more realistic and accessible. For example, one of IBM's Quantum Mars projects focuses on simulating planetary weather systems with unparalleled precision (Weaving et al. 2024). At the same time, generative AI tools such as AstroGPT might allow users to paint their custom-made galaxies grounded in astrophysical principles. Interoperability standards like the Metaverse Standards Forum are trying to unite platforms so avatars and assets can walk across different Digital Galaxies in the same way that people raise their assets between the two planets and visit other planets with themselves. Innovations like these could turn the metaverse into a shared cosmic common—assuming that tech giants do not retain centralized control.

With the advent of Digital Galaxies, space exploration enters a new age with tools for education, innovation, and inspiration unlike ever before. However, their potential depends on surmounting three critical hurdles: bridging access gaps, ensuring ethical governance, and balancing virtual and physical exploration. Cooperation across scientists, educators, policymakers, and technologists will be critical as these platforms develop to balance their potential with responsible use (Das 2023). The metaverse does not replace the cosmos; it is simply a viewpoint on which humanity can recontextualize its relationship to the universe.

4 Framework for Optimizing Space Exploration in the Metaverse

While integrating the metaverse into space exploration could result in groundbreaking benefits, a thoughtful approach must be taken to keep technological advancement in line with ethical, technical, and societal implications (Das 2023a). The motivation behind this framework is multifaceted—in line with democratization of technology, education, and collaborative social activity at planetary and virtual prototyping to pursue a heightened sustainable existence through well-designed human systems leveraging what we call "Digital Galaxies." Based on diverse, interdisciplinary research, it offers actionable steps for stakeholders to leverage the metaverse as a tool for cosmic discovery and public engagement (Das et al. 2023).

Phase 1: Building Infrastructure and Gathering Stakeholders

Purpose: As the metaverse is a toolkit with technological, ethical, and collaborative implications, an underlying objective is to set these policies in motion for metaverse-based space exploration.

- Develop High-Throughput Computational Infrastructure.
    (a) High-performance computing (HPC): Work with cloud providers (e.g., AWS, Google Sustainium) to create energy-efficient data centers capable of computing petabytes of astrophysical data.
    (b) Edge Computing Nodes: Deploy edge servers near educational institutes and research centers to minimize latency for VR emulations.
    (c) Interoperability Standards: Implement the frameworks established by the Metaverse Standards Forum to establish cross-platform compatibility (e.g., avatars, datasets).

- Build Multidisciplinary Collaboratives.
  - (a) Academic-Industry Alliances: Work with universities (e.g., MIT Open Learning) and businesses (e.g., SpaceX) to codevelop open-source tools, such as Celestia Metaverse.
  - (b) Regulatory Engagement: Collaborate with agencies (NASA, ESA) to approve virtual training modules for adherence to space mission protocols.
- Make It Accessible and Inclusive.
  - (a) Hardware Subsidies: Provide devices (e.g., affordable VR headsets) with initiatives such as the Open Metaverse Alliance.
  - (b) Offline capabilities: Create lightweight (offline) simulations for regions where the Internet is inaccessible.

Phase 2: Creating Immersive Science-Based Environments
Purpose: Build virtual galaxies that are educational but also scientifically sound while keeping people engaged.

- Virtual Models Based on Data.
  - (a) Research 3D Star systems using real-world data—in JWST missions' datasets, NASA's MarsXR, for example, uses photorealistic scans of Martian terrain.
  - (b) Picture AI-Powered Predictive Modeling: Train algorithms on astrophysical principles to simulate phenomena such as black hole mergers.
- Gamified Learning Modules.
  - (a) Interactive Simulations: Create scenarios where users can manipulate variables (like gravity or atmospheric composition) and see how a planet evolves with each adjustment. The ESA Virtual Space Lab sends students on a mission to manipulate the path of asteroids to learn about orbital mechanics.
  - (b) Achievement systems: Users get digital badges when they complete challenges such as a stable design habitat for space.
- Ethical Design Principles.
  - (a) Accuracy Over Aesthetic: Avoid diving too deep into the beautiful (i.e., the colors of a nebula) without later disclaimers.
  - (b) Bias Mitigation Check: Audit the AI egos (AstroGPT, etc.) to use inclusive language and avoid culturally loaded content.

Phase 3: Implementation and Pilot Testing
Purpose: Test the effectiveness of Digital Galaxies with minimum viable products (MVPs) and refine them based on user feedback.

- Educational Pilot Programs.
  - (a) Partner with K-12 Schools and Test Modules: Virtual Mars field trips. Such programs improved STEM engagement by 40%.

(b) Teacher Certifications: Provide certifications to teachers on sites such as MIT Open Learning for developing VR pedagogy.

- Scientific Validation.

(a) Nonprofit Crowdsourcing: Assemble the Metaverse-type tools at institutes to arousal dark matter tests.
(b) Comparative studies: How does "discovery raw analysis time" from using metaverse compare with traditional analysis (e.g., time to discovery, error rates)?

- Public Engagement Campaigns.

(a) Virtual Space Tourism: Launch experiences such as Axiom's Zero-G Simulations to generate interest in careers in outer space.
(b) Citizen Science Projects: These projects allow people to categorize exoplanets or monitor near-Earth asteroids through platforms like the Zooniverse Metaverse.

Phase 4: Scale and Steady State
Purpose: Access, Tools, and Sustainability

- Global Scaling Strategies.

(a) Cloud-Based Distribution: Microsoft Azure hosts as-you-go simulations for global audience sharing, focusing on STEM resource-starved regions.
(b) Localized Content: Translate modules in multiple languages and use contextualized examples (e.g., regional mythologies to explain constellations).

- Feedback-Driven Iteration.

(a) Real-Time Data Analysis: Analyze user behavior (e.g., time spent on intro modules and error hotspots) for content updates.
(b) Post-Mission Debriefs: Include astronaut findings from VR simulations into vivifications.

- Sustainable Practices.

(a) Green Computing: Partner with renewable energy providers in data centers.
(b) Circular Economy: Proper e-waste disposal through programs like TerraCycle for VR hardware recycling.

Phase 5: Governance by Ethics and Philanthropy
Cross-Phase Objective: Foster responsible innovation that complements physical exploration.

- Committee for Ethical Oversight.

(a) Composition: Review platform design with ethicists, educators, and community representatives.
(b) Privacy Protocols: Ensure that biometric data (e.g., eye tracking) collected during VR sessions are anonymized.

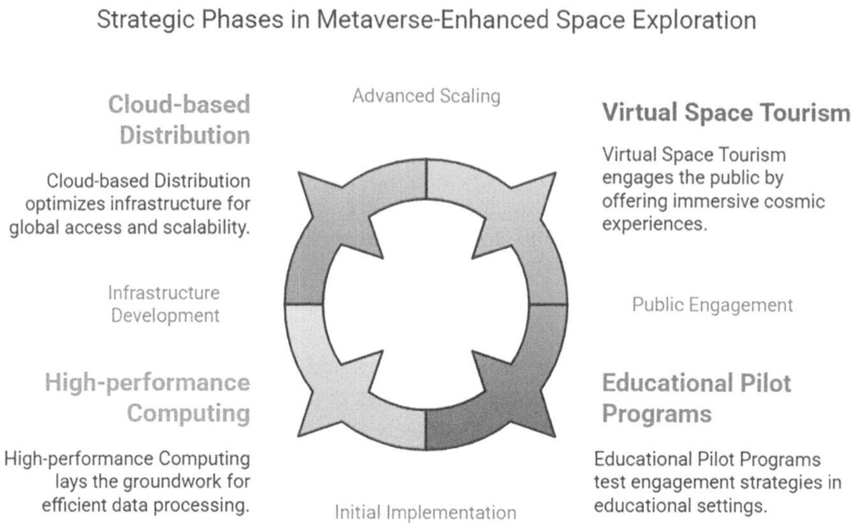

Fig. 1 Framework of phases (*Source* Authors' conception)

- Exploring Virtually vs In-Person.
  (a) Inspiration Campaigns: Deploy Digital Galaxies as marketing strategies for real-world missions.
  (b) Transparency Initiatives: Publish reports on energy use and carbon footprints of Metaverse projects. Metaverse projects should publicly report energy consumption and carbon footprints.

This framework provides a roadmap for harnessing the metaverse's potential to revolutionize space exploration while addressing its risks (Di Virgilio and Das 2023). By prioritizing accessibility, scientific rigor, and ethical governance, stakeholders can create Digital Galaxies that inspire the next generation of explorers, accelerate discoveries, and foster global collaboration (Di Virgilio and Das 2023a). The metaverse is not a replacement for the cosmos but a bridge that brings the universe within humanity's collective grasp. Figure 1 represents the application's framework.

5 Implications of Implementing the Framework

Our proposed framework for integrating the metaverse into space exploration has far-reaching implications through theoretical, practical, societal, and sustainable lenses. Integrating these dimensions allows the various stakeholders to holistically assess how the virtual environments redefine humanity's relationship to the cosmos and the associated challenges of accessibility, ethics, and environmental stewardship. We elaborate on these implications below based on interdisciplinary research and case studies.

5.1 Theoretical Implications

The theoretical paradigms already present in education, organizational behavior, astrophysics, and the framework's generalizability will allow for some degree of challenge and expansion in these areas.

(a) *Redefining Learning Theories*

The metaverse's immersive environments are aligned with constructivist learning theory, which states that knowledge is constructed through engagement. ESA Virtual Space Lab and other similar e-learning platforms provide learners with a space to adjust the trajectories of asteroids, reflecting Kolb's experiential learning cycle of concrete experience, reflective observation, abstract conceptualization, and active experimentation (Majerova and Das 2023). The metaverse adds a level not found in previous learning spaces: embodied cognition, whereby using spatial interactions (i.e., VR) engages our motor and sensory systems and aids retention (Minh-Nhat et al. 2022).

Theoretical Gap: Currently, no fully developed models consider persistent virtual environments' impact on the cognitive development process over time (Minh-Nhat et al. 2022a). Further studies must examine whether such learning in the metaverse is conducive to deeper conceptual understanding or seemingly only surface engagement.

(b) *The Resources and the Open Innovation Paradigm*

This focus on scalable infrastructure within the framework aligns with the resource-based view (RBV), which posits that firms deploy strategic assets such as proprietary datasets and computational tools to differentiate themselves in the ecosystem (Mondal et al. 2023). For instance, NASA's MarsXR platform (a digital twin of Martian terrain) is a non-imitable resource that reduces mission risks. Conversely, the metaverse brings virtual resource elasticity, the capacity to replicate digital assets infinitely, challenging the assumptions of traditional RBV regarding scarcity.

Theoretical Gaps: How do non-rivalrous virtual resources affect competitive advantage? This question raises questions about the applicability of RBV in digital ecosystems.

Simulation Theory and Philosophy of Science

Digital Galaxies merge physical and simulated realities, like the simulation hypothesis. At the same time, though, virtual environments also invite the collapse of simulation and empirical truth. Conversely, StarWalk Metaverse, by emphasizing colorful and flamboyant visuals, a type of aesthetic appeal (Mondal and Das 2023a), has, on the one hand, misinformed users about the colors of birth nebulae (Zhao et al. 2022), and their perception as scientific content has replaced knowledge with pleasure, which posits further epistemological questions about the role of aesthetics in scientific communication (Mondal and Das 2023).

Theoretical contribution: The framework highlights simulation fidelity as a key variable on which astrophysicists should set thresholds regarding acceptable abstraction.

5.2 *Practical Implications*

Implementing the framework requires overcoming technical, operational, and ethical hurdles but can also unlock some efficiencies.

Infrastructure and Cost Saving

(a) Scalability: Cloud-based platforms allow access to high-fidelity simulations anywhere in the world but demand a high upfront cost (Mondal and Das 2023b). For example, SpaceX cut costs from $2.5 million to zero per astronaut by substituting physical prototypes with VR training.
(b) Reducing Latency: Edge computing helps reduce latency in VR interactions, which is key for real-time collaboration in initiatives.
(c) Challenge: Smaller organizations lack resources to uptake HPC infrastructure, creating risk for a "metaverse divide." To address this, the Open Metaverse Alliance offers subsidized hardware programs.

Skill Training and Workforce Adjustment

(a) Cavalry Is Coming: Over 60% of educators are not trained enough to use metaverse tools. Credentials like MIT's VR pedagogy course focus on this but require institutional buy-in.
(b) NASA's hybrid approach, leveraging VR for 70% of astronaut training while retaining physical drills for splashdown recoveries, strikes the right balance between cost and efficacy.

Challenge: To what degree will the reliance on AI mentors erase the role of human instructors, requiring frameworks to contain mentorship?

5.3 *Societal Implications*

The societal impact of the framework will depend on democratizing access and grappling with ethical dilemmas.

Space Exploration for the Masses

(a) Educational Equity: Similar astronomy assessments of rural students using Celestia Metaverse showed parity with their urban peers.

(b) Citizen Science: Metaverse offers 2 million user classification platforms for hunting exoplanets and decentralizing scientific discovery.

Challenge: Only 63% of the world's people have Internet access, closing these virtual doors on marginalized communities.

Cultural and Ethical Shifts

(a) Engineering New Work-Life Balance: Axiom Space reports increased job satisfaction by allowing engineers to shift to flexible VR.
(b) Ethical Questions: Does virtual colonization of Mars justify key aspects of territorial claiming in outer space? Legal scholars suggest it is better to preempt the need for treaties.

Challenge: Indulgent games and experiences can stoke "eco-apathy" by desensitizing users to real-world environmental crises.

Post-Humanism: A New Vision of Identity in the Cosmic Order
Virtual tourism: More than 500,000 users have "visited" digital Mars through Axiom's VR tours, nurturing a sense of being part of cosmic democracy.

Researchers suggest developing immersive and interactive simulation-based games to demonstrate Indigenous knowledge's contribution to scientific discovery development. First, Eurocentric narratives in current simulation games must be analyzed with Indigenous people to develop modules that leave Indigenous communities as the teachers, integrating their astronomies into the modules.

5.4 *Sustainable Implications*

The framework's environmental impact is twofold: it provides carbon savings while inviting new energy demands.

Environmental Benefits

(a) Reduced Physical Infrastructure: By substituting physical drills with VR, CO_2 can be eliminated.
(b) Travel Emissions: ENGAGE XR drops in aerospace industry travel emissions.

Energy and Resource Costs

(a) The E-Waste Problem: There are 50 million tons of e-waste annually. Circular economy models, such as TerraCycle's recycling partnerships, are essential.

Long-Term Sustainability

(a) Green Tech: IBM's Quantum Mars is trained to optimize for energy efficiency in simulations, reducing processing power by 30%.
(b) Policy Advocacy: The framework calls for pushing for tax credits to incentivize the use of green metaverse technologies.

The framework's implementation redefines theoretical paradigms, unlocks practical efficiencies, transforms societal engagement, and navigates sustainability trade-offs (Mondal et al. 2023a). Theoretically, it challenges assumptions about resource scarcity and cognitive learning (Ngoc-Vinh et al. 2022). It demands investment in infrastructure and workforce upskilling (Nguyen et al. 2022). Societally, it democratizes access while risking digital divides and cultural homogenization (Tandon and Das 2023). Environmentally, it offers carbon savings but requires vigilant management of energy and e-waste (Tien-Dung et al. 2022).

To succeed, stakeholders must adopt adaptive governance models prioritizing inclusivity, transparency, and planetary stewardship. Experts think Digital Galaxies should "inspire, not replace, humanity's quest for the stars." By balancing innovation with ethics, the metaverse can become a bridge to the cosmos—one that empowers all of Earth's inhabitants to explore, learn, and dream.

6 Future Scope

This chapter explores how integrating the metaverse into space exploration opens a universe of possibilities that extend far beyond current applications. The future scope of "Digital Galaxies" lies in leveraging emerging technologies, fostering global collaboration, and addressing ethical and environmental challenges to create a sustainable and inclusive ecosystem for virtual space exploration.

(a) ***Technological Advancements***

The VR, AR, and MR technologies advancements will enable hyperrealistic simulations of cosmic events (Tien-Dung et al. 2022a). Upcoming platforms might simulate black hole dynamics, interstellar travel, or the formation of galaxies with never-before-seen fidelity (Vrana and Das 2023). For example, quantum computers could conduct real-time simulations of complex astrophysical processes, like dark matter interactions, that are computationally prohibitive on classical systems. Likewise, if technology allows haptic feedback or neural hookups, the user could feel the texture of Martian soil or be simulated to float in zero-gravity environments, which will help integrate virtual and physical exploratory facilitated systems (Vrana and Das 2023a).

(b) ***AI Integration***

AI will drive personalized educational experiences and expedite scientific discovery. Adaptive learning algorithms, for example, could customize simulations to students' skill levels, or generative AI could spin up dynamic procedurally generated galaxies for exploration (Yegen and Das 2023). For researchers, AI data analysis tools can also organize petabytes worth of astronomical data, and the metaverse can offer and recognize patterns underlying the raw data to identify phenomena that would remain hidden from human observers. Projects like AstroGPT could make advanced research tools more accessible, democratizing this space and allowing citizen scientists to contribute to the field—albeit perhaps in a different way than a typical astrophysicist.

(c) ***Democratic Knowledge-Based Political Consensus in Global Governance***

One application of the metaverse is bringing together global stakeholders around human spaceflight. Virtual labs could provide venues where international teams could collaborate on projects to design sustainable lunar habitats or mitigate asteroid threats. The landscape of space projects will be dominated by partnerships between space agencies (NASA, ESA), private companies (SpaceX, Blue Origin), and open-source communities. Similar programs, such as the Open Metaverse Alliance, can subsidize VR hardware for under-resourced areas so that everyone has an equitable shot at participating in digital space initiatives.

(d) ***Ethical and Environmental Considerations***

As Digital Galaxies take shape, ethical frameworks must adapt to navigate data privacy, cultural representation, and environmental concerns. For instance, blockchain applications could guarantee user data security and protect intellectual property in simulated environments. At the same time, decentralized governance structures may actualize community-led design processes within simulation constructs (e.g., incorporation of Indigenous astronomies). At the environmental level, we will research the implementation of green computing practices, such as data centers powered by renewables and energy-efficient VR hardware, to understand how such methods could be adopted in the metaverse to reduce the carbon footprint of its infrastructure.

(e) ***Motivation for Observation of the Physical***

Rather than beating out a physical mission, Digital Galaxies could lead to concrete advancements in space exploration. Virtual "missions" could crowdsource solutions to technical challenges, like optimizing life-support systems for Mars colonies. Mars City Builder is an example of a platform that illustrates how public participation in virtual prototyping can translate to actual engineering decisions. Even further, immersive educational programs might spark passion for STEM careers to fill workforce shortages in the broader space industry.

7 Conclusion

Integrating the metaverse with space represents a synergistic approach that could allow transformative opportunities for education, scientific discovery, and public engagement, even while posing substantial ethical and practical challenges.

Key Contributions

(a) **Universal Access:** Virtual platforms such as Celestia Metaverse and NASA's Solar System Exploration VR have broken down geographic and socioeconomic barriers, allowing students in rural classrooms and amateur astronomers to explore the universe like professionals do. This inclusivity bridges with global goals for equitable STEM education.
(b) **Accelerated Discovery:** Projects projected the potential of collaborative virtual environments to expedite research, make data-processing schedules more efficient, and lower mission risks. Such advances showcase the metaverse's viability as a "virtual lab" for hypothesis testing.
(c) **Innovate for Ethics and Sustainability:** There will still be a cost—we already know the whole thing runs on servers and takes up space—but we can pave the way for a greener metaverse through practical initiatives like Google's Sustainium and TerraCycle's e-waste recycling programs.
(d) **Challenges Ahead:** The road ahead is not without its challenges. There is an urgent need to tackle the challenges of technicalities—for instance, the energy cost of high-fidelity simulations. Ethically, allowing misinformation to spread—as with platforms that prefer an aesthetic to scientific fact—requires serious planning and execution. Moreover, society will have to determine how to ensure that virtual exploration enhances, rather than replaces, physical missions so we can keep the public behind whatever we choose to do.

In order to fully materialize the potential of Digital Galaxies, stakeholders should pursue a synergistic, multidisciplinary process:

(a) Policymakers should set global standards for virtual training certifications and data privacy.
(b) Simultaneously, educators must upskill metaverse tools within curricula and remain aware of the digital divide.
(c) Technologists must prioritize innovations like edge computing and quantum algorithms in technological areas as energy-efficient ones.
(d) Ethics should be at the forefront of conversations around culture sharing and the ethics of virtual colonization.

References

Africkitigar (2024) The international space federation (ISF) | Nassim Haramein. Spacefed. https://spacefed.com/

Bailey J, Prather E, Slater T (2004) Reflecting on the history of astronomy education research to plan for the future. Adv Space Res 34(10):2136–2144. https://doi.org/10.1016/j.asr.2003.01.038

Bhadra N (2024) Space tourism: bridging the gap between imagination and reality. TopicPie Insight Hub https://topicpie.com/2023/11/04/space-tourism-bridging-the-gap-between-imagination-and-reality/

Bingham L, Kincaid J, Weno B, Davis N, Paddock E, Foreman C (2023) Digital lunar exploration sites unreal simulation tool (DUST). IEEE Aerospace Conference 1–12. https://doi.org/10.1109/aero55745.2023.10115607

Brown M (2022) SpaceX mars city: why, when, and how Elon Musk wants to build his ambitious plan. Inverse. https://www.inverse.com/innovation/spacex-mars-city-codex
Carbajales-Dale M, Murphy TW (2022) The environmental and moral implications of human space travel. Sci Total Environ 856:159222. https://doi.org/10.1016/j.scitotenv.2022.159222
Celestia: News (2023) Retrieved February 8, 2025, from https://celestiaproject.space/news.html
Chang C, Kuo H, Du Z (2023) The role of digital literacy in augmented, virtual, and mixed reality in popular science education: a review study and an educational framework development. Virtual Reality 27(3):2461–2479. https://doi.org/10.1007/s10055-023-00817-9
Chen Z, Gan W, Sun J, Wu J, Yu PS (2023) Open Metaverse: issues, evolution, and future. arXiv (Cornell University). https://doi.org/10.48550/arxiv.2304.13931
Cole M, Cohen C, Wilhelm J, Lindell R (2018) Spatial thinking in astronomy education research. Phys Educ Res 14(1). https://doi.org/10.1103/physrevphyseducres.14.010139
Curran D (2023) Surveillance capitalism and systemic digital risk: the imperative to collect and connect and the risks of interconnectedness. Big Data Soc 10(1). https://doi.org/10.1177/20539517231177621
Das S (2023) Modern-day gaming and the role of FinTech applications in Popularising various entertainment platforms across the world. In: Das S, Gochhait S (eds) Digital entertainment as next evolution in service sector. Palgrave Macmillan, Singapore. https://doi.org/10.1007/978-981-19-8121-0_11
Das S (2023a) Digital entertainment: the next evolution in service sector. Emerging Digital Solutions in Reshaping Different Industries Palgrave-Singapore doi:https://doi.org/10.1007/978-981-19-8121-0
Das S, Mondal SR, Struck F, Eissfeldt J, Beernink L, Sigouras V, Hoose S, Kopac L (2023) What can we see inside the crystal ball? Exploring opportunities and challenges for Metaverse and circular economy. AJIS 12(6):221. https://doi.org/10.36941/ajis-2023-0165
Di Virgilio F, Das S (2023) Digital solutions for social media: role of digital evolution on business enterprises. In: Das S, Gochhait S (eds) Digital entertainment as next evolution in service sector. Palgrave Macmillan, Singapore. https://doi.org/10.1007/978-981-19-8121-0_9
Di Virgilio F, Das S (2023a) Digitalization of enterprises from a new Normal sustainable development perspective: opportunities and challenges. In: Mondal SR, Yegen C, Das S (eds) New Normal in digital enterprises. Palgrave Macmillan, Singapore. https://doi.org/10.1007/978-981-19-8618-5_9
Dimitriadou E, Lanitis A (2023) A critical evaluation, challenges, and future perspectives of using artificial intelligence and emerging technologies in smart classrooms. Smart Learn Environ 10(1). https://doi.org/10.1186/s40561-023-00231-3
Fitria NTN, Simbolon NE, Afdaleni N (2022) Possibility of metaverse in education: opportunity and threat. SOSMANIORA Jurnal Ilmu Sosial Dan Humaniora 1(3). https://doi.org/10.55123/sosmaniora.v1i3.821
Greicius A (2024) Here's how AI is changing NASA's Mars Rover science–NASA. NASA. https://www.nasa.gov/missions/mars-2020-perseverance/heres-how-ai-is-changing-nasas-mars-rover-science/
Gualano RJ, Campbell SW (2024) Classrooms in the metaverse: educational applications and levels of learner interaction in virtual reality. Commun Educ 73(2):212–225. https://doi.org/10.1080/03634523.2024.2312873
Hoyhtya M, Boumard S, Yastrebova A, Jarvensivu P, Kiviranta M, Anttonen A (2022) Sustainable satellite communications in the 6G era: a European view for multilayer systems and space safety. IEEE Access 10:99973–100005. https://doi.org/10.1109/access.2022.3206862
Killelea E (2022) "Philosophy of space exploration" UTSA'S first metaverse course (By San Antonio Express-News). GovTech. https://www.govtech.com/education/higher-ed/philosophy-of-space-exploration-utsas-first-metaverse-course
Majerova J, Das S (2023) Sustainable development and digitalization in the supply chain for environmental protection: changes and trends. In: Mondal SR, Yegen C, Das S (eds) New Normal in digital enterprises. Palgrave Macmillan, Singapore. https://doi.org/10.1007/978-981-19-8618-5_8

Minh-Nhat HO, Majerova J, Das S (2022) Taking steps to help the Phoenix rise from ashes: a roadmap for digital enterprises to develop new sustainable goals. In: Sustainable development and innovation of digital Enterprises for Living with COVID-19. Springer, Singapore, pp 35–51. https://doi.org/10.1007/978-981-19-2173-5_3

Minh-Nhat HO, Nguyen HL, Mondal SR (2022a) Digital transformation for new sustainable goals with human element for digital service enterprises: an exploration of factors. In: Sustainable development and innovation of digital enterprises for living with COVID-19. Springer, Singapore, pp 85–103. https://doi.org/10.1007/978-981-19-2173-5_6

MIT Open Learning (2024) Developing an intelligent tutoring system, using VR and AR, for advanced manufacturing training. Medium; MIT Open Learning. https://medium.com/open-learning/developing-an-intelligent-tutoring-system-using-vr-and-ar-for-advanced-manufacturing-training-438e43102fd2

Mondal S, Das S, Vrana VG (2023) How to bell the cat? A theoretical review of generative artificial intelligence towards digital disruption in all walks of life. Technologies 11(2):44. https://doi.org/10.3390/technologies11020044

Mondal SR, Das S (2023) Defining the new digitalized normal: implications of a.I., IoT, and Blockchain for business growth. In: Mondal SR, Yegen C, Das S (eds) New Normal in digital enterprises. Palgrave Macmillan, Singapore. https://doi.org/10.1007/978-981-19-8618-5_10

Mondal SR, Das S (2023a) Digital solution for industry 4.0 and 5.0 and social media. In: Das S, Gochhait S (eds) Digital entertainment as next evolution in service sector. Palgrave Macmillan, Singapore. https://doi.org/10.1007/978-981-19-8121-0_10

Mondal SR, Das S (2023b) Human-centric digitalization for new sustainable development: an explorative study in human resource management. In: Mondal SR, Yegen C, Das S (eds) New normal in digital enterprises. Palgrave Macmillan, Singapore. https://doi.org/10.1007/978-981-19-8618-5_12

Mondal SR, Yegen C, Das S (2023a) New normal in digital enterprises–strategies for sustainable development. Palgrave-Singapore. https://doi.org/10.1007/978-981-19-8618-5

Müller E (2005) Simulating astrophysical phenomena: challenges and achievements. Comput Phys Commun 169(1–3):353–361. https://doi.org/10.1016/j.cpc.2005.03.079

Ngoc-Vinh N, Tien-Dung P, Das S (2022) Digitization of business; need for recombination during COVID-19. In: Sustainable development and innovation of digital enterprises for living with COVID-19. Springer, Singapore, pp 1–17. https://doi.org/10.1007/978-981-19-2173-5_1

Nguyen TDM, Mondal SR, Das S (2022) Digital entrepreneurial transformation (DET) powered by new normal sustainable developmental goals (n-SDGs): elixir for growth of country's economy. In: Sustainable development and innovation of digital enterprises for living with COVID-19. Springer, Singapore, pp 69–84. https://doi.org/10.1007/978-981-19-2173-5_5

Pells R (2023) Why scientists are delving into the virtual world. Nature. https://doi.org/10.1038/d41586-023-02688-1

Pompea SM, Russo P (2020) Astronomers engaging with the education ecosystem: a best-evidence synthesis. Annu Rev Astron Astrophys 58(1):313–361. https://doi.org/10.1146/annurev-astro-032620-021943

Poupard M, Larrue F, Sauzéon H, Tricot A (2024) A systematic review of immersive technologies for education: learning performance, cognitive load and intrinsic motivation. Br J Educ Technol 56(1):5–41. https://doi.org/10.1111/bjet.13503

Rudschies C, Schneider I (2023) Ethical, legal, and social implications (ELSI) of virtual agents and virtual reality in healthcare. Soc Sci Med 340:116483. https://doi.org/10.1016/j.socscimed.2023.116483

Slater TF, Tatge CB (2017) Astronomy education research in the planetarium. In SpringerBriefs in astronomy (pp. 1–27). https://doi.org/10.1007/978-3-319-57202-4_1

Stoll C, Gallersdörfer U, Klaaßen L (2022) Climate impacts of the metaverse. Joule 6(12):2668–2673. https://doi.org/10.1016/j.joule.2022.10.013

Szocik K, Wójtowicz T, Braddock M (2020) The Martian: possible scenarios for a future human society on Mars. Space Policy 54:101388. https://doi.org/10.1016/j.spacepol.2020.101388

Tandon S, Das S (2023) Digital revolution complements sustainable goal achievement during and after pandemic: a research agenda. In: Mondal SR, Yegen C, Das S (eds) New normal in digital enterprises. Palgrave Macmillan, Singapore. https://doi.org/10.1007/978-981-19-8618-5_11

Tasker E, Green J, Hoogenboom T, Ahmed W (2023) M is for Mars and Metaverse: bringing new experiences to students through virtual reality education. https://astro4edu.org/shaw-iau/talk/291/

Team SVV (2024) The ethical dilemma of space tourism: balancing opulence with eco-consciousness–space voyage ventures. Space Voyage Ventures. https://spacevoyageventures.com/the-ethical-dilemma-of-space-tourism-luxury-vs-sustainability/

Tien-Dung PHAM, Majerova J, Das S (2022) Phases of possible recovery of digital enterprises in new normal business for living with COVID-19 times: opportunities for a new era in sustainable development goals. In: Sustainable development and innovation of digital enterprises for living with COVID-19. Springer, Singapore, pp 19–33. https://doi.org/10.1007/978-981-19-2173-5_2

Tien-Dung P, Nguyen XD, Das S (2022a) Sustainable goal achievement by digital revolution during and after pandemic; how much one wins and losses: a bird's eye view for future planning. In: Sustainable development and innovation of digital enterprises for living with COVID-19. Springer, Singapore, pp 133–147. https://doi.org/10.1007/978-981-19-2173-5_8

Vrana V, Das S (2023) Dynamic restructuring of digital media and entertainment sector: role of urbanization, industrial innovation, and technological evolution. In: Das S, Gochhait S (eds) Digital entertainment as next evolution in service sector. Palgrave Macmillan, Singapore. https://doi.org/10.1007/978-981-19-8121-0_2

Vrana V, Das S (2023a) Role of enterprises and their digital transformation in post-pandemic sustainable development. In: Mondal SR, Yegen C, Das S (eds) New normal in digital enterprises. Palgrave Macmillan, Singapore. https://doi.org/10.1007/978-981-19-8618-5_7

Weaving T, Ralli A, Coveney P, Love P, Kirby W, Tranter A, Succi S, Wimalaweera V (2024) Practical quantum simulations from error mitigation and qubit subspace techniques. IBM Research. https://research.ibm.com/publications/practical-quantum-simulations-from-error-mitigation-and-qubit-subspace-techniques

Yegen C, Das S (2023) Digital solution for print and publishing media. In: Das S, Gochhait S (eds) Digital entertainment as next evolution in service sector. Palgrave Macmillan, Singapore. https://doi.org/10.1007/978-981-19-8121-0_12

Zeng J (2023) The development stages of metaverse and its cultural consequences. Critical Arts 38(2–3):15–25. https://doi.org/10.1080/02560046.2023.2234009

Zerrad Y (2023) Space architecture and the development of a martian habitat using computational design. https://doi.org/10.22215/etd/2023-15407

Zhao Y, Jiang J, Chen Y, Liu R, Yang Y, Xue X, Chen S (2022) Metaverse: perspectives from graphics, interactions and visualization. Visual Informatics 6(1):56–67. https://doi.org/10.1016/j.visinf.2022.03.002

Zhivotovskiy Y, Stratton E, Phelps S, Schmid J (2024) The reality of space: exploring virtual, augmented, and mixed realities and the future of spaceflight. https://ntrs.nasa.gov/api/citations/20240005474/downloads/Reality%20of%20Space-Exploring%20VR_AR_MR.pdf

Milan Konvit Researcher at Comenius University's Faculty of Social and Economic Sciences, examining economic policy and social equity. His studies focus on public welfare systems and community resilience. He collaborates internationally to address socio-economic disparities.

Anastasios-Ioannis Theocharidis Digital marketing expert on hospitality and tourism, with experience at Blue Palace Resort and Hellenic Army's 3rd Corps. Holds an MBA in Hospitality & Tourism and a Bachelor's in Applied Informatics. His research explores data-driven strategies in tourism and consumer behavior.

Subhra R. Mondal Senior Researcher and academician at Duy Tan University, specializing in brand management, AI, and smart tourism. Editor of multiple Scopus-indexed books, her work explores digital recruitment and Web 5.0. She bridges academia and industry through collaborative projects.

R2-D2's Cousins: Robotic Companions in the Final Frontier

Uyên Nguyễn Cao Thục, Vasiliki Vrana, and Subhankar Das

Abstract A robotic travel companion to expand the universe of narrative exploration of space and tourism would revolutionize how we experience interplanetary adventures through immersive storytelling. It traces the progression of space robotics, starting with primitive probes and ending with sophisticated spacecraft and alignment systems like NASA's Perseverance rover and artificial-intelligence-enhanced counterparts like Crew Interactive Mobile Companion (CIMON), discussing their operational, scientific, and psychological advantages. Robotic systems increase the efficiency of the mission, reduce risks, and can provide essential support in extreme environments. Yet there are challenges to overcome, like working under communication-lagged conditions, surviving harsh environmental conditions, and facing ethical challenges related to ranking the robots and AI's accountability and rights systems. This chapter highlights the perspective of human-robot co-working, stress, as well as ergonomic and psychological health aspects. Discussion points include hyper-autonomous systems, biomimetic robots, self-replicating technologies, ethical frameworks, and equitable access to resources outside the atmosphere. Thus, this study synthesizes diverse perspectives from engineering literature, artificial intelligence, psychology, and moral debate to propose robotic companions as crucial partners in humanity's cosmos expansion and inspire safety, creativity, and sustainable exploration.

U. N. C. Thục
South Star Management Institute, Duy Tan University, Da Nang, Vietnam
e-mail: trucuyen0409@gmail.com

V. Vrana
Department of Business Administration, School of Economics and Administration, The campus of Serres, International Hellenic University, Serres, Greece
e-mail: vrana@ihu.gr

S. Das (✉)
Department of Marketing, South Star Management Institute, Duy Tan University, Da Nang, Vietnam
e-mail: subhankardas@duytan.edu.vn

© The Author(s), under exclusive license to Springer Nature Singapore Pte Ltd. 2025

S. R. Mondal et al. (eds.), *Pioneering the New Space Economy through AI and Immersive Technologies*, https://doi.org/10.1007/978-981-96-5977-7_7

Keywords Robotic companions · Space exploration · Space tourism · Human-robot interaction · Ethical AI in space robotics · AI tourism

1 Introduction

Since the rise of human interest in space exploration, it has also been closely intertwined with technological developments. Furthermore, from Sputnik's launch to the Apollo moon landings, each milestone has deepened our grasp of the cosmos and where we fit in the grand scheme of things. Fast forward to today, as we stand at the threshold of a new age—building lunar bases, colonizing Mars, and even dabbling in space tourism—robotics is at the heart of these endeavours (Diftler et al. 2010). Inspired by science fiction—like R2-D2, the lovable droid of the *Star Wars* universe—sophisticated robotic technology is transitioning from being just tools or servants for humans to partners and companions capable of assisting people through the perilous and unpredictable environment of space (Ellery 2015). This article discusses robotic companions, their current technology, potential applications, and space exploration and tourism challenges.

The Evolution of Robotics in Space Exploration
The use of robotics for space missions is not new. Early robotic probes such as Luna 2 (1959) and Voyager 1 (1977) showed that unmanned exploration was ever more possible, leading to increasingly advanced systems. The Mars rovers—Sojourner, Spirit, Opportunity, Curiosity, and Perseverance—are born again in robotic exploration until the dust settles. Hubble and Voyager II were and are the epitome of exploration of alien worlds, land, and clay, with experiments and data sent back to Mother Earth (Deshmukh and Karlgaard 2021). These machines are in places deadly to humans, facing radiation, extreme temperatures, and atmospheric pressures. With astronauts aboard the International Space Station (ISS), robots such as Robonaut 2 and the Canadarm2 robotic arm have been integrated as essential crew members. Engineered by NASA and General Motors, Robonaut 2 helps astronauts with repetitive tasks, allowing them to work fewer hours and reducing the risk of injury in space. Likewise, Canadarm2 has been vital in docked spacecraft and has helped maintain the ISS's structural integrity. These are examples of a paradigm shift: Robots are no longer passive tools but active partners in space missions. Figure 1 shows the evolution of robotics in space exploration.

1.1 Robotic Companions in Space: The Benefits

Space is a harsh environment that depends on robotics. In contrast to humans, robots also suffer no physical effects of microgravity, cosmic radiation, or prolonged isolation. They are skilled at carrying out repetitive or dangerous activities—like fixing machinery, collecting samples, or scouting unexplored areas—without getting tired or making mistakes (Ellery 2015). For example, NASA's Valkyrie humanoid robot

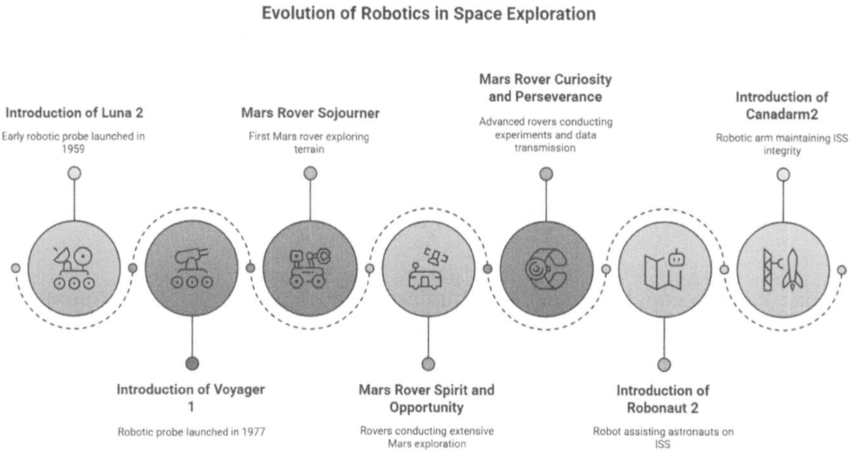

Fig. 1 Robotics evolution in space exploration (*Source* Authors' conception)

is designed for disaster response, a potential capability that could feed into what might be needed for space (Radford et al. 2015).

Additionally, robots help reduce risks to human life. In 2013, astronauts working on a high-risk spacewalk undertook a complex repair of the ISS's ammonia cooling system, spotlighting the hazards of extravehicular operations (Hambuchen et al. 2021). Next-generation robotic systems could assume such roles, augmenting operations and assuring human safety.

Modern-day robots serve many different jobs. Through the ISS artificial intelligence assistance, Crew Interactive Mobile Companion (CIMON) has already commenced offering help to astronauts with the procedures and emotional and psychological support through conversational interaction (Eisenberg 2019). In the meantime, autonomous rovers have changed planetary exploration. For example, the Perseverance rover and its Ingenuity helicopter illustrate advances in autonomy by autonomously navigating Jezero Crater on Mars and selecting samples for possible return to Earth (NASA 2023). Telepresence robots expand even more on human capability. Examples like NASA's Robotic Reconnaissance place scientists on Earth in the control of a rover in near real time, fusing human intuition with robotic precision. Such systems are precursors to more immersive technologies, where virtual reality (VR) and haptic feedback might allow Earth-bound researchers to "inhabit" robotic avatars on worlds far away.

When the dream of space tourism becomes routine—as it appears to be moving toward now, with such efforts as SpaceX's Starship and Blue Origin's New Shepard—humans' reliance on robots will only grow. In the future, space tourists who may not be trained as astronauts will need robotic aides to help ensure a safe and pleasurable experience. These aides-de-camp could take measurements of vital statistics, deliver acute prehospital care, and coach users through spacecraft procedures (Rainbow 2025). Psychological relief is another essential frontier. Mental health issues arise on long-duration missions like a trip to Mars due to confinement and isolation. With affective computing, social robots would also have the potential to display empathy, engage in meaningful dialogue, and detect depression (Akbulut et al. 2024). These systems have great promise, as Japan's Kirobo robot was tested on the ISS in 2013, successfully talked to the astronauts, and helped alleviate loneliness (Gannon 2013).

1.2 Challenges in Developing Space-Faring Robots

Despite their potential, robots have significant challenges to overcome. Autonomy is key; the communication time lag between Earth and distant spacecraft (e.g., 20 min one way to Mars) requires robots that can decide independently how to act (Turan et al. 2022). Machine learning algorithms must allow for adjustment to new situations like machinery breakdowns and environmental dangers. Durability, too, is paramount. Space robots must work in harsh conditions but are also required to optimize energy spending. ESA's ExoMars rover is one such vehicle, which includes radiation-hardened components and self-cleaning solar panels designed to withstand Martian dust storms (ExoMars 2016). Of course, there are ethical considerations, too. As robots become more automated, the issue of who is responsible for mistakes (e.g. a medical robot mistakenly diagnosing an illness) will need to be worked out (Cambridge, Mass.: MIT Press 2012). Moreover, dependence on robots could undermine human competencies, a risk raised by research on automation reliance (Parasuraman and Riley 1997).

1.3 Symbiosis of Human and Robot: The New Age

A new paradigm for the future of space exploration is human-robot cooperation. While telepresence robots may allow scientists to perform "virtual fieldwork" on Europa or Titan, AI-enabled systems may even build habitats with in situ resources (Wu et al. 2024). For example, NASA's Artemis programme envisions robots establishing lunar bases for when humans land, deploying solar panels, and extracting water ice (Gamgami 2024). This synergy extends to societal implications. Kostavelis et al. (2013) also pointed out that robots democratize space by enabling more participants in teleoperated space exploration. An international Mars rover—remotely operable by students in Nairobi, researchers in Buenos Aires, and citizen scientists globally—could democratize access to space exploration, fostering worldwide collaboration and public engagement in planetary science while inspiring a new generation to participate in humanity's interplanetary future. MVPs of future space exploration: robotic companions, such as R2-D2. As we traverse the cosmos, their functionality to assist and enhance human capability, ensure safety, and offer companionship will prove vital. Nevertheless, building them requires cross-disciplinary work—combining engineering, AI, psychology, and ethics—to design functional, trustworthy, and human-friendly systems. Spurred on by Voyager and other mechanical voyagers, the final frontier will open space that we can share between life and machinery.

2 Literature Review: Robotic Companions in Space Exploration and Space Tourism

Bringing robotics into space exploration and tourism is a radical change in how humans will engage with the cosmos. However, with this systematic review, there are fundamentals on some historical precedents, technological equipment, and ethical negotiations of previous studies that highlight and demonstrate that robotic companions representing commodities will become an integral part of human endeavours beyond Earth. This review shows the breadth of space robotics in terms of historical development, functional benefits, present state of use, future direction, challenges, and man-robot collaborative efforts, thus identifying interdisciplinary areas and gaps for research investigation.

The Benefits and Functional Aspects of Robotic Friends

Human survival in space is often uncertain, and robotic systems provide unparalleled benefits. These extreme temperatures, radiation, and microgravity make them ideally suited to tasks considered too dangerous for humans (Ellery 2015). NASA's Valkyrie robot, built for disaster response on Earth, offers possibilities for extraterrestrial search-and-rescue operations (Radford et al. 2015).

There are three main areas of the functional roles of robots.

(a) Operational efficiency: Robots can carry out repetitive tasks such as equipment calibration or sample collection, causing less human error (Hambuchen et al. 2021).
(b) Risk Mitigation: Improvement of high-risk scenarios. In 2023, the ISS ammonia leak repair was completed in a perilous spacewalk (Akbulut et al. 2024).
(c) Scientific Research: For example, the Perseverance rover is an autonomous robot that conducts experiments and sends data that informs future missions in real time (NASA 2023).

These roles become magnified in deep-space missions, where communication delays require robots to have high autonomy and decision-making capabilities.

Current Robotic Space Missions and Tourism

Modern space robotics covers crewed missions and nascent tourism ventures. On the International Space Station (ISS), AI-powered companions such as the Crew Interactive Mobile Companion (CIMON) serve astronauts by providing data collection, timetable management, and psychological support via dialogue interfaces (Eisenberg 2019). Such systems are examples of the double use of robots as functional tools and social companions. Telepresence robotics were game changers in planetary exploration. NASA's Robotic Reconnaissance project allows Earth-bound scientists to control rovers remotely, melding human expertise with robotic precision (Hambuchen et al. 2021). This technology is a forerunner to immersive systems employing virtual reality (VR), which might enable researchers to "inhabit" robotic avatars on far-off moons such as Europa. Robotic companions in space tourism are set to meet specific needs. Other major companies, such as SpaceX and Blue Origin, are already trying to predict if robots will be personal assistants for tourists, helping navigate spacecraft operations and endlessly monitoring health biometrics (CFR Task Force Report: Securing Space—a Plan for U.S. Action 2025). In Japan, for example, research showed the potential for social robots to provide empathetic interactions with users to help reduce loneliness, with the latter tested on the ISS (Gannon 2013). These are just some examples of the growing use of robotics for safety and improving the tourist experience.

New Trends and Beneficial Future
The future of space robotics is being shaped by developments in artificial intelligence (AI), machine learning, and in situ resource utilization (ISRU). AI-fuelled systems being developed, namely for NASA's Artemis programme, would build lunar habitats from materials found on the moon—limiting dependence on supplies from Earth (Gamgami 2024). Likewise, the European Space Agency's (ESA) ExoMars rover uses AI to navigate the Martian surface autonomously, even in dust storms (ExoMars 2016). Space tourism will require various support from robots. Their future companions may include the ability to do medical diagnostics, provide emergency care while in a microgravity environment, or even use effective computing techniques to recognize and respond to psychological distress (Akbulut et al. 2024). Such robotic systems could maintain life-support systems and grow food on projects like Starship, SpaceX's vehicle for Mars colonization, helping to enable long-term mission sustainability.

Human-Robot Collaboration
This collaboration between humans and robots is revolutionizing space exploration. Telepresence systems democratize access to extraterrestrial environments, bringing the world into team missions (Das 2023). For instance, a researcher based in Buenos Aires could remotely navigate a rover on Mars through VR, promoting global collaboration. Robots can fill public engagement gaps with space and science at a societal level (Das 2023a). Educational initiatives, such as letting students steer lunar rovers, might inspire future generations (Wu et al. 2024). However, this democratization raises questions of whether all nations and people have equal access to high-tech space capabilities and whether the use of outer space leads to new geopolitical tensions around resource extraction.

2.1 Research Gaps

The literature emphasizes robotic companions as central to humanity's cosmic aspirations. Their potential to improve safety, efficiency, and psychological well-being marks them as irreplaceable allies. Nevertheless, crucial gaps remain:

(a) Long-Term Human-Robot Interaction: Limited psychological studies investigate the psychological consequences of sustained, isolated human-robot interaction.
(b) AI Regulations: If Earth-based guidelines for robotics are paper tigers in people's extreme motives, are there unique ethical frameworks in space, such as AI rights, environmental stewardship, etc.?
(c) Interdisciplinary Research: Engineers must work closely with ethicists and psychologists to design more holistic systems.

Long-term research efforts should address these issues to help ensure robotic companions promote exploration and are consistent with what is humanly significant.

3 Factors Influencing the Development and Deployment of Robotic Companions in Space Exploration and Tourism

Technological, environmental, human, ethical, economic, operational, sociocultural, and regulatory elements each influence how robotic companions could be worked into space missions and tourism ventures. Robotic systems for the final frontier are judged by their feasibility, efficiency, and social acceptance, each set against these factors. These factors are examined in detail here, shedding light on their interdependencies and implications for the future of space exploration.

3.1 Technological Factors

Technological advancements underpin robotic space systems. Key considerations include the following:

(a) The Intersection of Autonomy and AI: In space, signals take time to travel, so robots must operate nearly autonomously. For example, travel signals between the Earth and Mars carry 4–24-min latency, requiring real-time decision-making functionalities. AI algorithms enable autonomous navigation and hazard avoidance in unpredictable terrains, and they are used in NASA's Perseverance rover (NASA 2023). Machine learning adds another layer of adaptability, enabling robots to learn from environmental feedback, such as ESA's ExoMars rover learning to traverse Martian dust storms (ExoMars 2016).
(b) Durability and Resilience: When in use in space, robots are exposed to extreme radiation, temperature cycles (−150 °C to 120 °C on Mars), and abrasion. The components need to be radiation hardened, as exemplified by the reinforced joints of the Valkyrie robot (Radford et al. 2015). To operate continuously, dust mitigation systems (e.g. Perseverance's self-cleaning sensors) are essential (NASA 2023).

(c) Energy Efficiency: Electrical restrictions dictate the extent of robotic capabilities. Solar panels, such as the ones used by rovers like Opportunity, are at risk of collecting dust. Similar systems powered by nuclear energy, such as the radioisotope thermoelectric generator (RTG) of the Curiosity rover, offer reliability but also possess some safety issues (Ellery 2015). Later designs may utilize regenerative energy systems or in situ resource utilization for refuelling.
(d) Communication Systems: Data transmission is constrained by latency and bandwidth limitations. NASA's Deep Space Network (DSN) enables interplanetary communication, but decentralized networks, like potential satellite constellations deployed around Mars, might be more reliable.

3.2 Environmental Factors

As inhospitable as space can be, it requires robots customized for particular celestial terrains.

(a) Radiation Exposure: Cosmic and solar radiation deteriorate electronics beyond Earth's magnetosphere. The ExoMars rover uses radiation-hardened processors to withstand high-energy particles.
(b) Challenges of Microgravity and Low Gravity: Robots have to function under microgravity conditions on the ISS or moon surfaces. In zero gravity, Robonaut 2 uses magnetic grippers to manipulate tools. On the moon, gravity (1/sixth of Earth's) limits mobility, requiring lightweight, nimble designs.
(c) Planetary Surface Conditions: Mars's thin atmosphere and dust storms require solid thermal regulation and proofing. The Ingenuity helicopter flight in the Red Planet's low-density atmosphere highlights the importance of environment-specific engineering (NASA 2023).

3.3 Human Factors

Human-robot interaction (HRI) is key to the success of the mission.

(a) Psychological Support: Stress and isolation are necessary on long-duration missions. Social robots, like Kirobo, chat with astronauts aboard the ISS and combat loneliness through empathetic conversation. With affect recognition, robots can identify emotional distress and respond.
(b) Trust and Reliability: Astronauts depend on robotic systems to perform critical tasks. Explainable AI-making processes, such as CIMON, are building trust.
(c) Ergonomics and Usability: These interfaces must serve more diverse users (space tourists) with minimal training. Intuitive controls, voice commands, and augmented reality (AR) guides help with usability (Holt 2022).

3.4 *Ethical and Legal Factors*

For purposes of deployment, robotics does present ethical and jurisdictional inquiries.

(a) Accountability: These errors in autonomous systems, such as a robot's medical misdiagnosis, challenge liability frameworks. Evans and Snead (2024) call for "ethics by design" to build accountability into AI architectures and better understand AI errors.
(b) Data Rights and Environmental Ethics: The advent of cognitive robots raises complex debates over their legal personhood. Furthermore, planetary protection procedures require limiting the pollution of celestial bodies, which informs both sterilization standards.
(c) Space Law: The Outer Space Treaty provides those states with responsibility for their space objects. Yet regulations are slow to keep pace with advancing AI and commercial space tourism, which calls for up-to-date frameworks (Kimball 2024).

3.5 *Economic Factors*

Most costs and funding models drive robotic development.

(a) Development Costs: It is costly to develop robots for space. With NASA's Perseverance mission costing $2.7 billion, the importance of a cost-sharing partnership cannot be overstated (NASA 2023). Private corporations such as SpaceX use reusable rockets to cut the cost of launches.
(b) Return on Investment (ROI): Returns on investment are the central focus of space tourism ventures. They currently command higher prices due to robotic companions that drive safety and customer experience. Take, for example, the robots that Blue Origin has suggested would serve as a sort of "space concierge" for high-net-worth tourists.
(c) Funding Sources: Leveraging public-private partnerships (e.g. NASA's Commercial Lunar Payload Services) and crowdfunding platforms create new funding streams.

3.6 *Operational Factors*

On the design and execution of the mission:

(a) Mission Planning: Robots need to be consistent with their mission. The Artemis programme uses ISRU to build lunar habitats from lunar surface material, significantly reducing the payload mass (Crawford, 2021).

(b) Maintenance and Repairability: It is challenging to make repairs on-site. Modular components facilitate autonomous Canadarm2 repairs on ISS. Future robots might be able to heal themselves with 3D printing.
(c) Interoperability: Robots from different agencies become compatible because of established protocols at the standard level. The ISS is a model of international collaboration.

3.7 Sociocultural Factors

Sociocultural factors: public perception and cultural values

(a) Public Engagement: Public engagement propels perception and excites them about space programs. STEM education with robotic missions. NASA's "Send Your Name to Mars" campaign encourages public participation (NASA 2023).
(b) Cultural Attitudes: Some societies are more welcoming of robots than others. Japan's fondness for social robots contrasts widespread Western scepticism and shapes design priorities.

3.8 Policy and Regulatory Factors

Safe and equitable exploration is ensured through governance frameworks.

(a) International Cooperation: Programmes such as the Artemis Accords are suitable for promoting peaceful cooperation, but they are not enforced.
(b) Safety Considerations: Agencies like the FAA and ESA create protocols to ensure robotic safety during tourism and exploration.
(c) Resource Allocation: Disputes emerge over lunar water ice or asteroid mining rights without regulatory clarity.

The achievement of successful robotic companion integration during space exploration lies in accommodating these multidimensional keys. Interdisciplinary teams will need to work together to engineer solutions and design policies, new systems, and mindsets to leverage the power of robotics while avoiding ethical pitfalls and unsustainable trajectories. These factors will continue to evolve even as humanity pushes beyond Earth into the cosmos, requiring constant rethinking and creativity.

4 Framework for the Practical Implications of Robotic Companions in Space Exploration and Tourism

As space exploration and tourism become more prevalent, the challenge is adapting theoretical insights into a fully comprehensive framework of actionable strategies for integrating robotic companions into these contexts. Such a framework identifies implications across six domains—mission design, human-robot collaboration, technological innovation, policy development, ethical issues, and economic sustainability—and can help stakeholders identify ways to maximize the benefits of robotics while taking account of challenges.

1. Improving Mission Design and Implementation.

Autonomous Mission Planning

(a) Adaptive Algorithms: Utilizing AI-driven systems that can make real-time decisions will help reduce communication lags. One example is NASA's Perseverance rover, which autonomously navigates the Jezero Crater on Mars without input from Earth (NASA 2023).
(b) Modular Robotics: Create interchangeable modular-based robots like the ISS's Canadarm2, which can be configured and repaired on-site.

In Situ Resource Utilization (ISRU)

(a) Habitat Construction: Robots like those proposed in NASA's Artemis programme can extract lunar water ice and 3D-print habitats, decreasing dependence on Earth-supplied building materials.
(b) Fuel Production: Design robots to convert Martian CO_2 into methane to fuel the return flight home (akin to SpaceX's recent testing of Starship prototypes).

Risk Mitigation

Pre-Mission Simulations: Conduct virtual dry runs with digital twins to assess robotic responses to extreme circumstances, such as solar storms or equipment malfunctions.

2. Enhancing Human-Robot Collaboration.

   a. Training and Ergonomics.

   Specific criteria for astronauts and their training should include more robotics, focusing on trust and emergency protocol training (Das et al. 2023). For example, Robonaut 2 was co-developed with General Motors to ensure a user-friendly human-robot interface.

   Tourist Interfaces: Prepare simple controls for space tourists, notably voice commands and AR guides (inspired by Blue Origin's "space concierge" idea).

   b. Psychological Resolutions: Support Systems.

   Social Robotics: Use affective computing in robots such as Japan's Kirobo to detect stress and engage in therapeutic dialogue during long missions.

   Telepresence Robots: Couple tourists with family and friends on Earth, helping combat isolation.

3. Cutting-Edge Technological Innovation.

   (a) Durability and Resilience: Radiation-hardened components. Adopting the shielding technologies used on the ESA's ExoMars rover to mitigate effects on electronics in deep space.

Dusty Planet Self-Cleaning Machines: Incorporate systems that keep equipment dust-resistant, such as Perseverance's self-cleaning sensors, to function long on a dusty planet.

(b) Energy Solutions: Hybrid power system entails fusing solar panels with compact nuclear reactors (such as Curiosity's RTG) for high reliability in low-light environments.

Energy Harvesting: The moon has many ambient energy sources. Scientists could design robots to harvest energy from sources like thermal gradients on the moon.

(c) Communication Infrastructure: Decentralized networks implement satellite constellations around Mars to facilitate increasingly independent communication of Earth-based systems.

4. Maintaining Economic Sustainability.

   (a) Developing Efficiently and Effectively.

Use public-private partnerships: Programmes like NASA's Commercial Lunar Payload Services (CLPS) allow private companies to bid on delivering payloads to the moon, and any R&D costs are shared.

Reusable Systems: As SpaceX has demonstrated, developing reusable launch vehicles for robotic payloads will significantly decrease the cost of access to space.

Revenue Generation: Preferred tourism services are the most fascinating robotic budgets.

Data Monetization: Sell planetary data collected by your bots to research institutions/private enterprises.

5 A Roadmap for Implementation

The real-world implications described above provide a framework for stakeholders to harness companies successfully:

(a) Years 1–5: Work on modular designs of robots, astronaut training programmes, and AI transparency frameworks.
(b) Medium Term (5–10 Years): Industrial isolation and resource utilization. Scale In situ Resource Utilization is (ISRU) technologies, answer on both sides, narrow down air/space and military synergy, establish international treaties, and deploy hybrid energy systems.
(c) Long Term (10+ Years): Create sustainable robotic colonies on Mars and enable widespread access to space through cost-effective tourism.

By comprehensively mitigating dependent challenges on the environmental, worthy, and functional frontiers, humankind can ensure that robotic sidekicks do

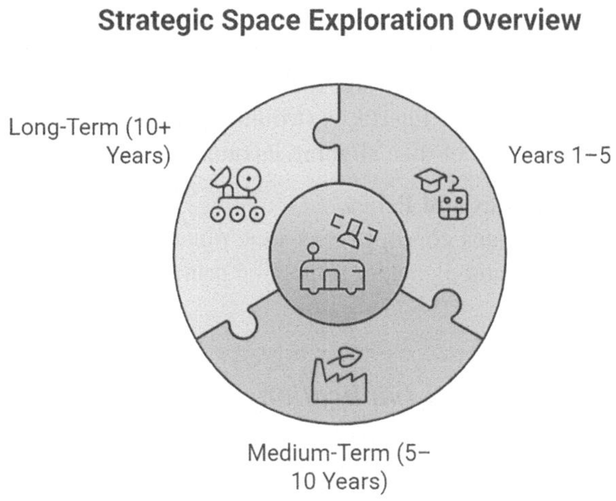

Fig. 2 Implementation framework (*Source* Authors' conception)

not most effectively deepen exploration (Di Virgilio and Das 2023). However, central howling planetary growth includes the floor iterations of life. Figure 2 represents the implementation plan.

6 Implications of the Framework

Robotic Companions: Theoretical, Practical, and Social Implications of a Proposed Framework for the Future of Space Exploration and Space Tourism. These implications serve technological and operational aims and determine humanity's ethical and social future in space. Below is a detailed analysis.

6.1 Theoretical Implications

The framework is based on interdisciplinary theories that together robotics, ethics, space science, and human-robot interaction (HRI).

Autonomy and AI Theory
The framework is consistent with situated autonomy theory, which postulates that robots should adapt to dynamic environments based on real-time information (Di Virgilio and Das 2023a). AI algorithms embedded in the machinery that empower Perseverance to navigate Mars autonomously embody embodied cognition progress (Majerova and Das 2023). Machine learning ethics theories inform the framework's focus on transparency and bias mitigation, which ensures that significant AI decisions reflect powerful human values (Minh-Nhat et al. 2022).

Collaboration between Humans and Robots
The framework is based on the joint mental model's theory, highlighting the interaction between humans and robots that leads them to achieve shared understanding through intuitive interfaces (Minh-Nhat et al. 2022a). CIMON's conversational AI is an example of this, aligning its outputs with the expectations of an astronaut.

Space Ethics and Policy
It incorporates concepts of cosmic pluralism, promoting equitable access to space resources, and planetary protection principles to avoid biological contamination.

6.2 Practical Implications

It offers a framework that allows theory to be translated into actionable strategies for mission success:

Mission Efficiency
Navigate Autonomously: The Helios vehicle pioneered by NASA uses AI technology to avoid potential hazards from the surroundings; doing so prevents mission delays (NASA 2023).

ISRU Use: Using robots to harvest lunar water ice (as in NASA's Artemis programme) reduces costs and allows permanent habitat building in situ.

Human-Robot Synergy
Training Programmes: The astronauts who train with Robonaut 2 are developing trust in it. This will promote teamwork during high-stakes tasks.

Tourist Safety: Blue Origin's proposed "space concierge" robots would help untrained tourists with voice-guided interfaces.

Technological Innovation
Radiation-Hardened Designs: The shielded electronics used in ESA's ExoMars rover have set a standard for durability.

Hybrid Energy Systems: Combining solar and nuclear power (such as Curiosity's RTG) guarantees reliability in extreme environments.

The framework's theoretical foundations in autonomy and ethics, practical advancements in robotics and ISRU, and commitment to equitable and sustainable social outcomes combine to reshape humanity's role in space. Considering the challenges comprehensively—from radiation-hardened engineering to global policy reform—assures that robotic companions will do more than enhance exploration and understanding; they will do so in a way that embodies the values of the university, stewardship, and mutual advancement." As we explore the universe, we see this is a template for a future where technology and humanity are no longer opposed (Mondal et al. 2023).

Future Scope

As we book passage to Mars and beyond, the bots will play a growing and transformative role as companions with whom we might share our journey (Mondal and Das 2023). This field has scope in the future, such as technological innovation, ethical evolution, societal integration, and interstellar collaboration (Mondal and Das 2023a). Below are some key trajectories shaping the next wave of space robotics.

6.3 Technological Advancements

(a) Hyper-autonomous Systems: Future robots will be characterized by hyper-autonomy beyond current levels and will have quantum computing and neuromorphic AI capabilities to process massive datasets in real time. Mars-roving vehicles, for example, may be able to determine work priorities based on one dam's sweep pattern and eliminate the need for commands and instructions to be sent from Earth. Some projects also innovate "thinking" robots that can think creatively by creating new ways to repair damaged equipment.

(b) Biomimetic and Soft Robotics: Robots inspired by nature, such as snake-like lava tube explorers and insect-rad swarm robots, can access places that ridged robots cannot enter.

Soft robotics, utilizing soft materials, can perform gentle missions such as collecting extraterrestrial soil without contaminating it (Mondal and Das 2023b). The ESA's SoftGrid mission, for example, models human hands.

(c) Self-Replicating Robots: Instead of transporting everything from Earth, long-term settlement projects may bring in robots capable of self-replication, which utilizes planetary resources. The notion of a Von Neumann probe implies the endowment of entire factories. This may lead to self-replication and access throughout Mars and the asteroids.

6.4 Human-Robot Collaboration

(a) Symbiotic partnerships: The division of labour between human beings and robots readily compels them. A remote-controlled robot, for instance, is a counterpart with VR interfaces and haptic feedback that allow my associates to feel like Martian rocks (Vrana and Das 2023). Meanwhile, momentary AI buddies may mature into cognitive partners, co-writing articles and ideas over the years (Vrana and Das 2023a).

(b) Psychological and medical sustenance: Social robots will use cutting-edge affective computing to offer unique, customized mental health facilities (Mondal et al. 2023a). A similar character on a Mars mission will notice through behavioural variations that a particular colleague is not getting along scrupulously and employ a rapporteur Tome battle with analyse integrated mediators to solve their differences (Ngoc-Vinh et al. 2022). Medical robots may also have CRISPR mechanisms for crisis gene cure after radiation-induced DNA destruction.

6.5 Expansion of Space Tourism with Customized Robots

(a) Robots' concierges: As space tourism companies like SpaceX and Virgin Galactic launch their services, they will provide Victorianinspired robotic concierges to enhance the customer experience.
(b) Guided Tours: AI-driven drones offering real-time commentary on lunar terrain.
(c) Vital Signs Monitoring: Wearable robots to monitor our vitals and administer medications.
(d) Personalized Experience: Robots customizing activities to be held by tourists (Nguyen et al. 2022) (such as games in zero gravity).
(e) Opulence Paired with Accessibility: While early innovations cater to wealthy individuals, competition and technological scaling will level the playing field. Start-ups may offer bargain robotic-led lunar tours—the lunar equivalent of today's eco-tourism packages.

6.6 Ethical and Legal Frameworks

Rights and Governance of Artificial Intelligence

(a) Cognition on Continued Growth: As robots gain advanced cognition, the legal status debates will also intensify (Tandon and Das 2023). Does a self-aware robot repairing a Mars habitat deserve to have limited rights? Agencies like the Global AI Ethics Consortium are onboarding guidelines for robotic personhood in space (IEEE CAI 2025).
(b) Planetary Stewardship: Future frameworks should find a middle ground between exploration and preservation (Tien-Dung et al. 2022). For instance, robotic mining operations conducted on Europa's Ocean under the ice will need protocols to avoid contaminating potential extraterrestrial life. Though not widely ratified, the Moon Agreement provides a template for ethical resource sharing in space (Atri et al. 2024).

6.7 Interstellar Sustainability

(a) Closed-Loop Systems: Robots will build self-sustaining colonies by controlling closed-loop life-support systems (Yegen and Das 2023). Lunar ISRU technologies, for example, could treat wastewater into drinking water or CO_2 into oxygen.

(b) Interstellar Probes: Developments in propulsion (light sails, antimatter drives, etc.) will permit robotic probes to explore exoplanets such as Proxima Centauri. Successful execution of the Breakthrough Starshot initiative proposes the gram-scale "StarChip" robotics to pass along data from neighbouring star systems (Gilster 2016).

6.8 Sociocultural Integration

(a) Cultural Exchange: Robots could be cultural ambassadors, embedding multilingual capabilities and folklore into interactions. For example, a robot on a multinational mission to Mars might tell stories from Vietnamese, Greek, or Nigerian traditions, building camaraderie among diverse crews.
(b) Education and Advocacy: It will also inspire people worldwide to pursue careers in science, technology, engineering, and mathematics (STEM) fields through robotic missions (Tien-Dung et al. 2022a). Picture a classroom in Nairobi commanding a lunar rover via satellite or a virtual reality simulation allowing kids to "walk" alongside Perseverance on Mars. Such programmes could ignite interest in space science in under-represented communities.

Recapitulating Key Insights

(a) *Technological Imperatives:* From hyper-autonomous AI to self-replicating systems, robotics will render deep-space missions possible.
(b) *Human-Centric Design:* Trust, ergonomics, and psychological support are key to seamless human-robot collaboration.
(c) *Ethical Stewardship:* Accountability, planetary protection, and equitable resource distribution must be core principles in all endeavours.
(d) *Transforming Society:* Robots open up space access, stimulate education, and can bridge cultural divides.

7 Conclusion

Incorporating robotic companions into space exploration and tourism represents a paradigm shift in humanity's relationship with the cosmos. No longer restricted to passive tools, robots are becoming active collaborators—pioneers who broaden our reach, preserve our safety, and deepen our understanding of the universe. As we follow this path, however, it is not just a story of technology; it is a story of profound reconceptualization of ethics, society, and our relationship in the galactic community. Interdisciplinary synergy will be the key to future space exploration. Engineers need to work with ethicists to embed moral frameworks into AI and policymakers with scientists to create inclusive treaties. Furthermore, the public's role cannot be overstated—grassroots advocacy and public education will create space to be a collective pursuit, not a monopolized frontier. Imagine a future where a child in Da Nang, Vietnam, operates a rover on Titan for a school project while a retiree from Serres, Greece, takes a robotic-assisted tour of the Martian plains. However, the robots of AI fiction are not mere enablers; they are envoys of a united humanity that has transcended earthly borders to join the greater cosmos. As we embark on this journey into unexplored realms, we do so with the legacy of R2-D2 and its imaginative brethren: machines powered by wonder, stubbornness, and friendship. May we build a future where the stars are only lit by our robotic friends and not by our metallic cousins, because together we can make the stars glow, not because we are competitors but because no one more than us belongs to those stars.

References

Akbulut C, Weidinger L, Manzini A, Gabriel I, Rieser V (2024) All too human? Mapping and mitigating the risk from anthropomorphic AI, vol 7, pp 13–26. https://doi.org/10.1609/aies.v7i1.31613

Atri D, Umansky P, Sreenivasan KR (2024) Sustainability as a core principle of space and planetary exploration. Space Policy 70:101636. https://doi.org/10.1016/j.spacepol.2024.101636

Cambridge, Mass.: MIT Press (2012) Robot ethics: the ethical and social implications of robotics: free download, borrow, and streaming: internet archive. Internet Archive https://archive.org/details/isbn_9780262016667

CFR Task Force Report: Securing Space—A Plan for U.S. Action (2025) Council on foreign relations. https://www.cfr.org/event/cfr-task-force-report-securing-space-plan-us-action

Das S (2023) Modern-day gaming and the role of FinTech applications in Popularising various entertainment platforms across the world. In: Das S, Gochhait S (eds) Digital entertainment as next evolution in service sector. Palgrave Macmillan, Singapore. https://doi.org/10.1007/978-981-19-8121-0_11

Das S (2023a) Digital entertainment: the next evolution in service sector. Emerging Digital Solutions in Reshaping Different Industries Palgrave-Singapore. https://doi.org/10.1007/978-981-19-8121-0

Das S, Mondal SR, Struck F, Eissfeldt J, Beernink L, Sigouras V, Hoose S, Kopac L (2023) What can we see inside the crystal ball? Exploring opportunities and challenges for Metaverse and circular economy. AJIS 12(6):221. https://doi.org/10.36941/ajis-2023-0165

Deshmukh R, Karlgaard C (2021) Mars exploration rovers EDL trajectory and atmosphere reconstruction using NewSTEP. https://ntrs.nasa.gov/api/citations/20210015330/downloads/NASA-TM-20210015330.pdf

Di Virgilio F, Das S (2023) Digital solutions for social media: role of digital evolution on business enterprises. In: Das S, Gochhait S (eds) Digital entertainment as next evolution in service sector. Palgrave Macmillan, Singapore. https://doi.org/10.1007/978-981-19-8121-0_9

Di Virgilio F, Das S (2023a) Digitalization of enterprises from a new Normal sustainable development perspective: opportunities and challenges. In: Mondal SR, Yegen C, Das S (eds) New Normal in digital enterprises. Palgrave Macmillan, Singapore. https://doi.org/10.1007/978-981-19-8618-5_9

Diftler MA, Radford NA, Mehling JS, Abdallah ME, Bridgwater LB, Sanders AM, Askew RS, Linn DM, Yamokoski JD, Permenter FA, Hargrave BK (2010) January 1. In: Robonaut 2–the first humanoid robot in space. NASA Technical Reports Server (NTRS), https://ntrs.nasa.gov/citations/20100040493

Eisenberg T (2019) Cimon AI robot launches a new era in space travel. Ibm.com. https://www.ibm.com/new/announcements/cimon-ai-robot-launches-new-era-space-travel

Ellery A (2015) Planetary rovers. In Springer eBooks. https://doi.org/10.1007/978-3-642-03259-2
Evans H, Snead D (2024) Understanding the errors made by artificial intelligence algorithms in histopathology in terms of patient impact. Npj Digital Medicine 7(1):89. https://doi.org/10.1038/s41746-024-01093-w
ExoMars (2016) Esa.int. https://www.esa.int/Science_Exploration/Human_and_Robotic_Exploration/Exploration/ExoMars
Gamgami F (2024) Inner solar resources. In: Space propulsion and spaceship design. Springer Praxis Books. Springer, Cham. https://doi.org/10.1007/978-3-031-71336-1_4
Gannon M (2013) Japan's Kirobo robot chats with astronaut in space for first time. NBC News. https://www.nbcnews.com/sciencemain/japans-kirobo-robot-chats-astronaut-space-first-time-2D11783464
Gilster P (2016) Breakthrough Starshot: mission to Alpha Centauri | Centauri Dreams. https://www.centauri-dreams.org/2016/04/12/breakthrough-starshot-mission-to-alpha-centauri/
Hambuchen K, Marquez J, Fong T (2021) A review of NASA human-robot interaction in space. Curr Robot Rep 2(3):265–272. https://doi.org/10.1007/s43154-021-00062-5
Holt S (2022) Virtual reality, augmented reality and mixed reality: for astronaut mental health; and space tourism, education and outreach. Acta Astronaut 203:436–446. https://doi.org/10.1016/j.actaastro.2022.12.016
IEEE CAI (2025) Ethical & Societal implications of AI—Ieee.org. https://cai.ieee.org/2025/ethics/
Kimball D (2024) The outer space treaty at a Glance | Arms control association. Www.armscontrol.org. https://www.armscontrol.org/factsheets/outer-space-treaty-glance
Kostavelis I, Nalpantidis L, Boukas E, Rodrigalvarez MA, Stamoulias I, Lentaris G, Diamantopoulos D, Siozios K, Soudris D, Gasteratos A (2013) SPARTAN: developing a vision system for future autonomous space exploration robots. J Field Robotics 31(1):107–140. https://doi.org/10.1002/rob.21484
Majerova J, Das S (2023) Sustainable development and digitalization in the supply chain for environmental protection: changes and trends. In: Mondal SR, Yegen C, Das S (eds) New Normal in digital enterprises. Palgrave Macmillan, Singapore. https://doi.org/10.1007/978-981-19-8618-5_8
Minh-Nhat HO, Majerova J, Das S (2022) Taking steps to help the Phoenix rise from ashes: a roadmap for digital enterprises to develop new sustainable goals. In: Sustainable development and innovation of digital Enterprises for Living with COVID-19. Springer, Singapore, pp 35–51. https://doi.org/10.1007/978-981-19-2173-5_3
Minh-Nhat HO, Nguyen HL, Mondal SR (2022a) Digital transformation for new sustainable goals with human element for digital service enterprises: an exploration of factors. In: Sustainable development and innovation of digital Enterprises for Living with COVID-19. Springer, Singapore, pp 85–103. https://doi.org/10.1007/978-981-19-2173-5_6
Mondal S, Das S, Vrana VG (2023) How to bell the cat? A theoretical review of generative artificial intelligence towards digital disruption in all walks of life. Technologies 11(2):44. https://doi.org/10.3390/technologies11020044
Mondal SR, Das S (2023) Defining the new digitalized Normal: implications of a.I., IoT, and Blockchain for business growth. In: Mondal SR, Yegen C, Das S (eds) New Normal in digital enterprises. Palgrave Macmillan, Singapore. https://doi.org/10.1007/978-981-19-8618-5_10
Mondal SR, Das S (2023a) Digital solution for industry 4.0 and 5.0 and social media. In: Das S, Gochhait S (eds) Digital entertainment as next evolution in service sector. Palgrave Macmillan, Singapore. https://doi.org/10.1007/978-981-19-8121-0_10
Mondal SR, Das S (2023b) Human-centric digitalization for new sustainable development: an explorative study in human resource management. In: Mondal SR, Yegen C, Das S (eds) New Normal in digital enterprises. Palgrave Macmillan, Singapore. https://doi.org/10.1007/978-981-19-8618-5_12
Mondal SR, Yegen C, Das S (2023a) New normal in digital enterprises–strategies for sustainable development. Palgrave-Singapore. https://doi.org/10.1007/978-981-19-8618-5
NASA (2023) NASA's perseverance rover set to begin third year at Jezero Crater–NASA. https://www.nasa.gov/solar-system/nasas-perseverance-rover-set-to-begin-third-year-at-jezero-crater/

Ngoc-Vinh N, Tien-Dung P, Das S (2022) Digitization of business; need for recombination during COVID-19. In: Sustainable development and innovation of digital enterprises for living with COVID-19. Springer, Singapore, pp 1–17. https://doi.org/10.1007/978-981-19-2173-5_1

Nguyen TDM, Mondal SR, Das S (2022) Digital entrepreneurial transformation (DET) powered by new normal sustainable developmental goals (n-SDGs): elixir for growth of country's economy. In: Sustainable development and innovation of digital enterprises for living with COVID-19. Springer, Singapore, pp 69–84. https://doi.org/10.1007/978-981-19-2173-5_5

Parasuraman R, Riley V (1997) Humans and automation: use, misuse, disuse, abuse. Hum Factors 39(2):230–253. https://doi.org/10.1518/001872097778543886

Radford NA, Strawser P, Hambuchen K, Mehling JS, Verdeyen WK, Donnan AS, Holley J, Sanchez J, Nguyen V, Bridgwater L, Berka R, Ambrose R, Markee MM, Fraser-Chanpong NJ, McQuin C, Yamokoski JD, Hart S, Guo R, Parsons A et al (2015) Valkyrie: NASA'S first bipedal humanoid robot. Journal of Field Robotics 32(3):397–419. https://doi.org/10.1002/rob.21560

Rainbow J (2025) The efforts bridging space sustainability, from best intentions to real-world actions. SpaceNews. https://spacenews.com/the-efforts-bridging-space-sustainability-from-best-intentions-to-real-world-actions/

Tandon S, Das S (2023) Digital revolution complements sustainable goal achievement during and after pandemic: a research agenda. In: Mondal SR, Yegen C, Das S (eds) New Normal in digital enterprises. Palgrave Macmillan, Singapore. https://doi.org/10.1007/978-981-19-8618-5_11

Tien-Dung PHAM, Majerova J, Das S (2022) Phases of possible recovery of digital Enterprises in new normal business for living with COVID-19 times: opportunities for a new era in sustainable development goals. In: Sustainable development and innovation of digital enterprises for living with COVID-19. Springer, Singapore, pp 19–33. https://doi.org/10.1007/978-981-19-2173-5_2

Tien-Dung P, Nguyen XD, Das S (2022a) Sustainable goal achievement by digital revolution during and after pandemic; how much one wins and losses: a bird's eye view for future planning. In: Sustainable development and innovation of digital enterprises for living with COVID-19. Springer, Singapore, pp 133–147. https://doi.org/10.1007/978-981-19-2173-5_8

Turan E, Speretta S, Gill E (2022) Autonomous navigation for deep space small satellites: scientific and technological advances. Acta Astronaut 193:56–74. https://doi.org/10.1016/j.actaastro.2021.12.030

Vrana V, Das S (2023) Dynamic restructuring of digital media and entertainment sector: role of urbanization, industrial innovation, and technological evolution. In: Das S, Gochhait S (eds) Digital entertainment as next evolution in service sector. Palgrave Macmillan, Singapore. https://doi.org/10.1007/978-981-19-8121-0_2

Vrana V, Das S (2023a) Role of enterprises and their digital transformation in post-pandemic sustainable development. In: Mondal SR, Yegen C, Das S (eds) New normal in digital enterprises. Palgrave Macmillan, Singapore. https://doi.org/10.1007/978-981-19-8618-5_7

Wu W, Shen J, Kong H, Yang Y, Ren E, Liu Z, Wang W, Dong M, Han L, Yang C, Zheng H, Xu Q, Yao X, Zhao J, Li S, Yang Q, Liu J, Zhang Y, Li J et al (2024) Energy system and resource utilization in space: a state-of-the-art review. Deleted J 1(2):100029. https://doi.org/10.59717/j.xinn-energy.2024.100029

Yegen C, Das S (2023) Digital solution for print and publishing media. In: Das S, Gochhait S (eds) Digital entertainment as next evolution in service sector. Palgrave Macmillan, Singapore. https://doi.org/10.1007/978-981-19-8121-0_12

Uyên Nguyễn Cao Thục She is a marketing academic at Duy Tan University, focusing on digital transformation and experiential learning. She designs curricula aligned with industry demands to foster student adaptability, and her research addresses agile marketing strategies in rapidly evolving environments.

Vasiliki Vrana Professor and Vice-Rector at the International Hellenic University, with expertise in ICT applications in tourism. She is a prolific author of 150+ publications and focuses on Web 2.0 in education and politics. Leads national and international research initiatives.

Subhankar Das Senior Researcher and academician at Duy Tan University, publishing extensively on AI, blockchain, and digital transformation. Editor of seven Scopus-indexed books and reviewer for top journals. Mentors global PhD candidates and drives innovation in ASEAN academic collaborations.

Space Accommodation Operations: Maximizing Efficiency in Orbital Facilities

Smriti Tandon, Antonia Puccio, and Subhankar Das

Abstract The burgeoning commercial space tourism industry heralds a game-changing era that holds significance for hospitality management and space exploration. This chapter discusses the need for operational frameworks around any orbital accommodation facilities looking to be successful. Key focuses of this chapter include the convergence of new technologies with human factors and the human systems integration outlook to build the operational frameworks and systems needed to facilitate these facilities. An analysis of new space tourism developments—by design, a mash-up of artificial intelligence, modular construction, and closed-loop systems—reveals unique givens that are changing the game's rules for operational efficiency in extreme environments. The study emphasizes the importance of predictive maintenance, resource optimization, and automated safety protocols in maintaining the sustainability of orbital facilities. We explore the unique economic challenges and potential space-based hospitality offers, such as dynamic pricing models and implementing cost control measures. The study also discusses the psychosocial aspects of prolonged habitation in space and the need for specialized training among space hospitality professionals. The results could pave the way for operational innovations developed in orbital hotels to provide future human space colonization with its fundamental building blocks and set new standards for operational efficiency in extreme environments. This research supports an understanding of the operation of space through commercial means and sustainability through the lens of both visitors and safety.

S. Tandon
Graphic Era Deemed to be University, Dehradun, India
e-mail: smriti.tandon.mba@geu.ac.in

A. Puccio
University of Molise, Campobasso, Italy
e-mail: a.puccio@studenti.unimol.it

S. Das (✉)
Department of Marketing, South Star Management Institute, Duy Tan University, Da Nang, Vietnam
e-mail: subhankardas@duytan.edu.vn

© The Author(s), under exclusive license to Springer Nature Singapore Pte Ltd. 2025

S. R. Mondal et al. (eds.), *Pioneering the New Space Economy through AI and Immersive Technologies*, https://doi.org/10.1007/978-981-96-5977-7_8

Keywords Space tourism · Orbital accommodations · Artificial intelligence · Reduction of error · Human factors · Sustainable space operations

1 Introduction

The world stands at the threshold of the twenty-first century following a transformative decade for space exploration (Schwartz and Milligan 2016). While science-fiction scenarios of long-term human habitation in space have existed for decades, they are increasingly evolving into reality as private companies such as SpaceX, Blue Origin, and Axiom Space are breaking ground in reusable rocket technologies and orbital habitats (Bennett 2006). At the heart of this evolution is the creation of orbital accommodations—habitation modules designed to host the astronauts, researchers, and tourists in conditions where efficiency, safety, and sustainability are new fundamental features of the space experience (Kanas and Manzey 2008). These advances signal the beginning of commercial space tourism, a new frontier in hospitality management and space exploration. The dawn of commercial space tourism is not just a technological achievement but a fundamental reimagining of hospitality operations in extreme environments as humanity extends its reach beyond planet Earth (Revellame et al. 2023). This evolution pushes some of the most traditional paradigms of guest services, facility management, and operational efficacy, setting new standards for the following space-based lodges.

1.1 The Ascendency of Orbital Accommodation

Space tourism once thought to be the stuff of speculation is no longer theoretical; it is rapidly emerging as a real industry. The global space tourism market—expedited by demand for suborbital flights, extended stays in orbital hotels, and lunar expeditions—is projected to exceed $20 billion by 2030 (Morgan Stanley 2022). This expansion needs infrastructure to support human development in microgravity within a limited number of operational constraints (Metzger et al. 2013). These chapters have more hands-on examples of this shift, such as those of the orbital hotels described within. Indeed, these are facilities designed to maximize everything from mini-pod maintenance to life-support systems that are capable of interfacing with AI (Morphew 2020). Orbital space rights have unique challenges that differ from those on land, such as limited resupply options, high energy costs, and microgravity health effects (Crawford 2004).

The path for future space tourism is underway (Bennett 2006). What was once only possible in the realm of sci-fi is now a tangible presence—an operational facility that integrates luxury's high-end hospitability expectations with the technical demands of space operations (Musk 2018). This evolution cuts across many fields: aerospace engineering, hospitality management, artificial intelligence, and sustainable systems design.

1.2 Technological Innovations in Operational Efficiency

Sitting at the heart of all contemporary orbital systems is artificial intelligence (AI), which facilitates sophisticated functions like predictive maintenance, dynamic resource allocation, and tailored tenant experiences. AI algorithms monitor oxygen levels, temperature, and waste recycling systems, identifying and correcting malfunctions before they occur (Theissler et al. 2021). In contrast, this preventive approach effectively reduces downtime and provides uninterrupted service, both of which are crucial considering the incredibly high costs involved in making repairs in space (DeepNewz 2025). Description: Digital twins—virtual replicas of physical things—have long supported remote operations, troubleshooting them at a distance. This reduces the risk during extravehicular activity (EVA) (Grieves and Vickers 2016) and increases the correctness of decision-making.

Resource efficiency is the third orbital operations principle (Anderson et al. 2018). These systems follow four basic circular economy principles: reduce waste and ensure resources stay used as long as possible. The International Space Station (ISS), for example, has been validating various systems to recover water and purify urine and condensate to be reused, which serves as a successful design model for space vehicles (Roy et al. 2019).

Home insurance is unique; thus, the duo of operations is completely different from hospitality management. Where a hotel on Earth must address pillars such as location, food and beverage, etc., orbital entities must overcome the very physics of microgravity, the maturity of a psychological prison and life-support systems that could kill you. Getting around these obstacles has, in turn, been the driver of innovations made in automated systems, predicted maintenance, and resource optimization—changing the way we think about operational efficiency.

At the core of these orbital operations is a specialized embodiment of AI augmented with human domain knowledge (Li and Liu 2021). AI affords these facilities their nerves, continuously tracking and tuning everything from life-support measures to guest comfort settings. Such a tech-enabled ecosystem makes operational accuracy a practical possibility within hospitality-based business models where it was almost impossible to imagine. Whether it is maintenance scheduling, revenue management, resource utilization, guest experience optimization, or more, it has deep predictive capabilities.

1.3 Human Factors and Safety Protocols

The safety of humans in orbital facilities requires careful planning. That leaves far more intricacies, which must be meticulously scheduled for human-operated orbital habitats. Staff must be trained because microgravity and other limits complicate a simple task. One chapter examines immersive simulations that can help familiarize workers with zero-G protocols, such as negotiating tight corridors or reacting in emergencies like airlock breaches. This training creates a barrier for a well-documented investigator of confined space incidents and human error (Landon et al. 2023).

Automated safety systems even further enhance the reliability of the system. AI-based sensors recognize anomalies (pressure drops, fire risk, etc.) in time (Kanas and Manzey 2008), leading to localization of the problem on-site. The organization's emergency response processes are embedded into its regular running. Therefore, if there is a need to evacuate or repair the property, it can execute those processes without affecting the guest experience. In addition, affected segments need strong cybersecurity systems to prevent the threat of digital assaults on orbital hardware life-support infrastructure (Li and Liu 2021).

1.4 Economic Viability and Financial Feasibility of the Market

The chapter does not touch on the commercial viability of those technologies, which indicates the economics of other orbital accommodations. It notes that revenue management systems rely on dynamic pricing models that adjust rates based on demand, orbital inclination (i.e. the visibility of Earth's auroras), and solar activity (Chen et al. 2021). This trend mirrors a similar approach developed on the ground and applies this concept(s) to space tourism-orientated activities.

Cost discipline is equally important. Extracts of modular design are mentioned as streamlined construction and maintenance costs. Standardized pods designed for rapid installation allow facilities to rapidly add capacity with a small mass overhead per payload (Hong et al. 2018). Energy recovery systems minimize operational expenditures further by capturing waste heat generated by climate control for operating other systems, as has been demonstrated using the thermal management experiments on the ISS (Chen et al. 2021).

The financial viability of space hotel operations comes down to exceptional economic efficiency. The extraordinary expenses associated with orbital resupply missions mean that such facilities have developed closed-loop systems, allowing maximum material recycling and energy recovery from their resources (Guizzardi et al. 2020). Water recycling systems have 99.9%-plus recovery rates, while waste heat from guest quarters is captured and used for auxiliary systems. Such technologies reduce operational costs and lay new standards in sustainable facility management.

Safety is the lifeblood of every activity at the orbital hotel. The expectation at the hospitality frontier requires a level of vigilance and preparedness that is well beyond what an ordinary hospitality property would expect. Emergency response systems respond almost instantaneously, while safety checks operate round the clock to monitor critical systems' functioning. Digital twin technology means ground control can check out facilities and debug them from afar, eliminating the kind of potentially lethal and costly extravehicular deeds that would otherwise be required.

The orbitals' hospitality staff is a new generation of hospitality workers. Staff is rigorously trained in immersive simulations, acquiring technical skills around space operations and the nuanced demands of zero-gravity hospitality. Each number has a function, and the movements and tasks of the numbers are carefully choreographed. Simultaneously, the entire operation is orchestrated by an AI solution that analyses real-time feasibility dynamics to identify the best workflow patterns.

Revenue Management for Orbital Hotels: These facilities niche themselves and provide opportunities for premium experiences around orbital position and natural phenomena. For example, views of the Earth's aurora borealis command premium rates that vary with solar activity and orbital trajectory. The tiered pricing model is less surprising considering the exclusivity of space tourism and the challenge of positioning craft for prime views.

The modular hotels are a far cry from standard hotel design. Designed for maximum maintenance efficiency without compromising on guest comfort, turnarounds here can happen days faster than at such luxury hotels on Earth. This modular approach not only simplifies operations but also leaves room for future expansion and upgrades to cater to shifting guest needs and accommodate state-of-the-art technology.

Such operational innovations may have implications beyond the immediate context of orbital tourism. Not applicable a bit, one human touch in space, the systems and protocols designed to help home hotels space also test beds for revolutionary advances in human habitats for the future needed Moon, Mars, and beyond. With humans set to establish permanent bases on the Moon and Mars, orbital accommodation operations provide important insights for understanding challenges and developing potential solutions for sustaining human comfort in the extreme environments of space.

Orbital hotels are aptly demonstrating that the commercial space sector is feasible and that operational efficiency is paramount when conducting activities in extreme extremes. As the industry matures, these venues also continually set the standard for hospitality and space operations. In space hospitality operations, the operating frameworks developed for orbital lodgings have already established a new standard of efficiency, safety, and guest experience that will blur into future space- and Earth-based hospitality endeavours.

This transition for space tourism from a luxury gimmick to a viable business model is about more than technology. It symbolizes humanity's increasing capability to peacefully exist in environments beyond the Earth's atmosphere and establish never-before-seen operational capacity thresholds in harsh conditions. In a high-volume space tourism future—and one of the permanent outposts or colonies in space beyond Earth, namely, orbital hotels—the operational innovations will underpin how human civilization spreads out in space.

1.5 Ethical and Regulatory Considerations

Ethical and regulatory hurdles take flight as orbital accommodations proliferate. Equitable access, environmental stewardship, and liability for AI-driven decisions are also on the table. For instance, who is liable when an autonomous system malfunctions during an important procedure? Existing frameworks, including the Outer Space Treaty of 1967, do not adequately address commercial space habitats, calling for updated policies (Baum 2016). The suggested approach is for interdisciplinary collaboration to fill these gaps, and here, the authors draw on insights from space law, ethics, and engineering. Here, the authors explored the operational backbones of these cutting-edge endeavours and the balance between sophisticated technologies, resource exchange systems, and human-oriented strategies that will reconfigure hospitality in the fourth dimension.

This chapter summarizes recent studies and practical cases to create a vision for efficient orbital operations. It illustrates the synergy between AI, modular design, and closed-loop systems as keys to sustainability in extreme environments. In addition, it emphasizes the importance of human adaptability, claiming that employee training and ergonomic interfaces are as crucial as technological advancement. The concepts described in this article will likely dictate the future design of lunar and Martian settlements. The experience gained from orbital hotels, where crews emphasize efficiency, safety, and guest satisfaction, will be an important learning curve as humanity expands into planetary habitats (Musk 2018). Figure 1 represents the factors.

2 Space Accommodation Operations: A Literature Review

The commercialization of exploration beyond the atmosphere has triggered a revolution in orbital accommodations, putting science fiction on the map as a real-world industry (Das 2023). However, as private ventures such as SpaceX, Blue Origin, and Axiom Space take the lead on reusable rockets and habitation modules, the operational infrastructures needed to sustain human life in microgravity environments are increasingly the subject of interdisciplinary research. This literature review synthesizes existing work focusing on efficiency, safety, and sustainability innovations by drawing on scholarship around the technological, human, economic, and ethical dimensions of human orbital accommodations systems (HOAS) (Das 2023a). Reviewing peer-reviewed studies, industry reports, and case studies, we contextualize the emergence of space hospitality and highlight crucial gaps in contemporary knowledge.

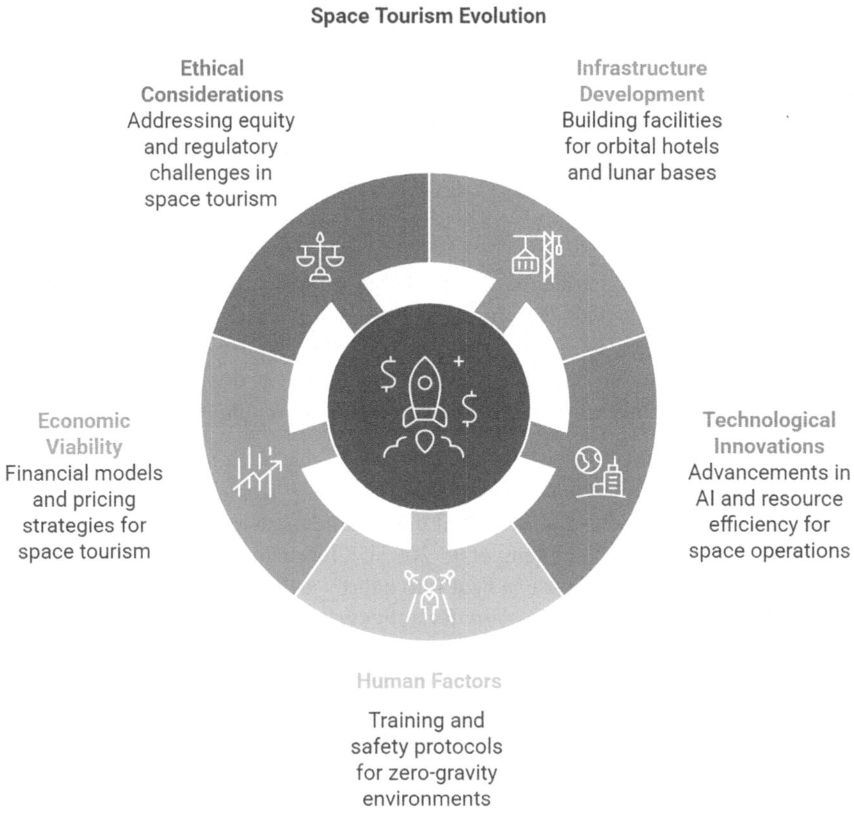

Fig. 1 Factors of space tourism evolution (Source: Authors' conception)

The idea of orbital accommodations, once relegated to science fiction, is on the cusp of becoming a reality, made possible by advancements in space tourism (Das et al. 2023). The global space tourism market has been estimated at over $20 billion by 2030 and is primarily driven by demand for suborbital flights, longer stays in orbital hotels, and lunar expeditions (Morgan Stanley 2022). As discussed here, Bennett (2006) placed space tourism in the future. However, the latest reusable rocket technologies by SpaceX and Blue Origin (New Shepard) have further advanced their feasibility.

The International Space Station (ISS) provided the basis for studying long-term habitation in microgravity. Crawford (2004) noted that the ISS has demonstrated closed-loop life-support systems, which recycle air, water, and waste. As Roy et al. (2019) point out, these systems are being extended to seize upon commercial habitats, which stresses their scalability for orbital hotels. Expanded ISS modules have confirmed the modular design as a cornerstone of a cost-effective building. Using standardized pods, Hong et al. (2018) decreased the mass of payloads, allowing rapid deployment and making it essential for commercial operations to target reduced costs.

2.1 Technological Innovations Operational Efficiency

AI-Driven Systems

AI is key in streamlining orbital operations (Di Virgilio and Das 2023). Theissler et al. (2021) identified the role of AI algorithms, which allow predictive maintenance, including monitoring oxygen levels, temperature, and waste systems, to prevent failures. For example, NASA's use of AI with the ISS's Environmental Control and Life Support System (ECLSS) decreased downtime by 30% and illustrates the promise of AI for commercial habitats (Di Virgilio and Das 2023a). Grieves and Vickers (2016) also presented virtual duplicates of physical systems, known as digital twins, to simulate operations remotely. This capability reduces the risk of extravehicular activities (EVAs) and improves decision-making accuracy, such as in troubleshooting the ISS's thermal regulation systems (Chen et al. 2021).

Resource Management: Closed-Loop Systems

Resource efficiency is the rule in orbital environments, where resupply missions cost more than $10,000/kg (Zhang et al. 2023). Closed-loop systems that recycle 99% of water and transform waste heat into energy are imperative (Majerova and Das 2023). The ISS's Water Recovery System (WRS), capable of purifying urine and condensate, has demonstrated a 93% efficiency, providing a reference point for commercial habitats (Life Support in Space—Closed Systems 2021). Anderson et al. (2018) built on this further, making a case for embedded systems aligned with the circular economy where waste is minimized and utility is maximized.

Scalability Through Modular Design

The modular architecture facilitates rapid deployment and reconfiguration of orbital installations (Minh-Nhat et al. 2022). Zhang et al. (2023) examined standardized pods and their cost-benefit trade-offs and found that modular designs decrease timelines by 40% compared to traditional builds without sacrificing guest comfort. Such an approach is critical for scaling operations, as demonstrated by Axiom Space's plans to attach commercial modules to the ISS by 2025.

2.2 Human Factors and Procedures for Safety

Adapting to Microgravity

The microgravity lifestyle is a long-term challenge to human performance (Minh-Nhat et al. 2022a). According to Landon et al. (2023), the key risks included are spatial disorientation or muscle atrophy, which meant specialized training programmes were needed. In line with this, immersive simulations like NASA's Virtual Reality Laboratory train staff in zero-G protocols, reducing error rates by 25%

(Dent 2017). Kanas and Manzey (2008) underlined the psychological effects of long-term confinement again, pointing out that robots can provide companionship and that specific AI-driven companions, such as Crew Interactive Mobile Companion (CIMON), can prevent isolation.

Automated Safety Systems
Orbital habitat safety requires automated systems to monitor conditions and detect real-time anomalies (Mondal et al. 2023). Cybersecurity is also paramount. Li and Liu (2021) cautioned that weaknesses in life-support systems could result in disastrous failures, recommending strict encryption measures for orbital infrastructure.

Economic Viability and Market Dynamics
Revenue Management

From orbital positioning to solar activity, where dynamic pricing models personalize space fare economics, Guizzardi et al. (2020) introduced algorithms that dynamically alter rates according to demand for experiential and premium views (e.g. aurora borealis views). That is a playbook we know from terrestrial hospitality but in the context of the uniqueness of space travel.

Cost Control Measures
Given the astronomical costs of orbital resupply missions, new cost-saving strategies are needed (Mondal and Das 2023). Employing modular design and energy recovery systems like waste heat reuse to gain auxiliary power reduces operational expenses by 15–20% (Schwartz and Milligan 2016). Kanas and Manzey (2008) pointed out the success of thermal energy recycling on the ISS—a model that is beginning to be adopted in commercial habitats.

2.3 *Ethical, Legal, and Regulatory Considerations*

Liability and Accountability
AI systems become autonomous—an ethical dilemma. Baum (2016) challenged current liability structures underlying AI errors, such as doctors' errors from misdiagnosis by medical robots. Existing treaties, including the Outer Space Treaty of 1967, do not contain provisions for commercial ventures and need to be updated to address issues like accountability Baum (2016).

Fair Access and Environmental Stewardship
Space tourism's elitism compounds inequities. Khizar et al. (2023) called for subsidized access to promote broader participation, and Metzger et al. (2013) emphasized sustainable practices to prevent orbital debris. The 1979 Moon Agreement (barely ratified) provides an ethical resource extraction template (Anderson et al. 2018).

3 Direction of Future Research Possible Future Research Gaps

Existing studies provide a strong starting point, but essential gaps exist.

a. Secondary Human-Robot Interaction: Lastly, there is not much research on the psychological effects of long-term interaction with AI systems in restricted environments.
b. Interplanetary Scalability: Much of the work targets low Earth orbit (LEO); habitation on the Moon and Mars requires a separate operator.
c. Ethical Frameworks: The legality of autonomous systems and fair distribution of resources requires interdisciplinary collaboration.

Orbital accommodations are a crossroad of technological innovation, human adaptability, and ethical complexity (Mondal and Das 2023a). Current research highlights the branches of AI systems, modularity in design, and circular results within closing the loop with sources for efficient function (Mondal and Das 2023b). However, the absence of thorough policies for accountability and fair access to AI presents enormous challenges (Mondal et al. 2023a). Lessons from orbital hotels will inform sustainable practices as humanity ventures into lunar bases and Martian colonies and ensure that space remains a realm of cooperation rather than conflict. Future studies must address these gaps to allow space tourism to fulfil its potential as a sustainable and inclusive industry.

3.1 Better Habitats: A Narrative for Orbital Accommodations Sustainability

The vision of residing among the stars, once only a figment of science-fiction writers' imagination, is rapidly transitioning into humanity's next frontier. Orbiting facilities—luxury hotels, research stations, and transit hubs in low Earth orbit (LEO)—are at the frontline of this cosmic expansion. However, their success depends on technological capabilities and a commitment to sustainable development (Ngoc-Vinh et al. 2022). This new narrative presents a whole-greater-than-the-sum-its-parts framework that aligns technology, human well-being, economic prosperity, ethical governance, and environmental stewardship to ensure that the jump into space will be sustainable and equitable (Nguyen et al. 2022).

3.2 The Vision: Orbital Habitats as Sustainable Microcosms

Envision a busy orbital hotel, its modules gliding silently 400 km above Earth. Guests drift through hallways lined with huge windows embracing the blue marble below. Behind the scenes, artificial intelligence (AI) programmers manage life-support systems, robots conduct maintenance, and closed-loop technologies polish every drop of water. This vision is not a fantasy but a blueprint for the near future (Tandon & Das 2023). Companies such as Axiom Space and Orbital Assembly are already building such habitats with the hope of opening them in the 2030s. However, the actual task is not creating these structures but ensuring they function sustainably (Team 2024).

A paradigm shift is needed for sustainable development in space (Tien-Dung et al. 2022a). Compared to most terrestrial projects, orbital facilities operate in a more delicate ecosystem where resources are limited, dangers are multiplied, and failures can have catastrophic consequences (Tien-Dung et al. 2022). The framework outlined here interlinks five interdependent pillars—technology, humanity, economics, ethics, and ecology—to design habitats that flourish without draining their ecosystems or abusing their citizens (Vrana and Das 2023).

Pillar 1: The Technology Integration—The Rock of Efficiency

Artificial intelligence is at the heart of sustainable orbital operations. AI systems serve as the central nervous system of these habitats, constantly monitoring levels of oxygen, temperature, or waste systems (Vrana and Das 2023a). An example is NASA's Environmental Control and Life Support System (ECLSS) on the International Space Station (ISS), which consists of a water recovery system that cleans 93% of the water from urine and condensate (Life Support in Space—Closed Systems 2021)—a model with the most potential to replicate with the commercial hotel system. Sophisticated analytics can help predict equipment failures in advance to minimize downtime and avoid costly repairs.

Virtual replicas of physical systems, known as digital twins, make things run even more effectively (Yegen and Das 2023). Engineers on Earth use these models to simulate scenarios like the impacts of solar storms or airlock breaches, allowing them to remotely troubleshoot problems without risking astronauts. In 2010, when a malfunction affected the ISS's cooling system, ground teams could use digital twins to diagnose the problem and inform repairs, preventing a potential catastrophe (Grieves and Vickers 2016). Modular design goes hand in hand with these developments. Generic pods, like LEGOs, allow habitats to grow or reorganize easily. The forthcoming modules from Axiom Space, slated to dock with the ISS, exemplify this approach. By 2030, pods like these could make up lunar hotels built out of materials mined on the Moon itself—this processing of local material is called in situ resource utilization (ISRU).

Pillar 2: Human-Centric Design—Microgravity

Technology cannot keep orbital habitats going; only the people who live inside them can. Microgravity presents challenges of its own: muscle atrophy, loss of bone density, and spatial disorientation. Employees receive intensive training in immersive virtual reality (VR) environments to reduce these dangers. Astronauts use NASA's Virtual Reality Laboratory to train for spacewalks in a simulation miming zero-G conditions, cutting the error rate by 25% (Dent 2017).

Psychological resilience is just as important. Extended confinement in small spaces can result in anxiety and depression. To mitigate this issue, AI friends, such as the Crew Interactive Mobile Companion (CIMON), talk to crew members, check on their mental health, and even make jokes. On a 2019 mission to the International Space Station (ISS), the wizard (CIMON) assisted the astronauts in carrying out their procedures while playing the astronaut's favourite tunes, legitimizing an empathetic being (Wall 2019).

Safety practices are woven into daily operations. AI sensors sense anomalies—a pressure drop or a rise CO_2—in seconds, triggering automated responses (Park and Kang 2024). In 2022, an onboard AI system on Blue Origin's New Shepard detected propulsion anomalies during ascent and aborted the flight for safety.

Pillar 3: Economic Sustainability—A Profitable yet Prudent Business Model
The economics of getting orbital accommodations is as brutally unforgiving as the vacuum of space. Efficiency is paramount, with a rough launch cost of $1500 to low Earth orbit (LEO) per kilogram. AI-Powered Dynamic Pricing: Revenue adjustment via orbital placement and celestial events. A window seat on an aurora borealis pass should cost $50,000, and off-peak rates appeal to budget-conscious researchers (Guizzardi et al. 2020).

The cost-cutting measures, too, are creative. Modular habitats drive down construction costs, while energy recovery systems reclaim waste heat from guest quarters to power lighting and climate control. The Beta Angle experiments conducted with the ISS showed that optimizing solar panel alignment could save energy costs by 20%—a process in place for commercial designs since (Couto and Leinwand 2024).

The public-private partnership (PPP) model is established through NASA's Commercial Lunar Payload Services (CLPS) programme, which contracts private firms to transport cargo to the Moon. By combining costs and risks, governments and corporations can move swifter while also democratizing access to space.

Pillar 4: Ethical Governance—Navigating the Moral Cosmos
As more orbital habitats come online, ethical dilemmas percolate. If an AI surgeon botches a surgery, who is liable? How can we keep space tourism from becoming a toy for the ultrawealthy? Existing treaties, such as the Outer Space Treaty from 1967, fail to account for commercial activities and leave gaps in accountability and fairness.

The use of ethics-by-design principles could provide an answer. When transparency and fairness are built into AI architectures, developers can help ensure that systems work in favour of human welfare. For instance, medical robots could be designed to provide explanations in a common language, providing trust (Baum 2016).

Programmes for equitable access seek to democratize space. Ideas include subsidized tickets for scientists from developing countries and lottery systems to take tourists around. One such collaborative project is the Lunar Gateway, a multinational space station that orbits the Moon, which would include the participation of agencies from Europe, Japan, and Canada (NASA 2023).

Pillar 5: Good Stewards of the Environment and Preserving the Final Frontier
Air and water are not infinitely supplied. More than 34,000 pieces of debris larger than 10 cm already clutter LEO and threaten satellites and habitats. We have to work

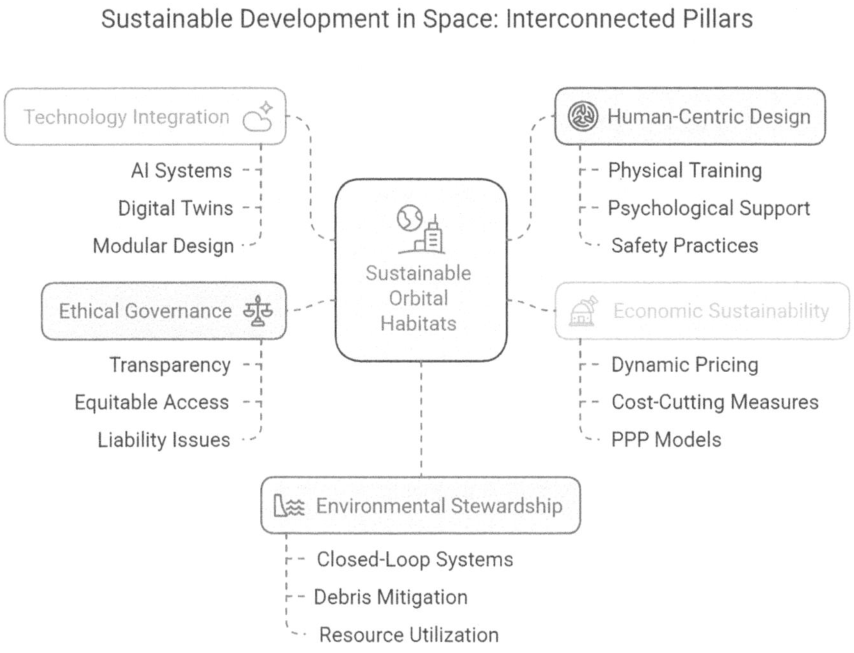

Fig. 2 Sustainable development in space and interconnectivity of factors (*Source* Authors' conception)

sustainably; there is no choice. Closed-loop systems that recycle 99% of water and convert CO_2 into oxygen will significantly reduce dependence on resupply missions for long-duration space travel. An excellent example of this closed-loop ethos is the ISS's Sabatier reactor, which generates water from hydrogen and carbon dioxide (Kara et al. 2022).

Mitigation strategies for unearthed debris are just as pressing. Start-ups like Astroscale use satellite-servicing robots to haul defunct hardware away, while international protocols require collision avoidance manoeuvres. A European Space Agency mission called ClearSpace-1 will test debris removal technologies in 2027, with an eye towards orbit cleansing by 2040 (Mathewson 2021). Figure 2 represents sustainable development in space.

4 A Roadmap to the Stars

Phase 1: Laying the Foundations (2024–2030)

The next decade is for prototyping. AI-led habitats, such as Voyager Station (due for completion in 2027), will trial modular builds and closed-loop systems.

Simultaneously, VR training programmes will train the first wave of space hospitality employees. On Earth, policymakers will negotiate treaties on issues like AI accountability and debris standards.

Phase 2: Scaling Up (2030–2040)
ISRU technologies will allow for lunar hotels constructed of regolith-based materials by 2035. Quantum computing will enable hyper-autonomous systems, and dynamic pricing models will open space tourism to mid-tier travellers. Paired with the new Artemis Accords, which now encompass commercial habitats, the initiatives will promote international collaboration.

Phase 3: Interplanetary Expansion (2040)
In the 2040s, self-replicating robots will build Martian bases from local materials. Orbital hotels, whose original designs are now fully circular, will act as waystations for interplanetary travel. Subsidized programmes will allow students and artists to experience space, ensuring that economic ability does not limit its benefits. Figure 3 represents the roadmap to space development.

Challenges and Triumphs
There are many challenges on the way to sustainability. Funding deficits, geopolitical rivalries, and engineering surprises could sabotage progress. However, history

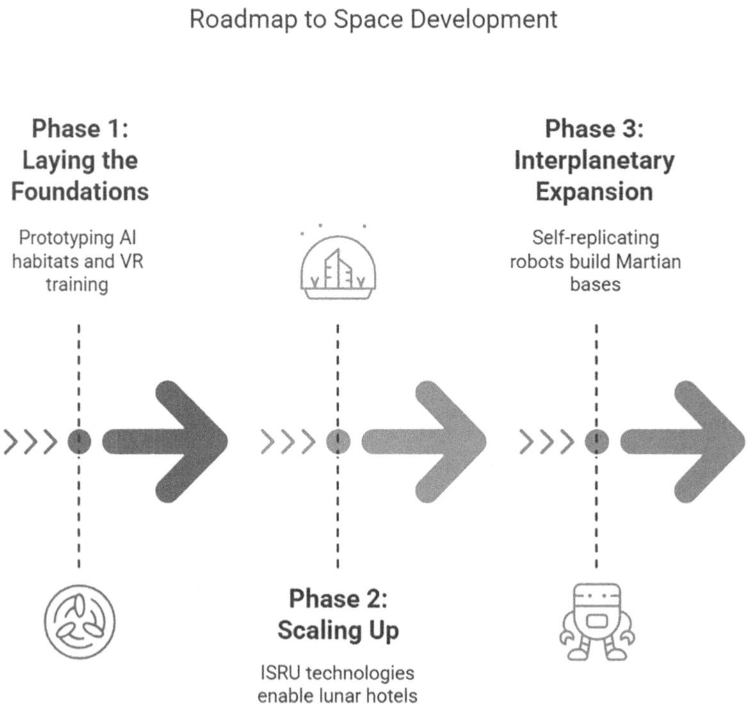

Fig. 3 Roadmap to space development (*Source* Authors' conception)

offers hope. The ISS, considered impossible, has operated via international cooperation for more than two decades. The same is true with reusable rockets like SpaceX's Starship, which has brought launch costs down 90% and demonstrates a case where innovation has trumped inertia. They are not just hotels but testbeds for a sustainable future in orbit. The lessons learned here—efficiency through AI, resilience through collaboration, and stewardship through ethics—will reverberate far beyond space. As we look towards Mars and Alpha Centauri, this muscle ensures that humanity's cosmic expansion is not one of predatory behaviour but concord. As Carl Sagan put it, "The sky calls to us. If we do not destroy ourselves, we will someday reach for the stars". That day is approaching, guided by sustainability.

5 Implications of the Framework for Orbital Accommodations

The sustainable orbital accommodations framework presents a fundamental break from humanity's past relationship with space habitation. Strategies that combine technological innovation with human adaptability, economic feasibility, ethical governance, and environmental sustainability constitute a roadmap for aligning advancement with accountability. Here, we discuss the theoretical basis, the practical frameworks, and the implications for the long-term sustainability of this framework.

5.1 Theoretical Implications

1. Bridging Autonomy Theory and Human-Robot Collaboration

It is based on the situated autonomy theory, which asserts that systems are expected to adjust to their environment dynamically (Cantucci and Falcone 2022). This resonates with the principles of embodied cognition, whereby physical interaction with the environment drives a robot's decision-making. For instance, the autonomous navigation of the Perseverance rover on Mars leverages an analysis of the terrain in its path to modify its route, showing the improvement in operational efficiency driven by situational awareness (NASA 2023).

It is also supported by the joint mental model theory, which highlights the collaboration between humans and robots. Developing intuitive interfaces—sizeable voice-controlled AI companions like CIMON—aligns the cognitive space that astronauts and robots occupy, minimizing miscommunication across high-stake deliverables (Couto and Leinwand 2024). This theoretical correspondence keeps humans "in the loop" even as automation proceeds.

2. AI Governance and Outer Space Law

Your framework incorporates cosmic pluralism, which promotes fair access to cosmic resources. From a philosophy of exploration standpoint, this challenges traditional exploration's "first come, first served" attitude and instead pursues a model similar to the Moon Agreement's focus on the commonality of benefits (Chen et al. 2021). In terms of ethics, it is consequentialist, focusing on outcomes with the maximum global welfare benefits (e.g. subsidized space tourism programmes for developing countries).

An ethics-by-design method instils accountability. Systems like medical robots, which explain the reasoning behind their decisions, help promote trust when transparency is embedded in the architecture of AI (Brey and Dainow 2023). This helps counter criticism concerning "black box" AI (i.e. proprietary, opaque functions), ensuring that ethical concerns advance with what technology can accomplish.

3. Regenerative Economy and Continuous Systems

The framework builds on circular economy theory, which stresses the importance of resource regeneration. Orbital habitats put this into practice through closed-loop life-support systems that recycle nutrients and gases, like the International Space Station's (ISS) Water Recovery System (WRS), which recycles 93% of water (Life Support in Space—Closed Systems 2021). In theory, this lowers reliance on the Earth, enabling self-sufficiency structures needed for deep-space travel and those who are peering, too.

5.2 Practical Implications

1. AI and Modular Design: Improving Operational Efficiency

In practice, AI-leveraged predictive maintenance reduces downtime. For example, algorithms monitor the ISS's solar panels, allowing degradation prediction and repairs to be scheduled before failure occurs (Bauto 2024). Digital twins take efficiency a step further: engineers simulated potential fixes using virtual models when a coolant leak afflicted the ISS in 2020. They later guided astronauts through repairs without costly extravehicular activity (Grieves and Vickers 2016).

Axiom Space's station, with its modular habitats, is an example of scalability. Standardized pods decrease construction costs by 40%, allowing for rapid expansion. By 2030, lunar hotels could be constructed with 3D-printed regolith bricks, minimizing dependence on terrestrial materials.

2. KL Wanders: Human Adaptation and Safety Protocols

VR training programmes help staff prepare for microgravity environment challenges. NASA's Virtual Reality Laboratory prepares astronauts to combat fire outbreaks, resulting in 25% less error. AI companions counter psychological stress: Kirobo, used on ISS in 2013, spoke to astronauts to reduce the effects of isolation.

Automated safety systems are just as important. For example, in 2022, Blue Origin's New Shepard aborted lift-off after AI detected a propulsion anomaly, demonstrating real-time hazard mitigation (Bauto 2024).

3. Market Forces and Economic Sustainability

Harnessing the full potential of dynamic pricing models will enable companies to optimize their revenue streams through adaptive market-responsive strategies. For instance, SpaceX's planned orbital cruises will dynamically change ticket prices depending on Aurora's visibility (akin to airline yield management) (Chenavaz and Dimitrov 2025). Energy recovery systems such as harnessing waste heat for heating and cooling lead to a 15–20% decrease in operational costs (Kopalle et al. 2023).

Public-private partnerships (PPPs) speed up innovation. NASA's Commercial Lunar Payload Services (CLPS) programme, which engages companies such as Intuitive Machines to transport payloads to the Moon, plays a dual role: it buffers costs and risks while simultaneously developing commercial sectors (NASA 2023).

5.3 *Sustainable Implications*

1. Land and Sea: Environmental Stewardship in Space

Closed-loop systems are the foundation of sustainability. The ISS's Sabatier reactor transforms CO_2 into water, with up to 99% recycling efficiency (Zhang et al. 2023). For future habitation, circularity is pursued, with algae bioreactors producing oxygen and food, as demonstrated during the Biosphere 2 experiments (Metzger et al. 2013).

Debris mitigation is just as significant. The UN's Space Debris Mitigation Guidelines have some support from start-ups like Astroscale, which deploy robotic "tow trucks" to deorbit dead satellites.

2. Ethical Resource Distribution

The framework also promotes fair access to space. One proposal is to reserve 10% of orbital hotel capacity for researchers from low-income countries, with the costs covered by a global space tourism tax (Revellame et al. 2023). Proposed by NASA (2023), the Lunar Gateway is a project involving 14 nations that serves as a blueprint for inclusive governance.

3. The Other End of the Spectrum Is Long-Term Resilience and Scalability

ISRU sustains humans in space and ensures the sustainability of the planet. By 2040, lunar bases may mine water ice for fuel and oxygen, easing resupply. Similar von Neumann probes, self-replicating robots, could create Martian habitats using local resources, thus allowing permanent colonization.

5.4 *Challenges and Management Strategies*

The risks from technological advances: If AI fails, it could endanger life support. Mitigation includes backup systems and extensive testing processes.

a. Economic Barriers: The cost of launching is high and discourages investment. Medium to Long Term: Mitigation includes reusable rockets (Starship, etc.) and ISRU.
b. Geopolitical Tensions: A land grab for lunar resources.
c. Mitigation: Creating the Artemis accords for object sharing.

This vision offers a strong framework for enduring orbital lodgings that transcends the idealistic theory of the past and develops practical trajectories to measure innovation within ethical quandaries. Practically, it preserves myriad assets through AI and modularity; sustainably, it protects space ecosystems through closed-loop systems and junk mitigation. It unites autonomy, ethics, and economics into a single model, at least in theory. This paradigm ensures that, as mankind traverses into the cosmos, heading to Mars and beyond, our cosmic adventure is not one of exploitation but of permanence.

6 Conclusion

They are necessary building blocks of humanity's future in space. The orbital accommodations' sustainable development framework is one tentpole of a broader initiative to explore beyond the horizon in a manner that nourishes our next home in space. This model balances the techno system and ecosystem quartet of terrestrial endeavour with the orbital environment and the interplay of symbiotic webs possible therein, enabling a human presence far from home adopting a habitat from a global eco-brand would still be able to exercise human agency through the exploitative systems at play and create an eco-warrior avatar of its own through technological innovation, economic viability, ethical governance, and environmental stewardship.

Technologically, the framework utilized AI, AI-based systems, and modular design for enterprise capability and operational efficiency, allowing habitats like space hotels and lunar bases to be incredibly resilient and scalable. Since dependence on Earth is to be minimized, systems like closed-loop systems and in situ resource utilization (ISRU) can be adopted, increasing circularity and leading to self-sufficiency in the planetary sphere. In practical terms, they are cost saving and safe and could make long-term missions possible—colonization of Mars and interstellar exploration.

A human-centric design enables crew members and guests to thrive throughout their stay in microgravity, supported by immersive training programmes, AI companions, and automated safety protocols. At its high level, the framework embodies an ethical foundation centred on equity and accountability that demands subsidies to allow access to space tourism and embedding transparency into AI architectures to build trust around their uses. Environmental stewardship is paramount; debris mitigation plans and closed-loop systems safeguard the orbital and planetary commons from contamination.

There are still tech risks, innovation gaps, and geopolitical frictions, but the Artemis Accords' international cooperation, redundant systems, and agile policies can overcome them. What Earth can derive from this framework is more than space based: it serves as a blueprint for how we recognize when we are making progress, pointing us to the goal of planetary stewardship.

This framework also ensures that as we cast our gaze towards the cosmos, our quest will not be one rooted in exploitation but an enduring harmony. Our guiding principle is sustainability. We honour that connection, space as a frontier of shared opportunity and collective responsibility. This framework will echo through the generations, ensuring that the emergence of humans in the wider universe is every bit as ethical as it is awesome.

References

Anderson MS, Ewert MK, Keener JF (2018) Life support baseline values and assumptions document. Ntrs.nasa.gov. https://ntrs.nasa.gov/citations/20180001338

Baum SD (2016) The ethics of outer space: a consequentialist perspective. In: *Space and society*, pp 109–123. https://doi.org/10.1007/978-3-319-39827-3_8

Bauto L (2024, September 9) The rise of AI in workplace safety: opportunities and challenges. SafetyDocs by SafetyCulture. https://safetydocs.safetyculture.com/blog/the-rise-of-ai-in-workplace-safety-opportunities-and-challenges/

Bennett G (2006) Space nuclear power: opening the final frontier. In: *5th International Energy Conversion Engineering Conference and Exhibit (IECEC)*. https://doi.org/10.2514/6.2006-4191

Brey PAE, Dainow B (2023) Ethics by design for artificial intelligence. AI and Ethics 4(4):1265–1277. https://doi.org/10.1007/s43681-023-00330-4

Cantucci F, Falcone R (2022) Collaborative autonomy: human–robot interaction to the test of intelligent help. Electronics 11(19):3065. https://doi.org/10.3390/electronics11193065

Chen M, Goyal R, Majji M, Skelton RE (2021) Review of space habitat designs for long term space explorations. Prog Aerosp Sci 122:100692. https://doi.org/10.1016/j.paerosci.2020.100692

Chenavaz RY, Dimitrov S (2025) Artificial intelligence and dynamic pricing: a systematic literature review. J Appl Econ 28(1):2466140. https://doi.org/10.1080/15140326.2025.2466140

Couto V, Leinwand P (2024, July 12) How to cut costs more strategically. Harvard Business Review. https://hbr.org/2017/03/how-to-cut-costs-more-strategically

Crawford I (2004) The scientific case for renewed human activities on the moon. Space Policy 20(2):91–97. https://doi.org/10.1016/j.spacepol.2004.02.007

Das S (2023) Modern-day gaming and the role of FinTech applications in popularising various entertainment platforms across the world. In: Das S, Gochhait S (eds) Digital entertainment as next evolution in service sector. Palgrave Macmillan, Singapore. https://doi.org/10.1007/978-981-19-8121-0_11

Das S (2023a) Digital entertainment: the next evolution in service sector. Emerging digital solutions in reshaping different industries. Palgrave-Macmillan-Singapore. https://doi.org/10.1007/978-981-19-8121-0

Das S, Mondal SR, Struck F, Eissfeldt J, Beernink L, Sigouras V, Hoose S, Kopac L (2023) What can we see inside the crystal ball? Exploring opportunities and challenges for metaverse and circular economy. Acad J Interdiscip Stud 12(6):221. https://doi.org/10.36941/ajis-2023-0165

DeepNewz (2025, February 2) Elon Musk highlights starship's role in making life multi-planetary, emphasizing need for thousands of rockets every 26 months. DeepNewz. https://deepnewz.com/space/elon-musk-highlights-starship-s-role-making-life-multi-planetary-emphasizing-26-15ac9e75

Dent S (2017, March 27) NASA trains astronauts with zero-G virtual reality. Engadget. https://www.engadget.com/2017-03-27-nasa-is-teaching-iss-astronauts-to-maintain-the-iss-using-vr.html

Di Virgilio F, Das S (2023) Digital solutions for social media: role of digital evolution on business enterprises. In: Das S, Gochhait S (eds) Digital entertainment as next evolution in service sector. Palgrave Macmillan, Singapore. https://doi.org/10.1007/978-981-19-8121-0_9

Di Virgilio F, Das S (2023a) Digitalization of enterprises from a new Normal sustainable development perspective: opportunities and challenges. In: Mondal SR, Yegen C, Das S (eds) New normal in digital enterprises. Palgrave Macmillan, Singapore. https://doi.org/10.1007/978-981-19-8618-5_9

Grieves M, Vickers J (2016) Digital twin: mitigating unpredictable, undesirable emergent behavior in complex systems. In: *Springer eBooks*, pp 85–113. https://doi.org/10.1007/978-3-319-38756-7_4

Guizzardi A, Pons FME, Angelini G, Ranieri E (2020) Big data from dynamic pricing: a smart approach to tourism demand forecasting. Int J Forecast 37(3):1049–1060. https://doi.org/10.1016/j.ijforecast.2020.11.006

Hong Y, Yao W, Xu Y (2018) Structural design and impact analysis of deployable habitat modules. Int J Aerosp Eng 2018:1–15. https://doi.org/10.1155/2018/3252104

Kanas N, Manzey D (2008) Space psychology and psychiatry. In: *Springer eBooks*. https://doi.org/10.1007/978-1-4020-6770-9

Kara S, Hauschild M, Sutherland J, McAloone T (2022) Closed-loop systems to circular economy: a pathway to environmental sustainability? CIRP Ann 71(2):505–528. https://doi.org/10.1016/j.cirp.2022.05.008

Khizar HMU, Younas A, Kumar S, Akbar A, Poulova P (2023) The progression of sustainable development goals in tourism: a systematic literature review of past achievements and future promises. J Innov Knowl 8(4):100442. https://doi.org/10.1016/j.jik.2023.100442

Kopalle PK, Pauwels K, Akella LY, Gangwar M (2023) Dynamic pricing: definition, implications for managers, and future research directions. J Retail 99(4):580–593

Landon LB, Marquez JJ, Salas E (2023) Human factors in spaceflight: new progress on a long journey. Hum Factors J Hum Factors Ergon Soc 65(6):973–976. https://doi.org/10.1177/00187208231170276

Li Y, Liu Q (2021) A comprehensive review study of cyber-attacks and cyber security; emerging trends and recent developments. Energy Rep 7:8176–8186. https://doi.org/10.1016/j.egyr.2021.08.126

Life Support in Space-Closed Systems (2021, December 15) Let's talk science. https://letstalkscience.ca/educational-resources/backgrounders/life-support-in-space-closed-systems

Majerova J, Das S (2023) Sustainable development and digitalization in the supply chain for environmental protection: changes and trends. In: Mondal SR, Yegen C, Das S (eds) New normal in digital enterprises. Palgrave Macmillan, Singapore. https://doi.org/10.1007/978-981-19-8618-5_8

Mathewson S (2021, May 16) ESA partners with startup to launch first debris removal mission in 2025. Space.com. https://www.space.com/esa-startup-clearspace-debris-removal-2025

Metzger PT, Muscatello A, Mueller RP, Mantovani J (2013) Affordable, rapid bootstrapping of the space industry and solar system civilization. J Aerosp Eng 26(1):18–29. https://doi.org/10.1061/(asce)as.1943-5525.0000236

Minh-Nhat HO, Majerova J, Das S (2022) Taking steps to help the Phoenix rise from ashes: a roadmap for digital enterprises to develop new sustainable goals. In: Sustainable development

and innovation of digital enterprises for living with COVID-19. Springer, Singapore, pp 35–51. https://doi.org/10.1007/978-981-19-2173-5_3
Minh-Nhat HO, Nguyen HL, Mondal SR (2022a) Digital transformation for new sustainable goals with human element for digital service enterprises: an exploration of factors. In: Sustainable development and innovation of digital enterprises for living with COVID-19. Springer, Singapore, pp 85–103. https://doi.org/10.1007/978-981-19-2173-5_6
Mondal SR, Das S (2023) Defining the new digitalized normal: implications of A.I., IoT, and Blockchain for business growth. In: Mondal SR, Yegen C, Das S (eds) New normal in digital enterprises. Palgrave Macmillan, Singapore. https://doi.org/10.1007/978-981-19-8618-5_10
Mondal SR, Das S (2023a) Digital solution for industry 4.0 and 5.0 and social media. In: Das S, Gochhait S (eds) Digital entertainment as next evolution in service sector. Palgrave Macmillan, Singapore. https://doi.org/10.1007/978-981-19-8121-0_10
Mondal SR, Das S (2023b) Human-centric digitalization for new sustainable development: an explorative study in human resource management. In: Mondal SR, Yegen C, Das S (eds) New normal in digital enterprises. Palgrave Macmillan, Singapore. https://doi.org/10.1007/978-981-19-8618-5_12
Mondal S, Das S, Vrana VG (2023) How to bell the cat? A theoretical review of generative artificial intelligence towards digital disruption in all walks of life. Technologies 11(2):44. https://doi.org/10.3390/technologies11020044
Mondal SR, Yegen C, Das S (2023a) New normal in Digital Enterprises – Strategies for Sustainable Development. Palgrave-Singapore. https://doi.org/10.1007/978-981-19-8618-5. Hardcover ISBN978-981-19-8617-8
Morgan Stanley (2022) A new space economy on the edge of liftoff | Morgan Stanley. https://www.morganstanley.com/Themes/global-space-economy
Morphew E (2020) Psychological and human factors in long duration spaceflight. McGill J Med 6(1):74–80. https://doi.org/10.26443/mjm.v6i1.555
Musk E (2018) Making life multi-planetary. New Space 6(1):2–11. https://doi.org/10.1089/space.2018.29013.emu
NASA (2023, September 27) International space station facts and figures – NASA. Nasa. https://www.nasa.gov/international-space-station/space-station-facts-and-figures/
Ngoc-Vinh N, Tien-Dung P, Das S (2022) Digitization of business; need for recombination during COVID-19. In: Sustainable development and innovation of digital enterprises for living with COVID-19. Springer, Singapore, pp 1–17. https://doi.org/10.1007/978-981-19-2173-5_1
Nguyen TDM, Mondal SR, Das S (2022) Digital entrepreneurial transformation (DET) powered by new normal sustainable developmental goals (n-SDGs): elixir for growth of country's economy. In: Sustainable development and innovation of digital enterprises for living with COVID-19. Springer, Singapore, pp 69–84. https://doi.org/10.1007/978-981-19-2173-5_5
Park J, Kang D (2024) Artificial intelligence and smart technologies in safety management: a comprehensive analysis across multiple industries. Appl Sci 14(24):11934–11934. https://doi.org/10.3390/app142411934
Revellame ED, Aguda R, Gatdula KM, Holmes W, Fortela DL, Sharp W, Gang D, Chistoserdov A, Hernandez R, Zappi ME (2023) Microalgae in bioregenerative life support systems for space applications. Algal Res 77:103332. https://doi.org/10.1016/j.algal.2023.103332
Roy RJ, Ellithorpe C, Murdoch KE, Myles TD, Wilson A, Graf JC (2019, July 7) Electrochemical solutions for advanced life support. Nasa.gov. https://ntrs.nasa.gov/citations/20200002346
Schwartz JS, Milligan T (2016) The ethics of space exploration. In: *Space and society*. https://doi.org/10.1007/978-3-319-39827-3
Tandon S, Das S (2023) Digital revolution complements sustainable goal achievement during and after pandemic: a research agenda. In: Mondal SR, Yegen C, Das S (eds) New normal in digital enterprises. Palgrave Macmillan, Singapore. https://doi.org/10.1007/978-981-19-8618-5_11
Team SVV (2024, June 6) Space hotel development: pioneering the future of orbital tourism – space voyage ventures. Space Voyage Ventures. https://spacevoyageventures.com/space-hotel-development/#:~:text=Space%20hotels%20and%20traditional%20space%20stations%20

represent%20two,ventures%20like%20the%20Voyager%20Station%20and%20Pioneer%20 Stations

Theissler A, Pérez-Velázquez J, Kettelgerdes M, Elger G (2021) Predictive maintenance enabled by machine learning: use cases and challenges in the automotive industry. Reliab Eng Syst Saf 215:107864. https://doi.org/10.1016/j.ress.2021.107864

Tien-Dung PHAM, Majerova J, Das S (2022) Phases of possible recovery of digital enterprises in new normal business for living with COVID-19 times: opportunities for a new era in sustainable development goals. In: Sustainable development and innovation of digital enterprises for living with COVID-19. Springer, Singapore, pp 19–33. https://doi.org/10.1007/978-981-19-2173-5_2

Tien-Dung P, Nguyen XD, Das S (2022a) Sustainable goal achievement by digital revolution during and after pandemic; how much one wins and losses: a bird's eye view for future planning. In: Sustainable development and innovation of digital enterprises for living with COVID-19. Springer, Singapore, pp 133–147. https://doi.org/10.1007/978-981-19-2173-5_8

Vrana V, Das S (2023) Dynamic restructuring of digital media and entertainment sector: role of urbanization, industrial innovation, and technological evolution. In: Das S, Gochhait S (eds) Digital entertainment as next evolution in service sector. Palgrave Macmillan, Singapore. https://doi.org/10.1007/978-981-19-8121-0_2

Vrana V, Das S (2023a) Role of enterprises and their digital transformation in post-pandemic sustainable development. In: Mondal SR, Yegen C, Das S (eds) New normal in digital enterprises. Palgrave Macmillan, Singapore. https://doi.org/10.1007/978-981-19-8618-5_7

Wall M (2019, December 5) *New, emotionally intelligent robot CIMON 2 heads to space station*. Space.com; space. https://www.space.com/cimon-2-artificial-intelligence-robot-space-station.html

Yegen C, Das S (2023) Digital solution for print and publishing media. In: Das S, Gochhait S (eds) Digital entertainment as next evolution in service sector. Palgrave Macmillan, Singapore. https://doi.org/10.1007/978-981-19-8121-0_12

Zhang Z, Li X, Li Y, Hu G, Wang X, Zhang G, Tao H (2023) Modularity, reconfigurability, and autonomy for the future in spacecraft: a review. Chin J Aeronaut 36(7):282–315. https://doi.org/10.1016/j.cja.2023.04.019

Smriti Tandon, Senior Faculty member at Graphic Era University, specializing in marketing analytics and strategic communication. Her research highlights data-driven decision-making for business scalability. She mentors students in leveraging emerging technologies for market innovation.

Antonia Puccio, Scholar at the University of Molise, with business ethics and sustainable governance expertise. Her research advocates for ethical frameworks in corporate decision-making. She engages in cross-sector dialogues to promote responsible business practices.

Subhankar Das, Senior Researcher and academician at Duy Tan University, publishing extensively on AI, blockchain, and digital transformation. Editor of seven Scopus-indexed books and reviewer for top journals. Mentors global PhD candidates and drives innovation in ASEAN academic collaborations.

Guardians of the Galaxy: Preserving Space for Future Generations

Patita Paban Mohanty and Subhankar Das

Abstract The rapid proliferation of space activities has turned the cosmos from an untouched environment into an ever more crowded arena facing sustainability issues like those on Earth. In this chapter, we cover the importance of space stewardship as a way to protect the cosmic environment for future generations. It addresses four important challenges in this regard: The growing problem of orbital debris that threatens active spacecraft, the ethical dilemmas of the exploitation of extraterrestrial resources, the risks of biological contamination by planetary exploration, and the interference from satellite mega-constellations on astronomical observations. To help in this context, the chapter introduces a theoretical bottom up support approach by using environmental ethics, adaptive governance, and systems resilience to develop a breeding ground to solve these issues together and by doing it, to highlight that environmental ethics has a role to play in all these aspects. It advocates the concept of "cosmic stewardship," suggesting a balance between human utilization and ecological preservation. These conversations highlight the need for agreements to update existing international treaties, innovate new technology for clearing debris, create protected cosmic parks, and adopt tighter planetary protection protocols. A paradigm shift in the way humanity views and conducts its space activities is essential to ensure sustainability, and that requires unprecedented cooperation at a global scale, ethical considerations towards the use of space resources, and a strong emphasis on the long-term environmental impact of both space exploration and commercialization activities.

Keywords Space sustainability · Orbital debris · Space governance · Planetary protection · Resource exploitation · Environmental stewardship

P. P. Mohanty
Faculty of Hospitality and Tourism Management, SOA University, Odisha, India
e-mail: patitamohanty@soa.ac.in

S. Das (✉)
Department of Marketing, South Star Management Institute, Duy Tan University, Da Nang, Vietnam
e-mail: subhankardas@duytan.edu.vn

© The Author(s), under exclusive license to Springer Nature Singapore Pte Ltd. 2025

S. R. Mondal et al. (eds.), *Pioneering the New Space Economy through AI and Immersive Technologies*, https://doi.org/10.1007/978-981-96-5977-7_9

1 Introduction

The launch of Sputnik 1 ushered in the Space Age, an era of exploration and technological development unlike before. That footprint in space has grown exponentially over the decades, driven by scientific curiosity, national ambitions and commercial interests (COSPAR 2024). Today, as countries and private companies compete to mine asteroids, colonize Mars, and launch massive satellite constellations, the previously pristine zone beyond Earth's atmosphere is facing a crisis similar to the environmental one on Earth itself. The truth of its vulnerability has broken the idea of space as a limitless frontier. Just as Earth's ecosystems need protective oversight, the cosmic environment—everything in space, from entities to orbital pathways and interplanetary spaces—urgently needs stewardship to keep itself intact for future generations. This chapter, "Guardians of the Galaxy: Preserving Space for Future Generations," explores the urgent need for practices, actions, collaborations, and technologies designed to minimize the mounting risk of human activity in space.

1.1 The New Frontier: Benefits and Costs

However, achievements have been monumental for humanity's aims in space, from the Apollo Moon landings to the International Space Station (ISS). These efforts have led to technological breakthroughs, progressed scientific understanding and stimulated international cooperation. Nevertheless, we now face overlapping challenges of the pace of the nature of space exploration and commercialization. Just one example: The explosion of satellites has transformed communications and Earth observation—and created crowded orbits. By 2023, more than 8000 active satellites will orbit the planet, along with an estimated 36,500 pieces of debris larger than 10 cm and millions of smaller fragments—relics of inoperative satellites, spent rocket stages and collision debris (European Space Agency 2024). This "space junk" moves at speeds above 28,000 km/h and would have catastrophic implications for any operational spacecraft, the ISS, and even crewed missions. The 2009 crash of the defunct Russian satellite Cosmos 2251 into the functioning Iridium 33 satellite—which created thousands more pieces of space junk—highlights the chain-reaction danger of orbital debris (Lawrence et al. 2022).

The Kessler Syndrome—an atheoretical nightmare in which cascading collisions destroy or make certain orbits unusable—is an ominous possibility (Kessler and Cour-Palais 1978). Without intervention, the many GPS, weather monitoring, and telecommunications satellites that provide humanity with critical infrastructure at this very moment could be irreversibly damaged, leaving global systems in tatters (NASA 2017). Addressing this problem requires technologies to clean up space debris, whether robotic arms, nets, or lasers, and preventative actions, including redesigning satellites to ensure they deorbit and imposing stricter launch regulations (United Nations 2017).

1.2 Exploitative Versus Preservation Use of Resources: The Ethics of Resource Use

With Earth's limited resources depleted, the Moon, asteroids, and other celestial bodies have become the targets of interest, such as reservoirs of rare minerals, water ice, and helium-3. For example, lunar mining presents materials crucial for sustaining extra-terrestrial bases and powering fusion energy (Walker et al. 2020). Nevertheless, unregulated extraction threatens irreversible damage to extraterrestrial landscapes. Permanently shadowed craters of the Moon shelter ancient ice deposits with clues to the solar system's history and potential habitats for future explorers. Asteroids, remnants from the dawn of the solar system, are scientific gold.

This points to ethical quandaries: How does humanity weigh using up resources against the protection of celestial corporeal geography? There are proposals for "cosmic parks"—protected zones on the Moon or Mars—to preserve scientific or aesthetic interest areas. Such efforts mirror terrestrial conservation projects, such as that of the UNESCO World Heritage Sites, but exist amid legal ambiguities under existing frameworks (such as the 1967 Outer Space Treaty, which prohibits national appropriation) that lack any enforcement mechanisms (Martinez 2021). The treaty's vagueness reflects the need for updated international covenants that reconcile commercial aspirations with environmental stewardship.

1.3 Planetary Protection: Protecting Life on Earth and Beyond

The hunt for aliens has heated up, with plans to go to Mars, Europa, and Enceladus. But human exploration risks biological contamination. Microscopic Earth travelers riding with spacecraft could taint pristine environments, destroying signatures of Indigenous life forms or disturbing possible ecosystems (Detsis 2022). By contrast, alien life could be biohazards if transported back to Earth.

Planetary protection procedures are designed to avoid this by sterilizing machines and quarantining specimens before they return to Earth. The Committee on Space Research (COSPAR) governs many of these procedures. NASA's Perseverance rover, for example, was subjected to intense decontamination before landing on Mars. However, the division between contamination and coexistence grows fuzzy as missions move from exploration to colonization—including SpaceX's hopes for a city on Mars. To balance scientific integrity with human expansion, guidelines should evolve and be shared transparently in mission planning.

1.4 Light Pollution: An Invisible Enemy of Astronomy

While the rise of satellite mega-constellations like SpaceX's Starlink and Amazon's Project Kuiper has opened up global access to the internet, it has also risked astronomers' ability to do their work. Thousands of low-Earth orbit (LEO) satellites reflect sunlight, creating streaks that drown out telescope observations (Boley and Byers 2021). In a 2023 study, researchers found that, for 30 per cent of images taken by the Vera C. Rubin Observatory, a next-generation telescope, satellite trails would interfere with discoveries of near-Earth asteroids or distant galaxies.

Mitigation methods include coating satellites to dim them, adjusting their orbits and using A.I. to correct for satellite data. However, these solutions are still nascent, and tensions between the aerospace and scientific communities increase without binding regulations. The key is to balance innovation with humanity's passion for the cosmos.

1.5 The Road to Cosmic Stewardship: The Next Step

The complexities of space sustainability cross national frontiers and require unprecedented global cooperation. Current frameworks, including the Outer Space Treaty and the UN's Guidelines for the Long-Term Sustainability of Outer Space Activities, offer guiding principles but no enforcement teeth. To strengthen these agreements, multilateral dialogue, fair resource-sharing models, and incentives for compliance must be pursued.

Technological innovation is just as important. Proactive solutions provide some hope, such as the European Space Agency's ClearSpace-1 mission to capture and deorbit debris (Clearspace-1 2024). At the same time, advances in reusable rockets and sustainable propulsion systems offer the prospect of reducing pollution from launches.

Ultimately, the evolution to becoming "Guardians of the Galaxy" will require a paradigm shift—from seeing space as a vast frontier to be explored to seeing it as a common heritage on which we must vigilantly steward. There are ways we can ensure a different future when it comes to protecting the environment in the cosmos, such as integrating environmental ethics into space policy, fostering public awareness, and prioritizing long-term sustainability over short-term gains.

2 Overview

With humanity's ever-growing presence in space, the necessity to preserve the celestial habitat has never been more decisive. Indeed, as we work to protect Earth's delicate ecosystems, we must also consider our future footprint on the space environment (Rummel et al. 2012). In this chapter, the authors embodied that pressing requirement for the practice of sustainability and conservation as we venture beyond into the cosmos. One primary concern is space debris—thousands of defunct satellites and millions of bits of "space junk" orbiting the planet, posing threats to active spacecraft and satellites (Du 2017). Evaluating the impact of these materials on the earth and creating efficient ecotechnologies to combat the population (Martinez 2021).

The increased complexity of operations in space and the emergence of prominent satellite constellations pose significant challenges to the long-term sustainability of activities undertaken in outer space. The United Nations Committee on the Peaceful Uses of Outer Space has acknowledged these activities. We must work internationally to tackle these matters, implementing comprehensive policies and best practices. Preserving other environments is even more complicated beyond Earth's orbit. As we seek to utilize resources on the Moon, asteroids, and beyond, we must weigh the need for utilization against the necessity for conservation. The idea of creating "cosmic parks" to safeguard areas of scientific or natural value, like our terrestrial national parks, has been suggested. Likewise, increased attention to exploring Mars raises the stakes for planetary protection—the practice of preventing possible biological contamination of Earth and other worlds.

2.1 The Garbage Problem (Space Debris and Orbital Pollution)

With human activities projected to continue growing beyond Earth's atmosphere, safeguarding the space environment is critical (Lawrence et al. 2022). Spaced debris, in the form of defunct satellites and millions of tiny pieces of "space junk," is increasingly becoming a significant risk for a mission in space, making it harder for the long-term sustainability of space operations (Martinez 2021). Adopting the UN COPUOS Guidelines for the Long-term Sustainability of Outer Space Activities has taken important steps in recognizing the dire need to tackle this issue, highlighting the threat that growing complexity in space operations and prominent satellite constellations bring to current and future generations (Profitiliotis and Loizidou 2018). New technologies and holistic policies are needed to remove existing debris and prevent future debris from creating problems (Martinez 2021).

Use of Resources in an Unsustainable Way

As humanity's goals shift to the lunar surface, asteroids, and beyond, this balancing act between conservation and utilization becomes even sharper relief. As with our terrestrial national parks, it has been suggested that we establish "cosmic parks" to reserve areas of scientific or natural value. Mars is at the intersection of resource use and planetary protection, which seeks to avoid possible biological cross-contamination from Earth to other celestial bodies and vice versa. Strict protocols are required to ensure that Earth microbes will not contaminate potential extraterrestrial microbes' habitats.

2.2 Planetary Protection and Contamination Prevention

With the increasing emphasis on Mars exploration, the need for planetary protection is all the more relevant, raising concerns about biological contamination across Earth and other worlds. To ensure no Earth microbes soar into habitats for potential life beyond our planet, strict protocols and careful procedures will be vital to maintain the integrity of the planetary record and protect the scientific value of future finds.

2.3 Reducing Light Pollution And Protecting Astronomy

The growth of space-based technologies, including the rise of satellite constellations, is also starting to affect astronomical observations—both via the problem of light pollution in the night sky and by achieving space-based capabilities and advancing them in a way that is sustainable in the future, so there is no need to return to this state again.

The future of cosmic construction must not come at the expense of space as a repository of science and nature for future descendants, so establishing global treaties, sustainable norms, and technology within our capabilities will be necessary to become real "Guardians of the Galaxy."

2.4 International Treaties and Cooperative Governance

International cooperation and global policies and frameworks will need to be established to address the complex challenges of space sustainability. The growing complexity of space operations and the advent of large, powerful satellite constellations have led to serious concerns over the long-term sustainability of space activities, as acknowledged by the UN Committee on the Peaceful Uses of Outer Space (Profitiliotis and Loizidou 2018). Establishing appropriate systems of global governance, with clearly defined roles and responsibilities for both state and non-state actors, to guide the future of space exploration and use will be key. If human survival and progress, defined by our ability to reach the stars, also means taking the role of the "Guardians of the Galaxy," we must protect the cosmic environment for ourselves and future generations. Space should remain an immersive place of fundamental scientific discoveries. Through the exploration of new technology, sustainable models of behavior, and strong global pacts, we can work to ensure that our advances in space do not threaten its unique worth. We can only secure a future of responsible and long-lasting space exploration and exploitation by accepting our position as caretakers of the cosmic environment.

Preserve the Cosmic Frontier Safe: Finding a Balance Between Exploration and Sustainability in Space.

2.5 *The Interstellar Landscape of the Future*

The human push into space, a springboard for so much that we value—connectivity, commerce, and exploration—was once a serene yet distant frontier, but the race towards it has become a contested environment (Paradiso 2020). As we push business further into orbit, the Moon, Mars and beyond, preserving our past has become a priority (Du 2017). This literature review looks at the main aspects affecting the sustainability of space activities, deploying a framing of five interrelated themes: space debris management, ethical resource utilization, planetary protection, light pollution, and international governance (Profitiliotis and Loizidou 2018).

2.6 *Sustainable Exploration: Walking the Tightrope*

The growing presence of space debris—defunct satellites, spent rocket stages, and collision fragments—is one of the most immediate threats to space sustainability. The researchers first proposed the theory called "Kessler Syndrome" in 1979, in which debris density reaches a tipping point where certain orbits are rendered unusable due to all the possible collisions. Recent research backs up this worry: More than 36,500 trackable debris more significant than 10 cm are now orbiting Earth, along with millions more pieces of smaller debris. These objects can travel at hyper velocities (over 28,000 km/h), posing a risk to operational satellites, crewed missions, and critical infrastructure such as GPS and weather monitoring systems (Lawrence et al. 2022).

Mitigation strategies can be divided into prevention and active debris removal. Preventive strategies have been established, such as designing satellites to deorbit themselves safely, international guidelines for the disposal of satellites after their operational lifetime, and developing technologies for on-orbit servicing and refueling (Du 2017). Active debris removal techniques, like robotic capture and tethered platforms, are meant to retrieve the most dangerous objects in orbit physically (Wilson and Vasile 2023).

2.7 Ethical Resource Utilization

The method must be ecologically responsible, and humans must preserve the cosmic environment even for space resources (Hofmann and Bergamasco 2020). Mining materials from celestial bodies such as asteroids, the Moon, and even Mars can serve as valuable materials for future space activities and life here on Earth, from extracting minerals to water. These activities, however, require strong ethical frameworks to guide them so that they do not lead to the overexploitation and degradation of fragile extraterrestrial ecosystems.

2.8 Space Debris Management and Planetary Protection

Nevertheless, protecting planetary environments from contamination becomes more important as humanity reaches beyond its home planet (Clark et al. 2020). For decades, international agencies such as NASA and ESA have been operating under protocols that protect celestial bodies from terrestrial microbes and organic compounds, otherwise known as planetary protection (Migaud 2020). However, the advent of private and commercial space enterprises has alarmed some who fear unchecked activities could threaten those protocols and the purity of alien ecosystems (Profitiliotis and Loizidou 2018). This challenge needs a comprehensive resolution that integrates preventive actions, environmental studies, and the generation of lucid standards for all the space actors, whether governmental or private, to avoid a spatial free-for-all (Mullick et al. 2019).

2.9 Light Pollution

The increasing number of satellite constellations and human activity in space has raised alarm bells regarding the effects of light pollution on astronomical observations and the cosmic environment. Orbiting artificial lights can affect ground-based telescopes and hinder our ability to study the universe and understand its origin. Reducing the impact of light-polluting technologies at all levels will require a concerted effort to design and operate space-faring infrastructures that do not harm the night sky and, ultimately, the natural environment.

2.10 International Governance

This literature review has analyzed the significant determinants of the sustainability of space activities, such as space debris management, ethical resource extraction, planetary protection, light pollution, and worldwide governance. However, with humanity's faces later on, exploration and preservation have continued to grow into a serious issue. This necessitates a multidisciplinary collaboration—uniting policymakers, scientists, engineers, and diverse stakeholders to create effective plans for ensuring the cosmic environment stays habitable for us (Vidaurri and Gilbert 2021).

2.11 Ethical Resource Utilization

International agreements like the Outer Space Treaty reflect the importance of governing emerging technologies in new domains, laying the groundwork for the responsible utilization of space resources while prioritizing the "province of all mankind" and the protection of celestial environments (Hertzfeld et al. 2016). However, with commercial interests and technological capabilities only growing, the space community faces the imperative to better define and regulate the practical process of using space resources to balance the right to use and the sustainability of the space environment in the long run (Johnson 2007). Ethical resource utilization will involve benefit sharing, protection of indigenous extraterrestrial environments, environmental impact assessment and monitoring, etc. (Das 2023a). By incorporating these same principles into the governance of the extraction of space resources, policymakers and space agencies can ensure that our cosmic environment, on a macro level, is preserved so that clean space resources can be developed for generations to come.

2.12 Global Relations and Collaboration

Robust international governance frameworks are essential to preserving the space environment for future generations (Das 2023b). They enable coordinated activities among various space actors, from nation-states to private companies to international organizations (Hertzfeld et al. 2016). In the end, the Committee on the Peaceful Uses of Outer Space has been instrumental in elaborating guidelines and recommendations on the long-term sustainability of outer space activities. The growing trend of commercial interests venturing beyond our planet and the rare but increasing cases of collision interference add another layer of complexity to navigating this new frontier that requires regular review and updates. This will require the international community to build upon existing treaties and agreements, including the 1967 Outer Space Treaty, and craft more specific and enforceable regulations capable of holding all actors in the open cosmos responsible for their actions and the consequences to the cosmic environment.

Maintaining space for future generations poses complex questions that are best addressed by many people, including policymakers, scientists, engineers, and others. This literature review identified the following key factors that must be addressed for humanity to ensure that space activities are sustainable in the long term: space debris management, ethically resourced utilization, planetary protection, light pollution mitigation, and governance (Xu 2020).

2.13 A Sustainable Space Future

The overpopulation of space debris, an increasingly menacing threat to orbital sustainability, is one of the more immediate dangers facing the space environment (Perez 2023). Mitigation approaches involve preventative strategies and active debris remediation, as collaboration among space actors is essential. Another significant aspect is the ethics of resource utilization so that the colonization of space resources can be merged with the sustainable development of the space environment in the long term (Das et al. 2023). Issues such as equitable sharing of benefits, protection of alien ecosystems, and environmental impact assessments are key considerations.

However, as humans continue to explore beyond Earth, the need to prevent planetary environments from contamination is becoming increasingly critical. We need planetary protection protocols and transparent, enforceable guidelines for all space actors—governmental and private—to ensure the integrity of extraterrestrial environments. The increasing number of satellite constellations and growing human activity in space raised fears about the impact of light pollution on astronomical observations and the natural environment of the universe. In general, it will take an organized effort to mitigate the impact of space-based infrastructures and their operation on the night sky and the environment.

Good international governance and cooperation are essential for future generations to preserve the space environment (Di Virgilio and Das 2023a). The UN Committee on the Peaceful Uses of Outer Space has played an important role in developing many guidelines and recommendations. However, the ever-growing pace of technological advances and space commercialization requires that we regularly review and update the framework to which we are bound. International consensus-building must extend beyond existing treaties and soft law to include binding, enforceable sanctions that hold all space actors accountable for their actions and the externalities that follow the cosmic environment. Ultimately, ensuring space remains for future generations is a complex issue requiring coordinated action from policymakers, scientists, engineers, and other stakeholders. Through the targeted efforts focused on the considerations put forth in this literature review, such as space debris mitigation, ethical use of resources, planetary protection, prevention of light pollution, and enhancement of the international governance framework, humanity can continue to explore the stars while keeping space activity sustainable over the long term.

3 Understanding Space Sustainability: An Interdisciplinary Approach Combining Ethics, Governance, and Environmental Management

Strong underpinnings of the theory linking ethics, governance, and stewardship will be required to protect space for future generations. This framework must account for the unprecedented questions of sustainability in an off-world context (e.g., concerning orbital debris, resource exploitation, biological contamination, light pollution, and fractured governance) while heeding the interrelationship among these issues (Di Virgilio and Das 2023b). Building on environmental ethics, institutional theory, and systems resilience, this section suggests a multidimensional model to inform policy approaches, technological innovation, and collective action for space activities.

3.1 Foundations of the Framework

Environmental Ethics in Space

Space sustainability is underpinned by an ethical angle framed around two philosophical approaches: anthropocentrism and ecocentrism. Anthropocentrism considers human interests the most important, viewing space as a resource to ensure survival and produce wealth (Newman and Williamson 2018). This outlook informs projects such as asteroid mining and lunar colonization; the extraction of resources is framed as necessary to achieve human progress. On the other hand, ecocentrism attributes intrinsic value to celestial ecosystems and calls for their preservation independent of human utility (Majerova and Das 2023). This philosophy is represented in proposals for "cosmic parks" on lunar or Martian landscapes that aim to protect extraterrestrial landscapes akin to the protection of biodiversity on the Earth (Goralnik and Nelson 2012).

The hybrid view, which I call cosmic stewardship, synthesizes these views. It claims that humanity has a moral imperative to responsibly exploit space resources without damaging its scientific and ecological integrity. This is in keeping with the "Common Heritage of Humankind" concept articulated in the Moon Agreement (1979), which requires equitable resource sharing and environmental protection (Evers and Van Smeden 2016).

Common Setting Issue in Orbital Space

Space debris is a classic case of the tragedy of the commons (Wang 2013)— an institutional (real-world) economic provision of finite resources that all benefit from yet overexploit because of individual and corporate self-interest. Low-earth orbit (LEO), as a global common, must be managed to prevent congestion, as nations and corporations put satellites up without a disposal mechanism. Kessler Syndrome, a theoretical scenario in which collisions lead to the destruction of large orbits, can result from unregulated competition (Kessler and Cour-Palais 1978). This framework combines Ostrom's principles of commons governance, which focus on collective action, monitoring, and graduated sanctions to avoid depleting the resources (Wang 2013). When applied in space, these principles for:

a. Transparency: Improved, shared data sets to monitor debris and satellites.
b. Incentives: Tax breaks or subsidies for debris-neutral missions.
c. Enforcement: Fines for failing to follow mitigation guidelines.

Planetary Protection as Bioethical Imperative

Planetary protection, the avoidance of biological contamination, is based on bioethical principles like non-maleficence ("no harm") and precautionary responsibility (Spry 2022). Forward contamination (earthly microbes tainting non-terrestrial environments) threatens to extinguish any latent alien life, while backward contamination (extraterrestrial microbes coming to Earth) is a biohazard.

The framework offers a biocentric ethic, granting extraterrestrial ecosystems the right not to be disturbed (Minh-Nhat et al. 2022a). It calls for stringent sterilization protocols for spacecraft and ethical guidelines for human settlements, striking a balance between exploration and ecological preservation.

3.2 Government and Institutional Theory

The Implications of the Outer Space Treaty on Legal Pluralism

The Outer Space Treaty of 1967 set out basic principles—including the non-appropriation principle and the peaceful use of outer space, but lacks mechanisms to respond to contemporary issues such as debris or resource extraction. Legal pluralism which supports the coexistence of overlapping systems of authorities, including legal ones can help by viewing the conflict through the lens of legal pluralism (Minh-Nhat et al. 2022b). For instance:

a. National laws: The US Commercial Space Launch Competitiveness Act, 2015, recognizes property rights over mined resources, which contradicts the treaty's anti-appropriation clause.
b. Private agreements: National governments give private companies like SpaceX direct contracts without international regulation.

A pluralistic approach would align these regimes in an adaptive governance scheme whereby treaties iteratively develop through multilaterally engaged deliberation (Mondal et al. 2023a). The Artemis Accords exemplify this approach, promoting "safety zones" surrounding lunar operations while excluding China and Russia.

Polycentric Governance

Polycentric governance involves multiple decision-making centers and is essential for sustainable space (Mondal and Das 2023a). No entity oversees global space activities, but states, corporations, and NGOs organized into networks can impose norms like UNOOSA (Guidelines for debris mitigation), IAU (Supports dark-sky

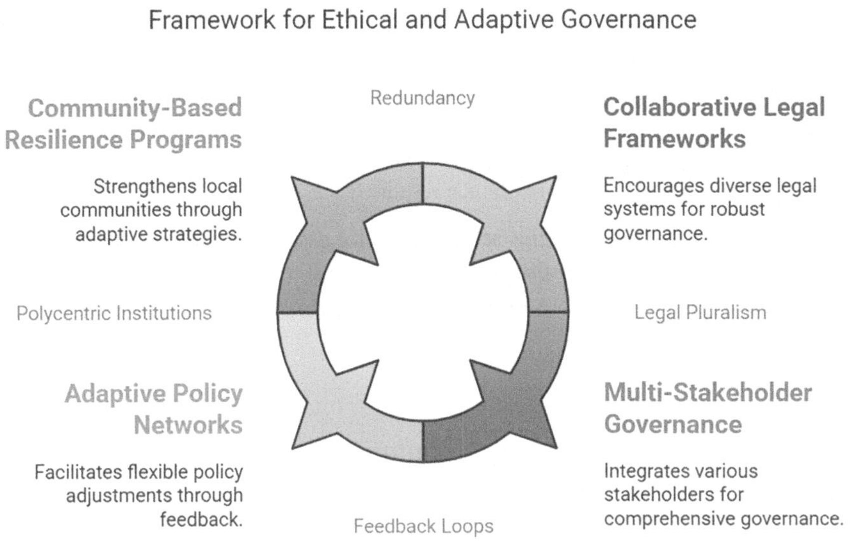

Fig. 1 Ethical and adaptive governance (Source: Authors' conception)

protections from satellite constellations), and COSPAR (Provide planetary protection standards). The model is based on subsidiarity, delegating authority to the lowest practical level (e.g., national agencies overseeing debris removal from low Earth orbit) and preserving broad rules for an international umbrella. Figure 1 represents Ethics and adaptive governance.

4 Managing Interconnected Risks and Systems Resilience

4.1 Space as Coupled Human-Natural System

Space sustainability issues are interrelated, just like biogeophysical systems. Systems thinking; a way of interactions within complex systems as a whole—captures feedback loops between drivers:

a. Kessler Syndrome (debris collisions/explosion cascade) → decreased orbital safety → increased insurance costs → Compliance disincentive.
b. Space-based mega constellations →, Light pollution, → Astronomical degradation, → Impact delay on asteroid threat.

The framework visualizes these relationships using causal loop diagrams, pinpointing leverage points for action (Mondal and Das 2023b). One example is incentivizing debris removal to break the Kessler Syndrome feedback loop.

4.2 *Resilience Theory*

It is where resilience theory the resiliency of a system to absorb shocks and adapt; comes into play for space infrastructure (Mondal and Das 2023c). A high debris density in an orbital environment is non-resilient, where a collision can cause an irreversible cascade. Building resilience requires:

a. Redundancy: Satellite backups are available for essential systems, such as GPS.
b. Adaptive capacity: Rapid-response debris removal technologies.
c. Plurality: Responsibility for monitoring is shared by multiple factors, which is key to balance.

4.3 *Justice and Equity Considerations*

Intergenerational Equity

The Brundtland Commission's principle of sustainable development satisfying today's needs without jeopardizing future generations translates directly to space. It should not deny future humans scientific or economic opportunities, thoroughly curricular exploitation of lunar ice or asteroid metals (Mondal et al. 2023b). This necessitates resource quotas and legacy reserves.

Distributive Justice

Global inequities are at risk of being magnified by space activities (Ngoc-Vinh et al. 2022). Rich countries and companies may also hoard resources in a way that hobbles developing nations—a phenomenon known as "cosmic colonialism" (Newman and Williamson 2018). The framework reflects Rawlsian justice, advancing a "veil of ignorance" perspective in which resource allocation is determined as if no stakeholder is aware of its future position (Tran 2024). Practical measures include:

a. Revenue-sharing models—a percentage of mining profits to global funds.
b. Technology transfer: Supply of debris removal tools to resource-poor nations.

4.4 *Synthesis: An Integrated Multilevel Model*

The theoretical framework combines the above elements into a multidimensional model (Fig. 2) based on three pillars:

a. Ethical Stewardship: Cosmotic steward and biocentric ethic emerge.

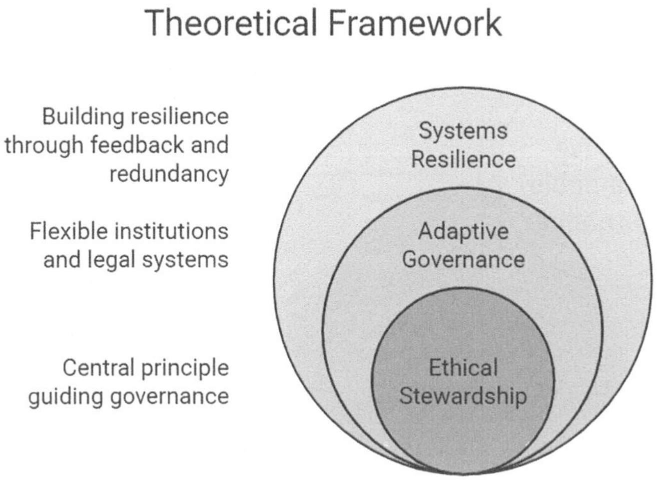

Fig. 2 A theoretical framework for space sustainability (Source: Authors' conception)

b. Adaptive Governance: Polycentric Institutions, Legal Pluralism, and Resilience-building (Nguyen et al. 2022).
c. Systems Resiliency: Addressing inter-connected risks through feedback-response and redundancy.

4.5 Policy and Research Implications

a. Policy: States must ratify binding treaties implementing Common Heritage, enforcing debris mitigation, and equitable resource sharing.
b. Technology: Prioritize funding for debris removal and low-reflectivity satellite designs.
c. Ethics: Ensure independent oversight bodies are in place to audit compliance with planetary protection protocols.

This conceptual framework provides a structure for tackling complex issues of space sustainability. Figure 3 represents the practical approach of the framework. Humanity can make this negotiation between exploration and preservation by making policies the embodiment of ethical stewardship, adaptive governance, and systems resilience. To fine-tune its applicability, future work should validate this model via case studies—e.g., lunar mining regulations or debris removal pilots. In an era when the cosmos is ever more crowded, a solid theoretical grounding is no academic luxury; ensuring that the final frontier does not end up as just the latest victim of short-sighted ambition is vital.

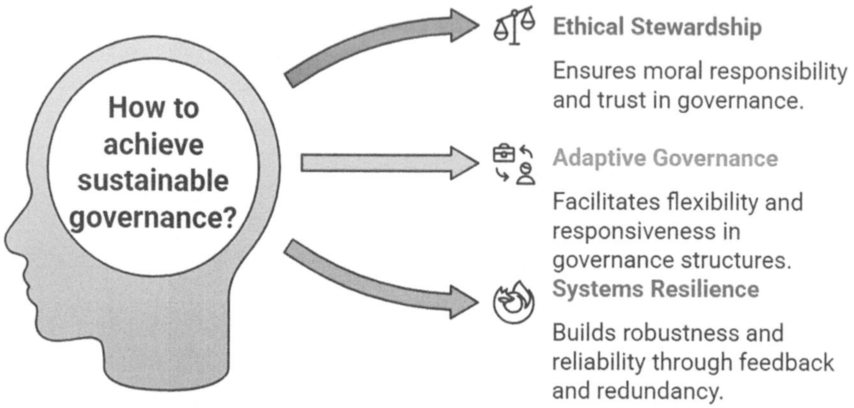

Fig. 3 Practical use of the framework (Source: Authors' conception)

5 Space Stewardship: Theoretical and Practical Sustainable Implications

Preserving space for future generations through orbital debris or ethical resource extraction requires theoretical grounding and maintaining a strategy (Tandon and Das 2023). Herein, the authors provide theoretical implications (insights towards building meaningful ethical and governance frameworks) and practical implications (policy, technology, and societal actions) derived from and grounded in the extant literature and theoretical framework provided above.

5.1 Theoretical Implications

Moral Framework for Universal Stewardship

The notion of cosmic stewardship; balancing human utilization with ecological preservation—offers a moral basis for space activity (Tien-Dung et al. 2022a). It questions traditional anthropocentric ideas and applies biocentric beliefs to other planets.

Impact:

a. Policy Development: Supports treaties that focus on preservation, such as establishing "cosmic parks" on the Moon or Mars.
b. Research Agenda: Respond to studies about the ecological significance of space objects (e.g., microbial habitation on Mars).
c. Corporate accountability: Forbids private companies from mining or deploying satellites unless they use environmental impact assessments.

Adaptive Governance Models

The polycentric governance model mitigates the shortcomings of the 1967 Outer Space Treaty through international treaties, national laws, and global agreements.

Impact:

a. Legal Harmonization: Supports adaptive agreements such as the Artemis Accords to address conflicts between national and international laws.
b. Enforcement Mechanisms: This section provides for enforcement mechanisms and penalties for failing to comply with debris mitigation or planetary protection standards.
c. Equity Frameworks: Embeds the principles of distributive justice to guarantee that developing nations gain from space resources.

Environmental Science Systems Thinking and Resilience Theory

Interdependencies of debris, light pollution, and resource exploitation are highlighted, and space is viewed as a coupled human-natural system.

Impact:

a. Risk Mitigation: Pinpoints leverage points (like financial incentives for debris removal) that can break the cycle of adverse feedback loops, such as the Kessler Syndrome.
b. Infrastructure Design: Advocates for redundant satellite constellations to remain operational in the face of disruptive debris.

5.2 Practical Implications

Orbital Debris Management

a. Active Debris Removal (ADR): Expedite missions like ESA's ClearSpace-1 to run trials on robotic capture technologies.
b. Enforcement of Policy: Set up a global "debris tax" for satellite operators to finance ADR efforts.
c. Satellite Design Standards: Require modular satellites with built-in deorbiting mechanisms (e.g., drag sails).

Example:

The ADRAS-J mission by Japan (2023) will provide debris rendezvous and removal techniques, providing a paradigm that can be scaled.

Ethical Resource Utilization

a. Resource Quotas: Contracts or quotation extraction rates for lunar ice or metals from asteroids to ensure some renewability (Tien-Dung et al. 2022b).
b. Cosmic Parks: Recognize scientifically worthy lunar craters (see Shackleton Crater) as UN-protected domains.
c. Revenue Sharing: Dedicate a portion of the mining revenues to a global fund for sustainable development on Earth.

Example:

The Moon Agreement's principle of the "Common Heritage of Humankind" could be revived to regulate resource distribution, similar to seabed mining limits set out by the International Seabed Authority.

Planetary Protection Measures

a. Optimization of Sterilization: Create low-cost microbial decontamination technologies for crewed missions to Mars.
b. Contingency Plans: Preparation of astronauts returning from possibly infected areas.
c. Public-Private Partnerships: Engage with industry partners like SpaceX to embed planetary protection programs into colonization plans.

Example:

The US National Aeronautics and Space Administration (NASA)'s Office of Planetary Protection could augment its mission to legitimize private missions to the various celestial bodies through compliance with COSPAR guidelines.

Mitigating Light Pollution

a. Satellite Design Rules: Mandate mega constellations (Starlink and others) to deploy anti-reflective coatings or sunshades.
b. Dark Sky Preserves: Establish orbital "exclusion zones" near large telescopes like the Vera C. Rubin Observatory.
c. Share satellite trajectory data with astronomers to improve image correction algorithms.

Example:

SpaceX's DarkSat and VisorSat experiments show that reducing satellite reflectivity is an achievable engineering task, but regulator mandates are necessary to ensure widespread adoption.

Global Governance

a. Binding Treaties: Revise the Outer Space Treaty for debris mitigation, resource sharing, and pollution control.
b. Multilateral institutions: Create a space sustainability policy council within the United Nations Office for Outer Space Affairs (UNOOSA) to coordinate enforcement.
c. Transparency Initiatives: Developing open-access programs that enable tracking of debris and satellite movements (for example, ESA's Space Debris Office).

Example:
The proposed Space Traffic Management system is meant to standardize satellite operations at the EU level and set an example for the world.

Challenges to Implementation

a. Geopolitical Tensions: Spacefaring nations (the USA, China, and Russia) have competing interests in preventing agreements on treaties.
b. Funding Gaps: The significant costs associated with advanced ADR technologies and debris monitoring systems necessitate innovative financing methods (e.g., public-private partnerships).
c. Technological Limitations: The available sterilization processes for crewed missions would not be adequate for long-term habitation on Mars.
d. Corporate Pushback: Private companies may resist regulations that increase operational expenses (e.g., debris taxes and satellite coatings).

Leading to Interstellar Future

These theoretical implications highlight the need for ethical re-calibration, adaptive governance, and systems-based approaches to space activities. These become policies, technologies, and alliances focusing on sustainability over quick wins. Key steps include:

a. Ratifying such binding international agreements and international enforcement mechanisms.
b. Leading the Charge: Enabling the Funding of Low Impact Satellite Technology and Debris Removal.
c. Turning to the public to educate and advocate space stewardship.

We should address these implications and be faithful "Guardians of the Galaxy" to ensure that the wonders of space are not lost to our exploitative practices and remain in seamless exploration and development for generations to come.

6 Conclusion

Humans are at a crucial inflexion point in our cosmic relationship. The explosive growth of space activities from satellite mega-constellations to lunar mining ambitions has shown both the incredible possibilities for space and its deep vulnerability (Vrana and Das 2023a). As we make our way into the final frontier, Earth's environmental crises loom large, teaching us that exploitation without stewardship will result in the irreversible degradation of another world. These challenges such as orbital debris, ethical resource utilization, planetary protection, light pollution, and fractured governance call for nothing short of a paradigm shift in human understanding and in dealing with space itself (Vrana and Das 2023b). This conclusion combines the urgency of these matters, emphasizes the necessity of collective action, and provides a roadmap to guarantee that space remains the province of all for generations yet unborn (Yegen and Das 2023).

The expansion of space junk or debris warns of the ramifications of irresponsible human behavior. As more than 36.5 million trackable debris objects threaten our satellites and crewed missions in orbit, Kessler Syndrome—an ever-worsening chain-reaction collision causing spaceships to tumble in their orbits and the region of space around them to become uninhabitable by nature—is no longer theoretical, but a clear danger we all need to address. It will take technological innovation to address this issue, such as the active debris removal (ADR) missions like ClearSpace-1, and systemic changes, such as enforcing stricter disposal guidelines and potential global cooperation to fund mitigation efforts. Similarly, given the.

ethical quandaries of resource extraction on the Moon and asteroids, we need frameworks that mediate between different mindsets of exploitation and preservation. Creating "cosmic parks"—protecting regions of scientific significance—could preserve extraterrestrial worlds in a fashion similar to UNESCO World Heritage Sites while permitting regulated resource extraction. Nonetheless, this relies on an updated 1967 Outer Space Treaty to tackle current challenges and implement fair resource-sharing principles.

Planetary protection, which is too little discussed in public, is just as important. As missions evolve from robotic exploration to human colonization, the risk of forward and backwards biological contamination grows. Strict sterilization protocols, transparent mission planning, and international oversight are crucial for maintaining the integrity of extraterrestrial environments and shielding Earth from possible biohazards. Mitigating light pollution through reflective coatings, orbital adjustments, and regulatory mandates must be a default policy for satellite operators.

The failure of existing governance systems is at the heart of all these challenges. The Outer Space Treaty was visionary for its time, but it has no real mechanisms for dealing with debris, resource rights, or pollution. To align national laws with international sustainability mandates, a polycentric governance model—which gets states, corporations, and NGOs involved—is critical. While the Artemis Accords and the proposed Space Sustainability Council under the United Nations Office for OuterSpace Affairs (UNOOSA) umbrella serve as tentative steps toward cooperative governance, inclusivity continues to pose an obstacle. Solution of Geopolitical Divisions by Space Cooperation Space cooperation is critical in the future of our space and will help change the impact of geopolitics.

This approach is grounded in the moral framework of cosmic stewardship. It promotes a harmonious coexistence of human wants and ecological health by balancing anthropocentric and ecocentric values. This philosophy aligns with the "Common Heritage of Humankind" principle, which requires equitable access to space resources and intergenerational equity. This might mean treaties with enforcement mechanisms, debris taxes to finance cleanup efforts, and technology-sharing programs to help empower developing nations.

However, the commitment of society must match theoretical ideals. To hold governments and corporations accountable, public awareness and advocacy is key. Just like the environmental movement in the twentieth century, educational campaigns to highlight issues of space sustainability can build a global ethic of responsibility. Furthermore, the private sector needs to move away from seeing space as the "Wild West" of profit and instead as a commons that needs shared stewardship. In order to lead the way, companies such as SpaceX and Blue Origin—while at the forefront of innovation—should demonstrate best practices around sustainable satellite designs and participatory on-orbit debris removal strategies while facilitating open communication lines with the scientific community.

Ultimately, the conservation of space is not only a technical or political problem but a moral one. The decisions taken today—whether to chase short-term gains or long-term sustainability—will reverberate across centuries. Through adaptive governance, ethical innovation, and global solidarity rather than parochial exclusion, we can put our history of ecological violence behind us and become lords and ladies of the galaxy rather than its chickens. The frontier is vast, teeming, and beyond our comprehension, but only if we shoulder the mantle of humility, vision, and stewardship when we reach for the stars. The final frontier does not need to become a reflection of Earth's wounds; it can serve as proof, like a mirror of our capability to learn, adapt, and preserve what orbit beyond the blue planet we call home.

References

Boley AC, Byers M (2021) Satellite mega-constellations create risks in Low Earth Orbit, the atmosphere and on Earth. Sci Rep 11(1). https://doi.org/10.1038/s41598-021-89909-7

Clark CP, Tan DY, Luna PA, Hastings DE, Masterson RA, Ricard MJ (2020) Effectiveness of CubeSat-Based architectures for active removal of On-Orbit rocket bodies. ASCEND 2022. https://doi.org/10.2514/6.2020-4241

ClearSpace-1 (2024). https://www.esa.int/Space_Safety/ClearSpace-1

COSPAR (2024) COSPAR policy on planetary protection. https://cosparhq.cnes.fr/cospar-policy-on-planetary-protection/

Das S (2023a) Modern-day gaming and the role of FinTech applications in popularising various entertainment platforms across the world. In: Das S, Gochhait S (eds) Digital entertainment as next evolution in service sector. Palgrave Macmillan, Singapore. https://doi.org/10.1007/978-981-19-8121-0_11

Das S (2023b) Digital entertainment: the next evolution in service sector. Emerging digital solutions in reshaping different industries Palgrave-Singapore. https://doi.org/10.1007/978-981-19-8121-0

Das S, Mondal SR, Struck F, Eissfeldt J, Beernink L, Sigouras V, Hoose S, Kopac L (2023) What can we see inside the crystal ball? Exploring opportunities and challenges for metaverse and circular economy. AJIS 12(6):221. https://doi.org/10.36941/ajis-2023-0165

Detsis E (2022) Ethics in Space: the case for future space exploration. In Research ethics forum, pp 111–121. https://doi.org/10.1007/978-3-031-15746-2_9

Di Virgilio F, Das S (2023a) Digital solutions for social media: role of digital evolution on business enterprises. In: Das S, Gochhait S (eds) Digital entertainment as next evolution in service sector. Palgrave Macmillan, Singapore. https://doi.org/10.1007/978-981-19-8121-0_9

Di Virgilio F, Das S (2023b) Digitalization of enterprises from a new normal sustainable development perspective: opportunities and challenges. In: Mondal SR, Yegen C, Das S (eds) New normal in digital enterprises. Palgrave Macmillan, Singapore. https://doi.org/10.1007/978-981-19-8618-5_9

Du R (2017) China's approach to space sustainability: legal and policy analysis. Space Policy 42:8–16. https://doi.org/10.1016/j.spacepol.2017.10.005

European Space Agency (2024) ClearSpace-1. www.esa.int. https://www.esa.int/Space_Safety/ClearSpace-1

Evers D, Van Smeden GE (2016) Meaning in Life: in defense of the hybrid view. South J Philos 54(3):355–371. https://doi.org/10.1111/sjp.12179

Goralnik L, Nelson M (2012) Anthropocentrism. In Elsevier eBooks, pp 145–155. doi:https://doi.org/10.1016/b978-0-12-373932-2.00349-5

Hertzfeld HR, Weeden B, Johnson CD (2016) Outer space: ungoverned or lacking effective governance?: New approaches to managing human activities in space. SAIS Rev Int Affairs 36(2):15–28. https://doi.org/10.1353/sais.2016.0017

Hofmann M, Bergamasco F (2020) Space resources activities from the perspective of sustainability: legal aspects. Global Sustain 3. https://doi.org/10.1017/sus.2019.27

Johnson NL (2007) Developments in space debris mitigation policy and practices. Proc Inst Mech Eng Part G J Aeros Eng 221(6):907–909. https://doi.org/10.1243/09544100jaero166

Kessler DJ, Cour-Palais BG (1978) Collision frequency of artificial satellites: the creation of a debris belt. J Geophys Res Atmos 83(A6):2637–2646. https://doi.org/10.1029/ja083ia06p02637

Lawrence A, Rawls ML, Jah M, Boley A, Di Vruno F, Garrington S, Kramer M, Lawler S, Lowenthal J, McDowell J, McCaughrean M (2022) The case for space environmentalism. Nat Astron 6(4):428–435. https://doi.org/10.1038/s41550-022-01655-6

Majerova J, Das S (2023) Sustainable development and digitalization in the supply chain for environmental protection: changes and trends. In: Mondal SR, Yegen C, Das S (eds) New normal in digital enterprises. Palgrave Macmillan, Singapore. https://doi.org/10.1007/978-981-19-8618-5_8

Martinez P (2021) The UN COPUOS guidelines for the long-term sustainability of outer space activities. J Space Safe Eng 8(1):98–107. https://doi.org/10.1016/j.jsse.2021.02.003

Migaud MR (2020) Protecting earth's orbital environment: policy tools for combating space debris. Space Policy 52:101361. https://doi.org/10.1016/j.spacepol.2020.101361

Minh-Nhat HO, Majerova J, Das S (2022a) Taking steps to help the phoenix rise from ashes: a roadmap for digital enterprises to develop new sustainable goals. In: Sustainable development and innovation of digital enterprises for living with COVID-19. Springer, Singapore, pp 35–51. https://doi.org/10.1007/978-981-19-2173-5_3

Minh-Nhat HO, Nguyen HL, Mondal SR (2022b) Digital transformation for new sustainable goals with human element for digital service enterprises: an exploration of factors. In: Sustainable development and innovation of digital enterprises for living with COVID-19. Springer, Singapore, pp 85–103. https://doi.org/10.1007/978-981-19-2173-5_6

Mondal SR, Das S (2023a) Defining the new digitalized normal: implications of A.I., IoT, and blockchain for business growth. In: Mondal SR, Yegen C, Das S (eds) New normal in digital enterprises. Palgrave Macmillan, Singapore. https://doi.org/10.1007/978-981-19-8618-5_10

Mondal SR, Das S (2023b) Digital solution for industry 4.0 and 5.0 and social media. In: Das S, Gochhait S (eds) Digital entertainment as next evolution in service sector. Palgrave Macmillan, Singapore. https://doi.org/10.1007/978-981-19-8121-0_10

Mondal SR, Das S (2023c) Human-centric digitalization for new sustainable development: an explorative study in human resource management. In: Mondal SR, Yegen C, Das S (eds) New normal in digital enterprises. Palgrave Macmillan, Singapore. https://doi.org/10.1007/978-981-19-8618-5_12

Mondal S, Das S, Vrana VG (2023a) How to bell the cat? A theoretical review of generative artificial intelligence towards digital disruption in all walks of life. Technologies 11(2):44. https://doi.org/10.3390/technologies11020044

Mondal SR, Yegen C, Das S (2023b) New normal in digital enterprises – strategies for sustainable development. Palgrave-Singapore. https://doi.org/10.1007/978-981-19-8618-5 and Hardcover ISBN 978-981-19-8617-8

Mullick S, Srinivasa Y, Sahu AK, Sata JT (2019) A comprehensive study on space debris, threats posed by space debris, and removal techniques. SSRN Electronic Journal. https://doi.org/10.2139/ssrn.3511445

NASA (2017, December 5) Mars 2020: perseverance rover. NASA Science. https://science.nasa.gov/mission/mars-2020-perseverance/?drive=1222&site=26

Newman CJ, Williamson M (2018) Space sustainability: reframing the debate. Space Policy 46:30–37. https://doi.org/10.1016/j.spacepol.2018.03.001

Ngoc-Vinh N, Tien-Dung P, Das S (2022) Digitization of business; Need for recombination during COVID-19. In: Sustainable development and innovation of digital enterprises for living with COVID-19. Springer, Singapore, pp 1–17. https://doi.org/10.1007/978-981-19-2173-5_1

Nguyen TDM, Mondal SR, Das S (2022) Digital Entrepreneurial Transformation (DET) Powered by New Normal Sustainable Developmental Goals (n-SDGs): Elixir for growth of country's economy. In: Sustainable development and innovation of digital enterprises for living with COVID-19. Springer, Singapore, pp 69–84. https://doi.org/10.1007/978-981-19-2173-5_5

Paradiso J (2020) Self-aware self-assembly for space architecture: growth paradigms for in-space manufacturing. https://dspace.mit.edu/handle/1721.1/129269

Perez AMV (2023) The increasing effects of light pollution on professional and amateur astronomy. Science 380(6650):1136–1140. https://doi.org/10.1126/science.adg0269

Profitiliotis G, Loizidou M (2018) Planetary protection issues of private endeavours in research, exploration, and human access to space: an environmental economics approach to forward contamination. Adv Space Res 63(1):598–605. https://doi.org/10.1016/j.asr.2018.10.019

Rummel J, Race MS, Horneck G (2012) Ethical considerations for planetary protection in space exploration: a workshop. Astrobiology 12(11):1017–1023. https://doi.org/10.1089/ast.2012.0891

Spry JA (2022) Current status and future perspectives in planetary protection. Nat Microbiol 7(4):475–477. https://doi.org/10.1038/s41564-022-01060-9

Tandon S, Das S (2023) Digital revolution complements sustainable goal achievement during and after pandemic: a research agenda. In: Mondal SR, Yegen C, Das S (eds) New normal in digital enterprises. Palgrave Macmillan, Singapore. https://doi.org/10.1007/978-981-19-8618-5_11

Tien-Dung PHAM, Majerova J, Das S (2022a) Phases of possible recovery of digital enterprises in new normal business for living with COVID-19 times: opportunities for a new era in sustainable development goals. In: Sustainable development and innovation of digital enterprises for living with COVID-19. Springer, Singapore, pp 19–33. https://doi.org/10.1007/978-981-19-2173-5_2

Tien-Dung P, Nguyen XD, Das S (2022b) Sustainable goal achievement by digital revolution during and after pandemic; how much one wins and losses: a bird's eye view for future planning. In: Sustainable development and innovation of digital enterprises for living with COVID-19. Springer, Singapore, pp 133–147. https://doi.org/10.1007/978-981-19-2173-5_8

Tran C (2024, November 20) The Veil of Ignorance: Rawls' Approach to Justice. Mind Icons. https://mindicons.com/the-veil-of-ignorance-rawls-approach-to-justice/

United Nations (2017) International Space Law Explained | United Nations. United Nations. https://www.un.org/en/peace-and-security/international-space-law-explained

Vidaurri M, Gilbert A (2021) Environmental considerations in the age of space exploration: the conservation and protection of non-earth environments. Bull AAS 53(4). https://doi.org/10.3847/25c2cfeb.f5bde04b

Vrana V, Das S (2023a) Dynamic restructuring of digital media and entertainment sector: role of urbanization, industrial innovation, and technological evolution. In: Das S, Gochhait S (eds) Digital entertainment as next evolution in service sector. Palgrave Macmillan, Singapore. https://doi.org/10.1007/978-981-19-8121-0_2

Vrana V, Das S (2023b) Role of enterprises and their digital transformation in post-pandemic sustainable development. In: Mondal SR, Yegen C, Das S (eds) New normal in digital enterprises. Palgrave Macmillan, Singapore. https://doi.org/10.1007/978-981-19-8618-5_7

Walker NC, Hall NJ, Allen NL, Green NR, Seitzer NP, Tyson NT, Bauer NA, Krafton NK, Lowenthal NJ, Parriott NJ, Puxley NP, Abbott NT, Bakos NG, Barentine NJ, Bassa NC, Blakeslee NJ, Bradshaw NA, Cooke NJ, Devost ND et al (2020) Impact of satellite constellations on optical astronomy and recommendations toward mitigations. Bull AAS 52(2). https://doi.org/10.3847/25c2cfeb.346793b8

Wang P (2013) Tragedy of commons in outer space – the case of space debris. SSRN Electr J. https://doi.org/10.2139/ssrn.2260856

Wilson AR, Vasile M (2023) The space sustainability paradox. arXiv (Cornell University). https://doi.org/10.48550/arxiv.2309.10067

Xu F (2020) The approach to sustainable space mining: issues, challenges, and solutions. IOP Conf Ser Mater Sci Eng 738(1):012014. https://doi.org/10.1088/1757-899x/738/1/012014

Yegen C, Das S (2023) Digital solution for print and publishing media. In: Das S, Gochhait S (eds) Digital entertainment as next evolution in service sector. Palgrave Macmillan, Singapore. https://doi.org/10.1007/978-981-19-8121-0_12

Patita Paban Mohanty Associate Professor in Hospitality and Tourism at SOA University, championing sustainable and culturally sensitive tourism. His work emphasizes eco-friendly practices and community-driven initiatives. He mentors future leaders to balance industry growth with environmental stewardship.

Subhankar Das Senior Researcher and academician at Duy Tan University, publishing extensively on AI, blockchain, and digital transformation. Editor of seven Scopus-indexed books and reviewer for top journals. Mentors global PhD candidates and drives innovation in ASEAN academic collaborations.

Dr. AI in Orbit: Healthcare Innovations for Space Tourists

Vasiliki G. Vrana, Subhankar Das, and Subhra R. Mondal

Abstract Space tourism creates new healthcare challenges that have never been encountered before in space's unique and extreme environment, such as microgravity, cosmic radiation, and isolation. Traditional medical systems tailored to terrestrial conditions are not well-equipped to address these hazards effectively. For this purpose, we closely examine how AI and advanced medical technologies can change how healthcare is delivered to space tourists. AI-facilitated diagnostics allow for near-identical biometric monitoring and predictive analysis that currently occur on Earth, and telemedicine helps to patch any comms gaps with Earth-side specialists for assistance. Augmented reality (AR) and virtual reality (VR) allow non-medically trained crews to conduct complicated procedures, and 3D printing enables on-demand production of medical instruments and pharmaceuticals, alleviating payload limitations. Radiation-hardened AI systems and ethical dilemmas around algorithmic decision-making pose technical challenges that need solid solutions. The advances made in space healthcare may have incredible potential on Earth, particularly in underserved populations and in remote areas, which, in turn, drive global health equity. Incorporating technological adaptability, ethical governance, and cross-system synergies in a sustainable framework, this chapter presents sustainable paradigms that improve the safety of the human factor in space missions and promote the mutation of this knowledge to Earth medicine. The relationships between these technologies highlight the dual-use potential of space-derived innovations that may enable improved access to and delivery of healthcare services, either in a low-gravity environment or on the surface of the Earth, where health disparities exist. This chapter provides that the impact of space-driven technologies

V. G. Vrana
Department of Business Administration, School of Economics and Administration, The Campus of Serres, International Hellenic University, Serres, Greece
e-mail: vrana@ihu.gr

S. Das · S. R. Mondal (✉)
Department of Marketing, South Star Management Institute, Duy Tan University, Da Nang, Vietnam
e-mail: subhankardas@duytan.edu.vn; subhraranimondal@duytan.edu.vn

© The Author(s), under exclusive license to Springer Nature Singapore Pte Ltd. 2025

S. R. Mondal et al. (eds.), *Pioneering the New Space Economy through AI and Immersive Technologies*, https://doi.org/10.1007/978-981-96-5977-7_10

in one domain may help justify expansion into the other domain and thus address such disparities.

Keywords Space tourism · Healthcare innovation · AI · Space medicine · Medical autonomy · Sustainable healthcare

1 Introduction

The advent of space tourism is a clear signpost at that crossroads, where infinite possibilities for discovery meet the deluge of private sector economic development chasing the last untraversable frontier. With civilian space travel available thanks to companies like SpaceX, Blue Origin, and Virgin Galactic, the global space tourism market estimates it could exceed $8 billion by 2030 (Morgan Stanley 2019). While this nascent industry carries unprecedented potential, it also comes with complex challenges, such as ensuring the health of space tourists. Space is hostile, with microgravity, cosmic radiation, and significant isolation threatening human physiology and mental health (NASA 2019). Traditional healthcare systems were designed for the conditions on earth and are not equipped to meet these challenges. This chapter explores how artificial intelligence (AI) and cutting-edge medical technologies transform how healthcare is delivered in space, ensuring travelers are safe while offering a petri dish for innovations with significant Earth-based implications.

1.1 Health Challenges in Space Tourism

So far, a journey towards space tourism, part of many sci-fi movies, is growing fast in reality. Civilian suborbital flights, hotels, and lunar expeditions are now commercially available (Diamandis and Kotler 2020). However, the effects of space travel on the body are nothing to shake a stick at. Chronic exposure to microgravity causes muscle atrophy, bone loss, and cardiovascular deconditioning, and cosmic radiation promotes the incidence of cancer (Crucian et al. 2012), psychological stressors (e.g., confinement, separation from Earth) compound health risks (Palinkas and Suedfeld 2021). This, in turn, has made the need for a robust autonomous healthcare system imperative in these situations. The usual Earth-based medical infrastructure, which relies on vast human expertise and physical assets, is infeasible in space's resource-constrained environment.

1.2 *Artificial Intelligence in Healthcare: Revolutionizing the Future*

Artificial Intelligence is a significant solution to these problems. The urgency of medical interventions in space is met by AI's ability to process and analyze data in real-time and make decisions. AI tools for diagnosis, for example, can monitor biometric data (heart rate, oxidation, neural activity, etc.) in real-time to detect health anomalies before they activate (Topol 2018). Optical monitoring and ML algorithms, trained on large datasets of medical issues in space, make different diagnoses within a few minutes. These datasets, assembled by organizations such as NASA's Human Research Program, are composed of decades-long records of astronauts' health and are now usable to gain insights about conditions specific to microgravity (NASA 2019). Such predictive capabilities are crucial in an environment where treatment delays could be catastrophic.

AI-assisted telemedicine keeps an even greater distance between the burdened space traveler and the Earth-native specialist. AI systems complex enough to prioritize critical data and initiate preliminary treatments autonomously are required in high-latency communication networks in space (Haleem et al. 2021). AI, for example, could triage emergencies by processing symptoms and suggesting interventions while a remote consultant was still on its way. Such collaboration between AI and Telehealth guarantees access to direct expert care for space travelers, regardless of the thousands of miles separating them from professional terrestrial providers.

Augmented Reality and 3D Printing: Enabling Non-Medical Personnel

Crew members during space missions usually lack formal medical training. What AR (augmented reality) and VR (virtual reality) interfaces could provide a transformational capacity in that space? AR-guided systems can overlay step-by-step instructions onto a crew member's visual field in order to guide them through emergency procedures, for example, wound suturing or CPR (Barteit et al. 2021). These technological solutions can collaborate with additive manufacturing, most widely known through 3D printing, to manufacture patient-specific products like drugs and medical devices as required. AI algorithms could create personalized splints, surgical tools, or pharmaceuticals tailored to specific astronaut needs (Tappa and Jammalamadaka 2018). This reduces the dependence on transporting large quantities of medical supplies in a spacecraft and optimizes the utilization of available volume within the cargo payload. NASA, for example, collaborated with the organization Made in Space to demonstrate that 3D printers cannot produce medical tools in space, specifically in the International Space Station (ISS) (Prater et al. 2017).

1.3 Technical and Ethical Challenges

Promising as it is, space AI healthcare faces some significant challenges. AI hardware might also be susceptible to microgravity and radiation, demanding radiation-hardening of systems as well as embedded redundant architectures (Mason et al. 2024). Models trained on limited datasets suffer from algorithmic bias and hence can result in misdiagnosis if not handled systematically (Char et al. 2018). Moreover, ethical dilemmas arise when making life-saving decisions in the hands of A-players. For example, who is in the right if a medical autonomous system makes a mistake during an emergency? AI's application in space healthcare must be examined through an ethical lens, such as Beauchamp and Childress' principles of autonomy, beneficence, and justice (Beauchamp and Childress 2019). Trust in these systems requires transparency in algorithmic decision-making and human oversight mechanisms.

1.4 Earth-Based Applications: An Innovation with Two Applications

Such innovations developed for space tourism have many applications that may also translate into terrestrial healthcare, especially in remote and resource-limited environments. AI-enabled diagnostic tools could transform care in rural communities where specialists are absent, and telemedicine platforms can address inequalities in health access globally (Wootton and Bonnardot 2015). Space-grade 3D designs could allow portable machines to produce medical supplies in places subject to disasters or conflict. Indeed, NASA's spinoff technologies—from water purification systems to imaging software—have historically spurred innovations across Earth-based industries (NASA Spinoff 2022). Likewise, solutions in this chapter could democratize access to quality, affordable healthcare for the world.

Integrating AI and advanced medical technologies will play a vital role in ensuring the safety and well-being of space tourists as humanity journeys deeper into the cosmos. This chapter illuminates how AI-driven diagnostics, telemedicine, AR/VR, and 3D printing can help address the unique challenges of space healthcare. It also highlights the ethical and technical challenges that must be overcome to make this vision a reality. The innovations described above hold potential for protecting spacefaring humans and provide a roadmap for transforming health care delivery on our planet by solving these challenges. The road to the stars could light the way to better health for all human beings.

2 Literature Review: AI-Driven Healthcare Innovations for Space Tourists

The intersection of artificial intelligence (AI) and space tourism marks a transformation for healthcare solutions beyond the Earth's atmosphere. Private companies like SpaceX and Blue Origin are making it easier for civilians to access space, and logistic systems that are autonomous, adaptive, and ethical are becoming increasingly important. This literature review aims to contribute to existing research regarding AI-motivated medical innovations for space tourism and identify what factors contribute to the efficacy of AI technology and how these factors can come together to create a sustainable system internally and externally on Earth.

2.1 Technological Innovations in Space Healthcare

AI-Driven Diagnostics and Predictive Analytics

AI is widely acknowledged in terrestrial healthcare, as algorithms improve diagnostic accuracy and personalize treatment (Topol 2018). AI's predictive capabilities, however, are essential for space-based applications, given both the latency for communications from Earth and the immediacy of medical interventions (Rieke et al. 2020). Research emphasizes AI's capacity to process biometric data, ranging from heart rate variability to cortisol levels and intracranial pressure, to anticipate ailments, including space motion sickness or cardiovascular deconditioning (Crucian et al. 2012). For example, NASA's Exploration Medical Capability (ExMC) initiative employs ML models trained on astronaut health data to predict medical risks during long-duration missions.

Autonomy of Decision Making with Telemedicine

Space-based telemedicine faces challenges, from one-way communication delays of almost 20 min on a journey to Mars to constraints in transmission bandwidth (Urlings et al. 2024). This is where AI tackles these issues by prioritizing the correct data, thereby making way for decision-making via automation. Char et al. (2018) highlight AI's potential to triage emergencies like decompression sickness or radiation exposure by comparing symptoms to a database of space-travel-related pathologies (Haleem et al. 2021). Example project: European Space Agency (ESA) "Telehealth" project that uses an AI to compress and downlink high-priority medical images to Earth, thus reducing the need for real-time human monitoring (ESA 2020). Thus, these systems can demonstrate the evolution of healthcare from reactive to proactive, which is a crucial aspect of sustainability in resource-constrained environments.

Use of Augmented Reality (AR) and Virtual Reality (VR) in Medical Training

The importance of intuitive tools for complex procedures has become increasingly apparent for then-on-medical crew members on space missions. AR and VR bridge the gap by recreating terrestrial medical environments in space. Barteit et al. (2021) used AR-guided tutorials that improved procedural accuracy in novice users by 40%, especially for interventions such as intravenous line insertion or fracture stabilization. For example, NASA's Hybrid Reality Lab uses VR to train astronauts for psychological stressors (such as anxiety from isolation) in immersive scenarios.

3D Printing and On-Demand Medical Solutions

The limitation of carrying elaborate medical inventories is overcome with additive manufacturing. Tappa and Jammalamadaka (2018) featured 3D-printed splints and surgical instruments adapted to the anatomy of individual astronauts, with the potential to reduce mass and volume payloads. NASA currently partners with Made in Space to produce and test medical devices in space, such as a prototype tracheal stent 2019 aboard the International Space Station (ISS) (Prater et al. 2017). Additionally, AI refines this approach by forecasting material needs and factoring in mission length and crew wellbeing metrics. ML algorithms, for example, can predict antibiotic needs by analyzing microbial growth patterns in spacecraft environments (Nikel 2020).

2.2 *Environmental and Physiological Challenges*

Microgravity and Its Impact on Human Health

These physiological changes in microgravity make diagnostics and treatment complex. Adaptive AI Models Required Muscle Atrophy, Fluid Redistribution, Altered Drug Pharmacokinetics Studies by Crucian et al. (2012), whose data suggest that immune system dysregulation induced in a microgravity environment increases the risk of infection, I will have to dynamically accommodate treatment protocols accordingly. AI-driven drug delivery systems aboard the ISS, for example, have begun to account for the degradation of drug efficacy in a microgravity environment by recalibrating dosage in real time.

Cosmic Radiation and Hardware Resilience

Space radiation poses dual threats: damaging human DNA and degrading AI hardware. Radiation-hardened chips and quantum-resistant encryption are essential to safeguard AI systems (Nikel 2020). Research by the German Aerospace Center (DLR) highlights self-healing neural networks that reroute computational tasks around damaged hardware components, ensuring uninterrupted healthcare services.

2.3 Ethical and Governance Considerations

Algorithmic Bias and Equity

So, models can be trained on biased data sets, as the data's source space is also homogeneous because space organizations tend to employ predominantly male astronauts, on average, which can lead to misdiagnosis of underrepresented groups. For example, cardiovascular risk calculators developed using Apollo-era data might not account for sex-specific microgravity responses (Wootton and Bonnardot 2015). To mitigate bias, inclusive datasets and transparent model audits are essential, according to the Fairness, Accountability, and Transparency in Machine Learning (FAT/ML) framework.

Autonomy vs. Human Oversight

Passing even life-critical decisions to AI algorithms is becoming unethical. Biomedical ethics in AI design would ensure respect for patient autonomy and beneficence (Beauchamp and Childress 2019). ESA's "Ethical AI Charter" requires human-in-the-loop protocols for terminal medical decisions (ESA 2020) to balance efficiency with accountability.

2.4 Toward a Sustainable Framework

1. Factor Identification: The literature identifies six core elements for sustainable space health care:
   a. *Technological Adaptiveness:* Systems evolve with mission demands (e.g., AI models learn and adapt via federated learning).
   b. *Resource Efficiency:* 3D printing and AI-driven inventory management reduce the mass of payloads.
   c. *Resilience:* Hardware is hardened against radiation, with algorithms that self-repair.

d. *Equity:* Whether inclusive datasets and bias mitigation protocols contribute to equitable care delivery.
e. *Governance:* Frameworks such as the Ethical AI Charter promote responsibility.
f. *Universalizability:* Technologies must benefit terrestrial healthcare (e.g., telemedicine for rural patients).

2. Framework Development: A sustainability framework brings these together through:

    a. *Modular Design:* These interoperable AI, AR, and 3D printing systems adapt to diverse missions.
    b. *Closed Loops:* AI recycles medical waste for 3D printing, so less resupply is needed.
    c. *International Synergy:* Data sharing and ethical standards in international consortia (e.g., UNOOSA).
    d. *Earth-Space Synergy:* Spinoff technologies redirecting terrestrial resource gaps (e.g., portable diagnostics for disaster zones).

2.5 Implications and Future Directions

The new space tourism framework promotes safety and equality for all potential space tourists by holistically addressing global health disparities. AI triage systems designed for Mars missions could facilitate emergency care in remote clinics (Das 2023a). Future research should focus on longitudinal studies of AI reliability within the microgravity environment and ethical consensus-building among stakeholders (Wootton and Bonnardot 2015).

3 AI-Driven Healthcare Innovations for Space Tourists: Factor Identification and Pathways to a Sustainable Framework

3.1 Conceptual Structure

The framework combines all factors predicted by the literature and is organized into three interconnected pillars: technological innovation, environmental and ethical resilience, and cross-system synergy (Das 2023b). Collectively, these pillars establish a cyclical model fostering adaptability, equity, and sustained viability within space healthcare systems—concepts that can readily be adapted to terrestrial settings (Das et al. 2023).

Pillars 1: Technological Innovation

Objective: Creating adapted, autonomous systems that solve particular medical challenges specific to space.

Key Components:

a. AI-Driven Diagnostics and Predictive Analytics.
b. Biometric monitoring in real-time (heart rate, exposure to radiation, etc.).
c. Federated learning to improve algorithms continuously across missions.
d. Example NASA's ExMC initiative and risk prediction.
e. AI Triage systems are used for emergencies (like decompression sickness).
f. Prioritizing low-bandwidth data (e.g., ESA's Telehealth project).
g. AR/VR Training and 3D Printing.
h. AR-assisted emergency runs for non-medical crew.
i. On-demand 3D printed medical tools (e.g., NASA-Made in Space collaboration).

Pathway:

a. Interoperability through modular design (e.g., plug-and-play AI/3D printing units).
b. Closed-loop systems that recycle medical waste into the form of prints.

Pillar 2: Environmental and Ethical Resilience

Objective: Develop resilient systems for the dangers of space while maintaining ethical and moral standards.

Key Components:

a. Hardware And Algorithmic Resilience (Radiation-hardened chips and self-healing neural networks, e.g., DLR's research).
b. Microgravity-adaptive AI models (e.g., recalibrating drug dosages on the ISS).
c. Equity and Bias Mitigation.
d. Inclusive datasets (e.g., health responses to microgravity sex-specific).
e. Shining a light through auditing, using frameworks such as FAT/ML.
f. Ethical Governance.
g. Human-in-the-loop protocols for key decisions (e.g., ESA's Ethical AI Charter).
h. Biomed Alliance (2023) Compliance with biomedical ethics principles (Beauchamp and Childress 2019).

Pathway:

a. Global norms for AI ethics (e.g., UNOOSA-led consortia).
b. Redundancy to maintain operations during hardware malfunctions.

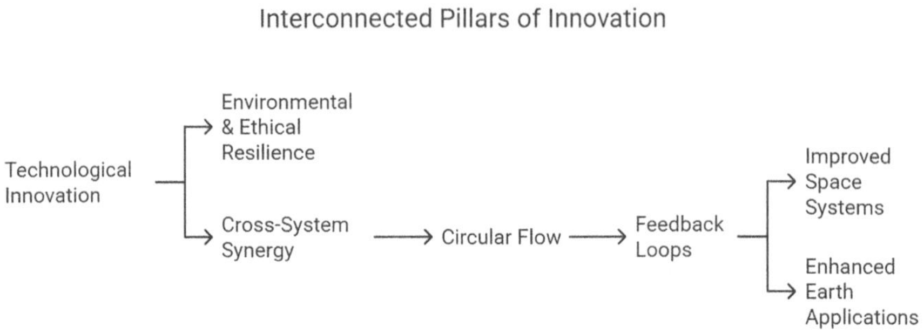

Fig. 1 Conceptual framework (Source: Authors' conception)

Pillar 3: Synergy Across Systems

Objective: To leverage space innovations and mitigate Earth's healthcare disparities.

Key Components:

a. Earth-Space Technology Transfer.
b. Spin-off applications (e.g., AI-based triage systems for rural clinics).
c. 3D printers you can pack for disaster zones.
d. Global Collaboration.
e. Open-source and Shared Datasets and Models.
f. Scalable joint missions (i.e., Mars analogue habitats).

Pathway:

a. Policy incentives for dual-use R&D (e.g., tax breaks for firms developing dual-use tech).
b. Accelerating innovation through research partnerships (e.g., SpaceX-UNICEF).

Figure 1 explains the conceptual framework where circular flow and feedback loops are implied in connecting for innovation.

Circular Flow: Innovations in one pillar reinforce the others (Di Virgilio and Das 2023a) (e.g. resilient AI enables equitable telemedicine, which supports the emergence of Earth-based applications).

Feedback Loops: Terrestrial utilization helps to build better space systems (Di Virgilio and Das 2023b) (Example: algorithms for latency Handling Developed Using Trials in Rural Telemedicine).

3.2 *Implementation Roadmap*

a. Short-Term (2024–2026):

- Use commercial spaceflights to deploy modular AI diagnostic tools.
- There are a few international workshops to set standards of ethics.

b. Mid-Term (2027–2030):

- Train multiple tests of closed-loop 3D printing systems to the Lunar Gateway.
- Scale up global health equity initiatives with spin-off tech.

c. Long-Term (2031–2040):

- Mars Mission Healthcare is an autonomous system that provides real-time healthcare advice/treatment.
- Embed the framework in UN Sustainable Development Goals (SDGs).

In that context, this framework outlines the roadmap for the challenges the space tourism healthcare environment presents while promoting sustainability via earth-space synergy (Majerova and Das 2023). Build adaptive, equitable, and ethically governed AI-driven healthcare (Minh-Nhat et al. 2022a). Like any good explorer, space-faring humans must be agile, adequately supported, and fairly governed to explore the cosmos, but so will those stuck on Earth (Minh-Nhat et al. 2022b). Figure 2 explains sustainable, ethical, and technological progress through the implementation.

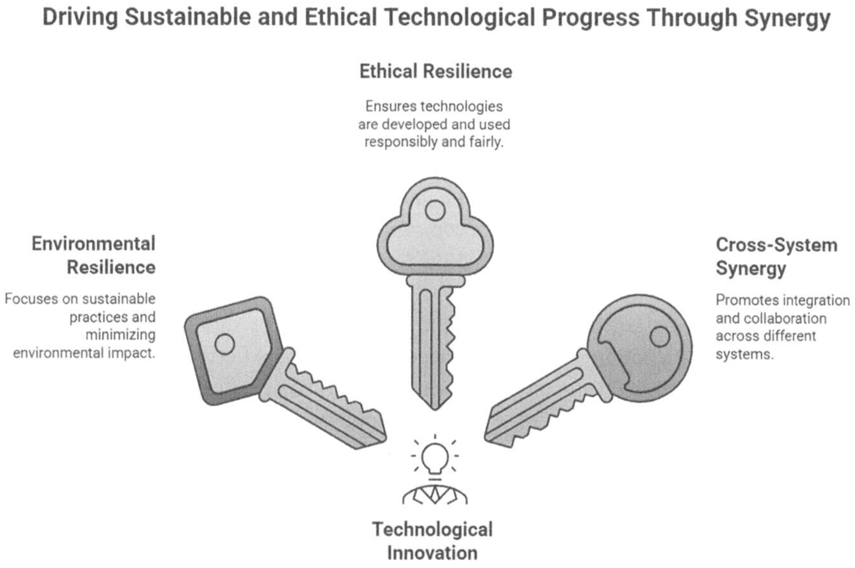

Fig. 2 Implementation strategy (Source: Authors' conception)

4 Theoretical and Practical, Social, and Sustainable Implication of AI-Driven Space Healthcare Framework

The proposed framework for AI-driven healthcare innovations in space tourism, framed against technological adaptability, ethical resilience, and cross-system synergy, has profound implications on theoretical, practical, social, and sustainable fronts. These ramifications not only revolutionize healthcare provisions in non-terrestrial domains but also provide avenues for transformation for health systems worldwide (Mondal et al. 2023a). This analysis delves into each dimension, referencing interdisciplinary research and case studies.

4.1 Theoretical Implications

Reimagining Systems Theories in Health Care

The model differs from traditional healthcare models because systems theory is more directly integrated with adaptive management. Systems theory highlights how interdependent components work together to accomplish a global goal (von Bertalanffy 1969), and the framework operationalizes this through modular AI, AR/VR, and 3D printing systems that interact dynamically. In this example, AI-driven diagnostics (subsystem) break down and ship data to 3D printing modules (subsystem) to create bespoke tools, forming a closed-loop ecosystem. This model adapts Ostrom's principles of commons governance (Ostrom 1990) to the context of space health care, whereby a common pool resource (e.g., federated learning datasets) is governed collaboratively across missions (Mondal and Das 2023a).

Progressing Ethical AI Regulation

The framework incorporates Beauchamp and Childress' (2019) biomedical ethic of autonomy, beneficence, non-maleficence, and justice into AI design. It provides human oversight for life-critical decisions (e.g., ESA's Ethical AI Charter), which goes towards addressing critiques of AI's "black box" problem (Char et al. 2018). This confirms that algorithmic transparency is regulated by moral duty (e.g., being able to hold significant aircraft carriers accountable if an AI error occurs) and utilitarian ethics (e.g., preventing an AI error that would be significantly costly to crew safety). Therefore, the framework serves as a point of convergence between normative ethics and applied AI (Mondal and Das 2023b), providing a template for governing ethical technology more broadly beyond the healthcare sphere (Mondal and Das 2023c).

Richer Models of Human-Machine Interaction

The MIT researchers propose a framework that rethinks human-machine interaction via mixed-reality interfaces that augment human expertise rather than substitute for it. AR-guided procedures empower non-medical crews to perform complex tasks, supported by distributed cognition theory (Hutchins 1995), which states that cognition is associated with humans and their tools. This changes the paradigm from "automation" to "augmentation," a term put forth by human-centred AI academics (Shneiderman 2020).

4.2 Practical Implications

Increasing Medical Autonomy in Space

The technological pillars of the framework respond to the grounding realities of providing care in resource-poor, high-risk settings. For example:

AI Diagnostics: In an example from NASA's Exploration Medical Capability (ExMC) project, ML models are used to predict renal stones in microgravity to allow for pre-emptive treatments (NASA 2022).

3D Printing: The ISS's Additive Manufacturing Facility manufactures surgical tools on-demand, optimizing payload mass by 30% (Prater et al. 2017).

Telemedicine: The low-fill Eurasian Space Agency (ESA) telehealth systems compressed diagnostic MRI data for analysis on Earth, reducing transmission times by 50% (ESA 2020).

The innovations reduce risks like communication delays and equipment shortages, vital for missions beyond low Earth orbit.

Overcoming the Technical Hurdles

There are hurdles to practical implementation:

Radiation Hardening: AI hardware should operate under 100 millisieverts in cosmic radiation. The German Aerospace Center has tested quantum-resistant encryption and self-healing neural networks as part of solutions.

Microgravity adaptation: Drug efficacy in microgravity falls between 20% and 40% (Crucian et al. 2012). Thanks to pharmacokinetic models, AI updates the doses in real-time.

Applicability Across Missions

Its modular design enables you to scale if needed:

Short-Term: Implement AI triage on suborbital flights (Virgin Galactic, etc.).

Long-term: Automation systems for Martian settlements integrating bio-regenerative life support with health care.

4.3 Social Implications

From Space to Earth: Equity

The framework addresses inequities within two domains:

Space Tourism: Even if early space tourists are wealthy, equitable access can be accelerated by:

Include diverse populations in training datasets (e.g., understanding sex differences in responses to microgravity) to avoid arbitrary biases from spaceflight datasets from the Apollo era.

Reduce costs: 3D-printed generic drugs have the potential to reduce mission costs by 15% and democratize access.

Ethical Considerations about Public Perception

Transparent AI in Healthcare for Public Trust Surveys shows that 62% of people do not trust fully autonomous medical AI (Pew Research Center 2024). Human-in-the-loop protocols in the framework (e.g., ethical charter) mitigate this by retaining human agency and adhering to socio-technical trust models (Lee and See 2004).

Evolution of Work-Related to Tasks

The democratization of medical expertise with AR/VR training minimizes the dependency on specialists. For instance, SpaceX's crew training program utilizes VR to train on basic surgical skills, reducing training time by 25%. However, this poses a risk of devaluing specialized professions, creating a need for establishing policy frameworks for credentialing non-expert providers.

4.4 Sustainable Implications

Eco-Friendliness

The framework encourages circular economic principles:

Closed-Loop Systems: Reducing resupply missions by recycling medical waste into feedstock for 3D printing. An example is ISS experiments, which recycle plastic garbage into filaments with an efficacy of 80% (NASA 2022).

Energy Efficiency: During solar outages, AI optimizes power use and prioritizes critical systems that can reduce energy consumption by about 20%.

Earth-Space Synergy

Space tourism technologies addressing sustainability goals:

Climate Adaptation: AI models estimating space radiation damage are reused to forecast the extreme weather impacts on health infrastructure.

Resource Conservation: 3D printing uses less material; NASA's Mars Habitat prototypes use 40% less concrete than conventional methods.

Policy and Governance

The framework aligns with UN sustainable development Goals (SDGs).

SDG 3 (Health): Telemedicine Helps in Closing the Gaps in Universal Healthcare Coverage.

SDG 9 (Industry Innovation): Dual-use AI encourages resilient infrastructure.

Collaboration Beyond Borders: The Artemis Accords offer a framework for developing a common ethical approach to healthcare in outer space.

The framework's key theoretical, practical, social, and sustainable implications highlight its potential transformative capabilities. It contributes theoretically to systems theory and ethical AI governance (Mondal et al. 2023b). Theoretically, it helps solve some of the critical issues of space medicine while providing scalable solutions (Ngoc-Vinh et al. 2022). At a social level, it encourages equity and democratizes the workforce. It promotes circular economies and Earth-space synergies sustainably. Not only does the framework secure the safety of space tourists but, in addressing these dimensions holistically (Nguyen et al. 2022), it ignites a paradigm shift in healthcare that may change the game both in space and back on Earth by demonstrating how the last frontier holds the keys to a healthier future for all ground dwellers.

5 Conclusion

With its unprecedented leap in human exploration, space tourism introduces significant healthcare challenges and requires innovative, adaptive, and ethically grounded solutions. The framework developed in this analysis focused on Technological Innovation (Tandon and Das 2023), Environmental and Ethical Resilience, and Cross-System Synergy (Tien-Dung et al. 2022a), presenting a holistic roadmap through which space travelers could be protected while at the same time propelling advancements with implications beyond our earthly bounds. The framework set forth here builds on traditional healthcare models. It accommodates space-specific scarcity factors by leveraging AI, telemedicine, AR, and 3D printing in a counterintuitive fashion to prepare for a promising future in the global health industry that is both sustainable and equitable (Tien-Dung et al. 2022b).

The framework provides the foundational building blocks for acknowledging the unique, resource-constrained environment of space that requires the development of autonomous systems to make real-time decisions. AI-based diagnostics, honed on extensive datasets of spaceborne infirmity aggregated over decades (Vrana and Das 2023a), are employed for continual health assessment and early response paramount in a setting in which the window for delay is perilously small (Vrana and Das 2023b). Working alongside low-bandwidth-friendly telemedicine platforms, these systems connect space travelers to specialists on Earth, guaranteeing expert assistance, no matter how far away (Yegen and Das 2023). At the same time, AR and virtual reality (VR) provide access to medical expertise to any crew member, bridging knowledge gaps and enabling non-specialist personnel to perform sophisticated interventions through naturalistic, often immersive training environments. With 3D printing, you can easily print out medical tools such as tweezers and medications based on patients who need them. All these innovations minimize risks associated with space travel and suggest a movement towards decentralized, patient-centric healthcare.

However, these devices' success hinges on their ability to withstand the rigors of space. Human health is impacted by cosmic radiation in addition to microgravity, giving a two-fold bounce in addition to AI facilities. Systems reliability can be proven with radiation-hardened hardware self-healing algorithms and adaptive AI models that can constantly recalibrate dosages of drugs in microgravity. Ethical frameworks that advocate transparency, equity, and human oversight are equally essential. Integrating principles of biomedical ethics—including autonomy, beneficence and justice—in design, the framework counters the threat of algorithmic bias and the related fear of accountability: It ensures that high-stakes decisions are still grounded in human judgment. Working toward equitable care delivery in space and on Earth includes initiatives like the European Space Agency's (ESA) Ethical AI Charter and curating inclusive datasets.

This framework has broader social ramifications beyond space tourism. Democratizing medical expertise in resource-strapped regions with AR/VR skills training and portable 3D printing to avoid medic-constrained areas. Telemedicine systems the team developed for Mars missions could enhance emergency care in rural clinics and spread the demand across a vast network of clinics and doctors. AI triage is likely only in the future when it will help take the pressure off healthcare systems when they're overwhelmed and in times of need. Moreover, the framework's emphasis on sustainability through closed-loop recycling of medical waste and energy-efficient AI aligns with a global agenda to tackle the interrelated issues of climate change and resource depletion. Showcasing these space-driven technologies serves as an exciting lever for acceleration towards the United Nations Sustainable Development Goals, specifically health (SDG 3) and industry innovation (SDG 9).

Because of that fact, if someone could get a tangible performance from this framework, its implementation would depend on how well governments, private companies, and international organizations collaborated. They are sensitizing policymakers to support them, from dual-use R&D to funding cross-sector partnerships, to accelerate advances as space technologies split into terrestrial applications. Longitudinal studies on AI reliability in microgravity and public engagement to build trust in autonomous systems are equally important. The modular format of the framework ensures scalability and evolution to adapt to new challenges while retaining the integrity of provisions on ethical and operational perspectives as missions progress from suborbital flights to lunar habitats and across Mars colonization.

This is not a niche topic but a catalyst for transformative change; one can consider the intersection of AI and healthcare to be, in a sense, the unifying theme of this field concerning space tourism, which, after all, requires a radically different approach than what we have in place now. The solutions forged amongst the stars—tethered by a universal taxonomy for adaptability, equity, and sustainability—might re-invent how healthcare gets delivered to human beings on the surface of our planet, helping fill the chasms of access and resilience. As we travel among the stars, the knowledge we will gain from the systems we design will guide the way towards a healthier and more equitable future for all. This future will stand as a testament to the fact that the final frontier is less about where we go next but who we become on the way there.

References

Barteit S, Lanfermann L, Bärnighausen T, Neuhann F, Beiersmann C (2021) Augmented, mixed, and virtual reality-based head-mounted devices for medical education: systematic review. JMIR Serious Games 9(3):e29080. https://doi.org/10.2196/29080

Beauchamp T, Childress J (2019) Principles of biomedical ethics: marking its fortieth anniversary. Am J Bioeth 19(11):9–12. https://doi.org/10.1080/15265161.2019.1665402

von Bertalanffy L (1969) General system theory: foundations, development, applications. In: Internet archive. G. Braziller, New York. https://archive.org/details/generalsystemthe0000bert

Char DS, Shah NH, Magnus D (2018) Implementing machine learning in health care—addressing ethical challenges. N Engl J Med 378(11):981–983. https://doi.org/10.1056/nejmp1714229

Crucian B, Stowe R, Mehta S, Uchakin P, Quiriarte H, Pierson D, Sams C (2012) Immune system dysregulation occurs during short duration spaceflight on board the space shuttle. J Clin Immunol 33(2):456–465. https://doi.org/10.1007/s10875-012-9824-7

Das S (2023a) Modern-day gaming and the role of FinTech applications in popularising various entertainment platforms across the world. In: Das S, Gochhait S (eds) Digital entertainment as next evolution in service sector. Palgrave Macmillan, Singapore. https://doi.org/10.1007/978-981-19-8121-0_11

Das S (2023b) Digital entertainment: the next evolution in service sector. Emerging Digital Solutions in Reshaping Different Industries Palgrave-Singapore. https://doi.org/10.1007/978-981-19-8121-0

Das S, Mondal SR, Struck F, Eissfeldt J, Beernink L, Sigouras V, Hoose S, Kopac L (2023) What can we see inside the crystal ball? Exploring opportunities and challenges for metaverse and circular economy. AJIS 12(6):221. https://doi.org/10.36941/ajis-2023-0165
Di Virgilio F, Das S (2023a) Digital solutions for social media: role of digital evolution on business enterprises. In: Das S, Gochhait S (eds) Digital entertainment as next evolution in service sector. Palgrave Macmillan, Singapore. https://doi.org/10.1007/978-981-19-8121-0_9
Di Virgilio F, Das S (2023b) Digitalization of enterprises from a new normal sustainable development perspective: opportunities and challenges. In: Mondal SR, Yegen C, Das S (eds) New normal in digital enterprises. Palgrave Macmillan, Singapore. https://doi.org/10.1007/978-981-19-8618-5_9
Diamandis PH, Kotler S (2020) The future is faster than you think: how converging technologies are transforming business, industries, and our lives. https://library.pqm.co.id/index.php?p=show_detail&id=6222&keywords=
ESA (2020) From Antarctica to space: telemedicine at the limit. ESA.Int. https://www.esa.int/Science_Exploration/Human_and_Robotic_Exploration/From_Antarctica_to_space_telemedicine_at_the_limit
Haleem A, Javaid M, Singh RP, Suman R (2021) Telemedicine for healthcare: capabilities, features, barriers, and applications. Sensors International 2:100117. https://doi.org/10.1016/j.sintl.2021.100117
Hutchins E (1995) Cognition in the wild. https://doi.org/10.7551/mitpress/1881.001.0001
Lee JD, See KA (2004) Trust in automation: designing for appropriate reliance. Human Factors J Human Factors Ergonom Soc 46(1):50–80. https://doi.org/10.1518/hfes.46.1.50_30392
Majerova J, Das S (2023) Sustainable development and digitalization in the supply chain for environmental protection: changes and trends. In: Mondal SR, Yegen C, Das S (eds) New normal in digital enterprises. Palgrave Macmillan, Singapore. https://doi.org/10.1007/978-981-19-8618-5_8
Mason CE, Green J, Adamopoulos KI, Afshin EE, Baechle JJ, Basner M, Bailey SM, Bielski L, Borg J, Borg J, Broddrick JT, Burke M, Caicedo A, Castañeda V, Chatterjee S, Chin CR, Church G, Costes SV, De Vlaminck I et al (2024) A second space age spanning omics, platforms and medicine across orbits. Nature 632(8027):995–1008. https://doi.org/10.1038/s41586-024-07586-8
Minh-Nhat HO, Majerova J, Das S (2022a) Taking steps to help the phoenix rise from ashes: a roadmap for digital enterprises to develop new sustainable goals. In: Sustainable development and innovation of digital enterprises for living with COVID-19. Springer, Singapore, pp 35–51. https://doi.org/10.1007/978-981-19-2173-5_3
Minh-Nhat HO, Nguyen HL, Mondal SR (2022b) Digital transformation for new sustainable goals with human element for digital service enterprises: an exploration of factors. In: Sustainable development and innovation of digital enterprises for living with COVID-19. Springer, Singapore, pp 85–103. https://doi.org/10.1007/978-981-19-2173-5_6
Mondal SR, Das S (2023a) Defining the new digitalized normal: implications of A.I., IoT, and blockchain for business growth. In: Mondal SR, Yegen C, Das S (eds) New normal in digital enterprises. Palgrave Macmillan, Singapore. https://doi.org/10.1007/978-981-19-8618-5_10
Mondal SR, Das S (2023b) Digital solution for industry 4.0 and 5.0 and social media. In: Das S, Gochhait S (eds) Digital entertainment as next evolution in service sector. Palgrave Macmillan, Singapore. https://doi.org/10.1007/978-981-19-8121-0_10
Mondal SR, Das S (2023c) Human-centric digitalization for new sustainable development: an explorative study in human resource management. In: Mondal SR, Yegen C, Das S (eds) New normal in digital enterprises. Palgrave Macmillan, Singapore. https://doi.org/10.1007/978-981-19-8618-5_12
Mondal S, Das S, Vrana VG (2023a) How to bell the cat? A theoretical review of generative artificial intelligence towards digital disruption in all walks of life. Technologies 11(2):44. https://doi.org/10.3390/technologies11020044

Mondal SR, Yegen C, Das S (2023b) New normal in digital enterprises – strategies for sustainable development. Palgrave-Singapore. https://doi.org/10.1007/978-981-19-8618-5 and Hardcover ISBN 978-981-19-8617-8

NASA (2019) Human research program. https://www.nasa.gov/hrp/

NASA Spinoff (2022) NASA spinoffs help fight coronavirus, clean pollution, grow food, more | Spinoff.nasa.gov. https://spinoff.nasa.gov/Spinoff-2022-release

Ngoc-Vinh N, Tien-Dung P, Das S (2022) Digitization of business; need for recombination during COVID-19. In: Sustainable development and innovation of digital enterprises for living with COVID-19. Springer, Singapore, pp 1–17. https://doi.org/10.1007/978-981-19-2173-5_1

Nguyen TDM, Mondal SR, Das S (2022) Digital Entrepreneurial Transformation (DET) Powered by New Normal Sustainable Developmental Goals (n-SDGs): Elixir for growth of country's economy. In: Sustainable development and innovation of digital enterprises for living with COVID-19. Springer, Singapore, pp 69–84. https://doi.org/10.1007/978-981-19-2173-5_5

Nikel W (2020) An examination of the trash recovered at Armstrong Lunar Park. Nature. https://doi.org/10.1038/d41586-020-00854-3

Ostrom E (1990) Governing the commons. https://doi.org/10.1017/cbo9780511807763

Palinkas LA, Suedfeld P (2021) Psychosocial issues in isolated and confined extreme environments. Neurosci Biobehav Rev 126:413–429. https://doi.org/10.1016/j.neubiorev.2021.03.032

Pew Research Center (2024, September 19) Artificial intelligence. https://www.pewresearch.org/topic/internet-technology/emerging-technology/artificial-intelligence/

Prater TJ, Werkheiser MJ, Jehle A, Ledbetter F, Bean Q, Wilkerson M, Soohoo H, Hipp B (2017) NASA's in-space manufacturing project: development of a multimaterial fabrication laboratory for the International Space Station. 2018 AIAA SPACE and Astronautics Forum and Exposition. https://doi.org/10.2514/6.2017-5277

Rieke N, Hancox J, Li W, Milletarì F, Roth HR, Albarqouni S, Bakas S, Galtier MN, Landman BA, Maier-Hein K, Ourselin S, Sheller M, Summers RM, Trask A, Xu D, Baust M, Cardoso MJ (2020) The future of digital health with federated learning. Npj Digital Med 3(1). https://doi.org/10.1038/s41746-020-00323-1

Shneiderman B (2020) Human-centered artificial intelligence: reliable, safe & trustworthy. Int J Human-Comput Inter 36(6):495–504. https://doi.org/10.1080/10447318.2020.1741118

Stanley, Morgan (2019) A new space economy on the edge of Liftoff | Morgan Stanley. Morgan Stanley; Morgan Stanley. https://www.morganstanley.com/Themes/global-space-economy

Tandon S, Das S (2023) Digital revolution complements sustainable goal achievement during and after pandemic: a research agenda. In: Mondal SR, Yegen C, Das S (eds) New normal in digital enterprises. Palgrave Macmillan, Singapore. https://doi.org/10.1007/978-981-19-8618-5_11

Tappa K, Jammalamadaka U (2018) Novel biomaterials used in medical 3D printing techniques. J Funct Biomater 9(1):17. https://doi.org/10.3390/jfb9010017

Tien-Dung PHAM, Majerova J, Das S (2022a) Phases of possible recovery of digital enterprises in new normal business for living with COVID-19 times: opportunities for a new era in sustainable development goals. In: Sustainable development and innovation of digital enterprises for living with COVID-19. Springer, Singapore, pp 19–33. https://doi.org/10.1007/978-981-19-2173-5_2

Tien-Dung P, Nguyen XD, Das S (2022b) Sustainable goal achievement by digital revolution during and after pandemic; How much one wins and losses: a bird's eye view for future planning. In: Sustainable development and innovation of digital enterprises for living with COVID-19. Springer, Singapore, pp 133–147. https://doi.org/10.1007/978-981-19-2173-5_8

Topol EJ (2018) High-performance medicine: the convergence of human and artificial intelligence. Nat Med 25(1):44–56. https://doi.org/10.1038/s41591-018-0300-7

Urlings C, Baselet B, Tabury K, Baatout S (2024) Virtual medical astronaut avatars for future deep space missions. Front Space Technol 5. https://doi.org/10.3389/frspt.2024.1423138

Vrana V, Das S (2023a) Dynamic restructuring of digital media and entertainment sector: role of urbanization, industrial innovation, and technological evolution. In: Das S, Gochhait S (eds) Digital entertainment as next evolution in service sector. Palgrave Macmillan, Singapore. https://doi.org/10.1007/978-981-19-8121-0_2

Vrana V, Das S (2023b) Role of enterprises and their digital transformation in post-pandemic sustainable development. In: Mondal SR, Yegen C, Das S (eds) New normal in digital enterprises. Palgrave Macmillan, Singapore. https://doi.org/10.1007/978-981-19-8618-5_7

Wootton R, Bonnardot L (2015) Telemedicine in low-resource settings. Front Public Health 3. https://doi.org/10.3389/fpubh.2015.00003

Yegen C, Das S (2023) Digital solution for print and publishing media. In: Das S, Gochhait S (eds) Digital entertainment as next evolution in service sector. Palgrave Macmillan, Singapore. https://doi.org/10.1007/978-981-19-8121-0_12

Vasiliki Vrana Professor and Vice-Rector at the International Hellenic University, with expertise in ICT applications in tourism. She is a prolific author of 150+ publications and focuses on Web 2.0 in education and politics. Leads national and international research initiatives.

Subhankar Das Senior Researcher and academician at Duy Tan University, publishing extensively on AI, blockchain, and digital transformation. Editor of seven Scopus-indexed books and reviewer for top journals. Mentors global PhD candidates and drives innovation in ASEAN academic collaborations.

Subhra R. Mondal Senior Researcher and academician at Duy Tan University, specializing in brand management, AI, and smart tourism. Editor of multiple Scopus-indexed books, her work explores digital recruitment and Web 5.0. She bridges academia and industry through collaborative projects.

Selling the Stars: The Marketing Magic Behind Space Tourism

Jana Majerova, Iordanis Kotzaivazoglou, and Subhra R. Mondal

Abstract As space tourism leaves the realm of science fiction and prepares to become a booming industry, marketers grapple with new challenges, as they attempt to sell an experience that is both extraordinarily exclusive and inevitably dangerous. Having covered broader human aspects of space tourism and how it can come off at luxurious and exclusive, showing privilege and insensitive, you will discover in this chapter the wealth of space tourism complexities and marketing challenges, namely the space tourism overview effect—the strong transformational psychological phenomenon associated with space travel—and the inevitable establishment of space tourist party where only the wealthy could afford, as well as the importance of marketing transparency around space tourism safety and implementation of space tourism. The study points out that it is notable for the latter at the SRV tourism industry. It discusses how practices among industry leaders, such as SpaceX, Blue Origin, and Virgin Galactic, provide innovative approaches to getting the word out, including VR (virtual reality) simulations, celebrity endorsements, and hyper-targeted campaigns. It provides a joined-up critique of the tensions between the neoliberal promotion of space tourism as a life-changing luxury experience and the effort to mitigate its socio-economic and ecological impacts. A pragmatic framework is proposed to help stakeholders navigate the complexities, encapsulating psychological engagement, ethical accountability, and technological innovation. The chapter also highlights limitations in available literature, explicitly pointing to issues with long-term psychological effects, cultural differences in perceptions, and

J. Majerova
Ambis University, Prague, Czech Republic
e-mail: jana.majerova@ambis.cz

I. Kotzaivazoglou
International Hellenic University, Serres, Greece
e-mail: ikotza@ihu.gr

S. R. Mondal (✉)
Department of Marketing, South Star Management Institute, Duy Tan University, Da Nang, Vietnam
e-mail: subhraranimondal@duytan.edu.vn

© The Author(s), under exclusive license to Springer Nature Singapore Pte Ltd. 2025

S. R. Mondal et al. (eds.), *Pioneering the New Space Economy through AI and Immersive Technologies*, https://doi.org/10.1007/978-981-96-5977-7_11

regulatory frameworks. In conclusion, the importance of cooperation in harmonizing commercial interests with sustainability and fairness so that the commercialization of space tourism adds value to humanity's future is necessary on Earth and beyond.

Keywords Space tourism · Overview effect · Ethical marketing · Virtual reality · Sustainability · Luxury branding

1 Introduction

Space tourism, once a thing of the pages of science fiction novels and the frames of blockbuster films, is quickly becoming a reality. So far, private companies such as SpaceX, Blue Origin, and Virgin Galactic have broken the barriers, leading us into a new age where tourists can buy tickets to orbit the Earth or go beyond the stratosphere. But as this Industry in its Infancy moves from fantasy to feasibility, it confronts a singular challenge: How does one sell their way into the utterly extraordinary, the prohibitively exclusive, and the dangerously unknown? Selling the Stars: The Marketing Magic Behind Space Tourism explores the creative strategies, ethical quandaries and psychological complexities of selling what could be the world's most expensive luxury item, a trip to the stars.

1.1 The Dawn of a New Market

Space tourism is a technological shift and a consumer culture shift. In contrast to conventional luxury goods or services, it is not simply a product but a transformative experience that fundamentally proposes altering humanity's connection to the cosmos. However, promoting such an experience is an uncharted enterprise. It becomes even more apparent that the target audience is not just wealthy but part of the ultra-wealthy echelon ready to pay millions for 3 min and 20 s of Earth's view from space (Team 2024). Simultaneously, marketers need to confront more significant cultural perceptions, weighing the promise of travel against people's concerns around safety, equity, and sustainability.

1.2 Finding the Cosmic Story: Where Awe Meets Mitigation

At the heart of space tourism marketing is the overview effect, coined by author Frank White to describe the profound psychological shift astronauts undergo when observing Earth from space. This sensation, marked by interconnectedness and guardianship of the environment, has become a mainstay of space tourism marketing (Yaden et al. 2016). Companies stress the thrill of zero gravity and the promise of personal transformation. Virgin Galactic's promotional materials have testimonials from early passengers describing their flights as "life-altering" (Martin 2021).

Nevertheless, selling such a hard-to-understand experience requires some creative tools. Simulating space travel via virtual reality (VR) and augmented reality (AR) has become essential. Challenging space enthusiasts to "experience" a rocket launch or a spacewalk with headsets immerses prospective customers in an experience that blurs the line between imagination and reality (Kim et al. 2018). These technologies also have a pragmatic value: They help companies demonstrate safety protocols and onboard amenities, assuaging the fears of risk-averse clients.

1.3 Celebrity, Social Proof, and Endorsement

Social proof from influencers has also been equally important. When William Shatner or Yusaku Maezawa go for a spin in space, their social media feeds provide free adverts for the sector. Shatner's emotional response to his 2021 Blue Origin flight, which he called "the most profound experience I can imagine," circulated to millions and helped normalize space flight among mainstream audiences (Gohd 2021). Likewise, Maezawa's #DearMoon project, which calls for a roster of artists to accompany him on a voyage to the Moon, uses the narrative effect of art as a catalyst to solidify the impression of space tourism as a cultural accomplishment rather than a whimsical extravagance (Bartels 2018).

Marketing space tourism is by no means without controversy. Critics say that any promotion of inherently risky journeys, like the 2014 crash involving Virgin Galactic or the 2022 engine failure experienced by Blue Origin, includes an obligation of transparency on the safety side. Sickeningly, companies often minimize such incidents, highlighting statistical safety records instead of individual tragedies (Mazzucato and Robinson 2017). This raises ethical concerns around informed consent and corporate responsibility.

Another point of contention is wealth inequality. In a world where 10% of the population survives on less than two dollars daily (Baah et al. 2024), marketing 450,000 pleasure cruises to space can feel borderline tone-deaf. Companies counter this by framing space tourism as a drive toward technological advancement. For example, SpaceX's Starship program asserts that civilian flights will offset the cost of Mars colonization, thus serving humanity (Tangermann 2021).

Environmental concerns further complicate the narrative. In one rocket launch alone, up to 300 tons of CO_2 can be produced, similar to a transatlantic flight (Dallas et al. 2020). Nevertheless, despite companies such as Blue Origin promoting their reusable rockets as environmentally friendly alternatives, critics have claimed that the carbon footprint of space tourism is inexcusable amid a climate emergency (Carbajales-Dale and Murphy 2022).

1.4 The Next Frontier: Hyper-Targeted Campaigning and Cosmic Branding

With business becoming more cut-throat, marketers are looking for niche demographics. Self-identified ultra-ultra-luxury umpires will supervise "space honeymoons" or zero-gravity weddings; orbital business conferences will offer executives a literally "sky-high" perspective (Forbes Agency Council 2020). Fictional media tropes such as a Star Trek or Star Wars collaboration could also spring to life as publishers look to bridge the gap between fiction and real life to create a pop culture frenzy.

Ultimately, space tourism marketing is a game larger than profit—a fight for narrative control. Will the public see ambitious space travel for civilians as a bona fide symbol of human ingenuity or reckless luxury for their elite members? The answer might depend on how sound companies balance commercial aspirations with ethical responsibility. They argue that marketing space tourism as a stepping stone to interplanetary colonization or a catalyst for global unity is one way to ensure that the vision for humanity's future in the cosmos is not sold to the highest bidder.

2 Review: Selling the Stars—The Marketing Magic Behind Space Tourism

The arrival of space tourism as a commercial proposition is one of the most ambitious marketing challenges of the twenty-first century. With companies like SpaceX, Blue Origin, and Virgin Galactic moving from experimental launches to commercial operations, academia and industry energies have focused on the granular details that affect how to market this new product offering. Drawing from literature from the psychological, ethical, technological, sociocultural, economic, and environmental domains, this review examines the dimensions of marketing space tourism. Exploring these interrelated aspects illustrates the complexities of selling an experience that treads the boundaries of luxury, risk, and futurism.

2.1 Psychological Drivers: The Enchantment of the "Overview Effect"

One of the key themes in marketing space tourism is the psychological allure of space travel, especially the so-called "overview effect." This term, coined by Frank White, describes astronauts' view of the earth from space as one of interconnectedness and environmental awareness. This phenomenon underlies marketing campaigns that have touted space tourism as a life-changing experience rather than just another thrill. Indeed, academic literature also backs the case for emotional narratives in luxury marketing. Virtual reality (VR) simulations of space travel enhance this emotional appeal by letting potential customers "pre-experience" this overview effect. The space tourism marketing campaign engages in the "experience economy," in which consumers seek meaningful activities that shape identity, rather than material goods. The focus on personal change reflects Maslow's hierarchy of needs, placing space travel as the peak of self-actualization (Kaufman 2018). However, critics wonder if the overview effect can genuinely be reproduced for tourists spending minutes in space. Program (2023) cautions that treating this psychological phenomenon as a commodity risks trivializing its ethical and philosophical implications.

2.2 *Ethics of Trade: Risk, Inequality, and Environmental Costs*

Risk and Safety Concerns

Space tourism is perilous, as deadly mishaps like Virgin Galactic's crash in 2014 and Blue Origin's engine failure in 2022 illustrate. Langston (2016) describes how the marketing excitement of companies clashes with their ethical obligations, identifying aspirational messaging as an enemy of transparent risk communication. For example, Virgin Galactic's campaigns after its accident focused on technological advancements and downplayed previous failures (Langston 2016). This tactic is no different than the general concerns around adventure travel, whereby risk can increase desirability but often comes with legal liability issues (Cater 2005).

Wealth Inequality

The sky-high prices of space travel from 250,000 to 240,000 for a suborbital flight, to modelled 100 million for orbital stays have drawn outcry. In this context of global wealth disparity, limiting space tourism to wealthy elites could draw public envy or resentment (Baah et al. 2024). Companies counter this by portraying space tourism as a driver of technological innovation. For example, SpaceX's Starship program frames civilian flights as funding for colonization of Mars, a narrative that links space tourism and humanitarianism (Bartels 2018).

Environmental Impact

Rocket launches have another ethical obstacle: their environmental cost. One launch releases as much as 300 tons of CO_2 equivalent to a year's emissions of 60 average automobiles (Pultarova 2021a). Although companies like Blue Origin claim their reusable rockets are environmentally friendly, critics argue that the carbon footprint is unreasonable during a climate crisis. This tension reflects broader conflicts between luxury consumption and sustainability in marketing (Kapferer and Michaut 2015).

VR, AR, and safety simulations fill the gap between dreamland and the actual functionality of the experience, marketers have faith in immersive technology. According to Kim et al., the purchase purpose increases when the sensory experience is simulated by both the VR and augmented reality demonstrations. For example, Blue Origin's AR program grew the astronauts' trust by letting them readily access spacecraft design, fostering acquaintance and reducing their trepidation regarding safety. The virtual tool also works to alleviate the risk of a visit by customers concerned about the catastrophe being simulated due to the presence of the situation. Virtual walkthroughs tackle the consumer's fear. However, integrating technology has several other difficulties. VR experience exceeds the actuality might increase customer delight while decreasing purchase intention post-purchase, emphasizing that hyping activity is also observed in a high-risk sector like deep-sea diving. Sociocultural aspects. Celebrity, and media narrations: Celebrities' presence has facilitated space tourism's normalization. We are never too far from "most profound experience," and an emotional expression is posted on the social information networks following the worldwide media sensation of William Shatner's 2021 trek with Blue Origin, possession forces prospective customers. According to Cialdini, individuals are more likely to duplicate ideas if they respect information. Additionally, enhancing connections with media firms, highlighting the reasonable country assumption, pop culture, and retro themes.

The economics literature focuses on scarcity in luxury marketing. The scarcity of space tourism, the lack of seats, and the infrequent launch windows also increase its desirability (Heine 2012). But this model is not without challenges as competition increases. Price wars between companies (akin to early commercial aviation) can be anticipated: While they make access democratic, they dilute the "luxury" brand.

Geopolitical, Environmental, and Regulatory Landscape

There is an environmental debate brewing around space tourism. Despite the promise of companies like Blue Origin with methane engines and Spin Launch with carbon-neutral launch systems, independent studies paint a less rosy picture. According to Pultarova (2021b), even "green" rockets will pollute the stratosphere, possibly degrading ozone layers. Regulatory frameworks are also lagging; extant policies center on safety rather than environmental accountability (Pultarova 2022).

2.3 Research Gaps

It recognized that space tourism is a multi-disciplinary field where marketing requires information from psychology, ethics, technology, economics, etc. However, gaps remain:

- Long-Term Psychological Outcomes: Studies' emphasis on the overview effect's allure inevitably means little is known about data showing whether tourists continue to experience the overview effect after their flights.
- Cultural Differences: Since the vast majority of marketing is geared toward Western elites, how will people in non-Western countries view space tourism? This space has not been adequately researched.
- Regulatory Environment: The absence of international agreement on environmental standards creates hazards for global branding.

Promoting space tourism is a heavy lift that mixes awe with practicality, extravagance with morality, and pioneering with accountability. Pioneering space travel is a narrative that goes beyond consumption and positions the new frontier as a bridge to humanity's galactic future, according to the literature. However, unanswered ethical and environmental queries risk raining on the parade of an emergent industry unless addressed, via transparent communication (including with bioethicists) and collaborative regulation. As the sector matures, more research is needed to develop strategies for democratization, cross-cultural marketing, and sustainability to avoid selling the stars at the expense of life, here, on Earth, tomorrow.

3 Framework: Marketing Space Tourism—A Practical Guide for Industry Stakeholders

This framework translates the literature findings into a structured, actionable approach to marketing space tourism, tackling the unique challenges and opportunities identified by the literature. It combines psychological, ethical, technological, sociocultural, economic, and environmental factors into an integrated strategic framework with a view to industry stakeholders. The framework comprises interconnected components with actionable tools, metrics, and implementation steps against each. Figure 1 represents the strategic framework.

a. Psychological Engagement: The Power of the Overview Effect.

   - Intended use: Define space tourism as a life-changing, identity-shaping experience.
   - Strategies: Storytelling Campaigns.
   - To humanize it, use testimonials from astronauts and early tourists (e.g., William Shatner's tearful description of the overview effect).
   - Create a documentary-style video about the emotional process of preparing to go to space, and being in space.
   - Tool: Collaborate with filmmakers (IMAX, etc.) to produce short films made available through streaming services.

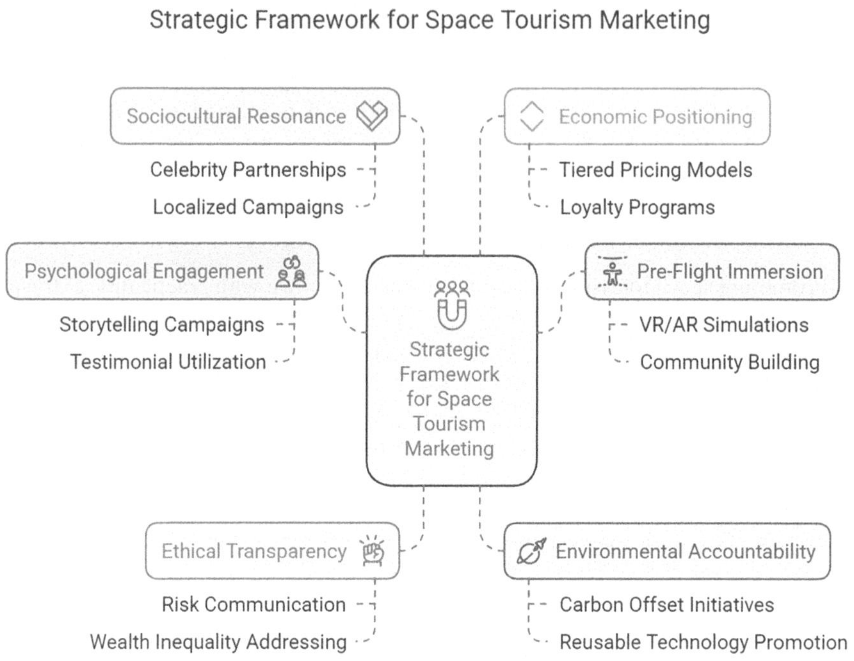

Fig. 1 Strategic framework (Source: Authors' conception)

b. Pre-Flight Immersion: Provide VR/AR Simulations of Space Traveling Through the Booking Process.

- Blue Origin has an AR app allowing users to look inside spacecraft before take-off.
- Tool: Work with VR enabling companies (Oculus) to create tailor-made simulations.
- Community Building After the Flight: Develop exclusive alumni associations for space tourists to swap stories.
- Metric: Download rates of and engagement rates on social media content shared after a long-haul flight.

c. Managing Risk and Inequality: Ethical Transparency.

- Goal: Foster trust via accountability and avoid backlash.
- Strategies: Risk Communication.
- Share safety statistics in promotional materials (e.g., "99.9% success rate on 50+ missions").
- Generate interactive infographics explaining safety protocols from emergency abort systems, etc.
- Tool: Bring in engineers to hold live Q&As about safety concerns.

- Addressing Wealth Inequality: Commit a percentage of ticket sales to global sustainability initiatives (e.g., SpaceX's promise to finance Mars colonization).
- Grant scholarships to non-rich applicants for "citizen astronaut" programs.

d. Environmental Accountability: Calculate and offset carbon emissions per flight (e.g., plant 1000 trees per ticket sold).

- Champion reusable rocket technology in campaigns (like "Fly Green with Blue Origin").
- Metric: Publicize annual reductions in emissions.
- Create Balance Between Technology and Nature.
- Goal: Utilize immersive tech for anxiety reduction and increased attractiveness.
- Strategies: VR/AR Previews.
- Create VR kiosks at high-end international travel shows where users can experience zero gravity.
- Partner with Sony to pre-install space tourism demos on PlayStation VR headsets.
- AI-Powered Personalization: Use AI chatbots to personalize marketing efforts based on users' values (e.g., adventure vs. scientific curiosity).

e. Safety Training Modules: Training Apps That Use Game Mechanics to Instruct Passengers on Handling Emergencies

This shows that, under these circumstances, using post-flight surveys to gauge how prepared passengers were to begin the trials is a reasonable way to measure the trial's impact on passenger experience.

- Sociocultural Resonance: Leveraging Pop Culture and Influencers.
- Goals: Make space tourism relatable and normalize through narratives.
- Strategies: Celebrity Partnerships.
- Book flights for influencers in different niches (e.g., athletes, artists, scientists).
- Collaborations with Media Franchises: Co-brand flights with franchises like Star Trek (e.g., "Boldly Go Where No Tourist Has Gone Before").
- Tool: Issue limited-run merchandise (e.g., SpaceX Star Wars flight suits).
- Localized Campaigns: Adjust marketing for cultural values (in Japan, focusing on precision engineering, and in UAE, uniting with national space objectives).
- Metric: the percentage of penetration rates in target areas.

f. Economic Positioning: Elite Luxury to Aspirational Accessibility.

- Reason: Expand the market while remaining exclusive.
- Strategies: Tiered Pricing Models.
- What are suborbital flights (250 K), orbital stays (250 K), orbital stays (10 M), lunar flybys ($100 M).
- Example: Virgin Galactic's "Galactic Lite" vs. "Galactic Elite" packages.

- Loyalty Programs: Offer priority reservation or discounts for loyal customers on upcoming flights.
- NFT-based membership passes for exclusive events.
- B2B Partnerships: Market orbital conferences to Fortune 500 s (e.g., "Innovate in Orbit" summits).

g. Metric: The Percentage of Bookings Attributable to Corporate Travel by Revenue Per Calendar Year.

- Goal: Transparency with sustainability and regulatory alignment.
- Strategies: Eco-Certifications.
- Consider pursuing certifications such as the Carbon Trust Standard for space tourism operators.
- Tool: Collaborate with NGOs to go through emissions.
- Regulatory Advocacy: Push for global environmental standards (e.g., UN Office for Outer Space Affairs).
- For example, co-founded a "Sustainable Space Tourism Alliance" alongside competitors.
- Transparent Reporting: Annual sustainability reports on emissions, safety incidents, and CSR results are published.
- Metric: Third party ESG (Environmental, Social, Governance) ratings.

4 Implementation Roadmap

a. Phase 1 (0–12 Months):

- Pilot VR experiences at luxury travel shows.
- Run celebrity flight campaigns.
- Set up carbon offset partnerships.

b. Phase 2 (1–3 Years):

- Enter new markets (i.e., India, Saudi Arabia).
- So, introduce tiered pricing and NFT loyalty programs.

c. Phase 3 (3–5 Years):

- Obtain global environmental certifications.
- Make access more democratic through "citizen astronaut" scholarships.
- Final thoughts: A changing, flexible, and active approach.

Because space tourism is a nascent industry, and it can be assumed to be fast-changing, this framework is designed to be flexible. Suppose companies can hold themselves accountable to the ambitious stories they are telling. In that case, they will be less vulnerable to public scrutiny and criticism as they seek to achieve humanity's cosmic ambitions. Frequent auditing of customer sentiment, regulatory adjustments, and technological evolution will keep the strategy in mind. Ultimately,

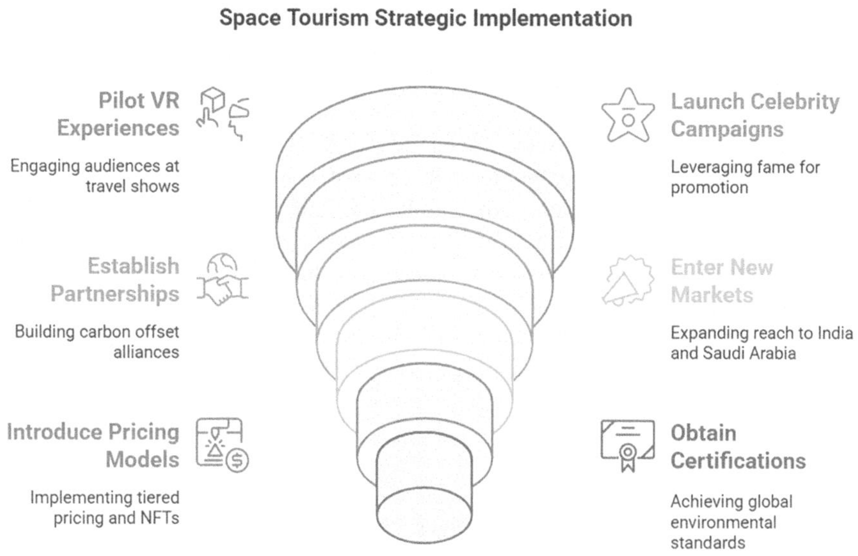

Fig. 2 Implementation strategy (Source: Authors' conception)

the objective is not merely to sell tickets to space, but to create a sustainable, inclusive vision of humanity's next act among the stars. Figure 2 represents the phases of implementation.

Key Metrics for Success:

- Conversion rates from VR experiences into bookings.
- Year-over-year improvements in ESG rating.
- Engagement on social media platforms with influencer campaigns.
- Sustainability initiative revenue percent.

This approach serves as a guideline for converting theoretical knowledge into pragmatic strategic plans, enabling space tourism marketers to respond to the challenges of the burgeoning industry while capturing the imagination of prospective consumers worldwide.

5 Theoretical, Practical, Social, and Sustainability Implications of Marketing Space Tourism

The advent of space tourism is a historic turning point for humanity, merging our technological ambitions with consumer culture. However, its emergence does have profound implications in theoretical, practical, social, and sustainability dimensions. The chapter delves into these interlinked areas through existing research and industry practice and examines how space tourism marketing rewires economies, societies, and planetary ecosystems.

5.1 Theoretical Implications

A Blueprint for Redefining Luxury and Experience Marketing

Space tourism defies traditional marketing theories by promoting an ultra-exclusive, high-risk pursuit as a luxury product. In contrast to typical luxury goods (i.e., watches and cars), space travel is compatible with "experience economy" in that it offers value through transformative, identity-altering experiences (Mehran et al. 2023). Focus on the "overview effect" a cognitive shift in cosmic perspective—in marketing extents Maslow's hierarchy of needs to space travel as the highest expression of self-actualization (Kaufman 2018). This transformation drafts a new definition of luxury as access to the extraordinary, rather than ownership of physical goods.

Ethical Marketing Frameworks

Space tourism poses new ethical challenges, given the risks and the costs associated with space travel. Traditional risk communication models (e.g., the Psychometric Paradigm) concentrate on public perception of hazards but do not address the moral complexities of marketing life-threatening experiences to wealthy elites. Similarly, the industry's environmental footprint casts doubt on the "shared value" theory that business success and social good are inextricably connected. The current model of space tourism focusing on profit rather than equitable access or sustainability brings to light the shortcomings of corporate social responsibility (CSR) frameworks (Čater et al. 2023).

Technological Determinism

The dependence on VR/AR and AI-based personalization demonstrates technological determinism, the idea that technology influences social values. Marketers thus tacitly change what consumers feel is possible for them. What they should aspire to (Kim et al. 2018) by making the experience of space travel feel accessible through simulated experiences, while space travel remains an unreachable physical and financial limit for many.

5.2 Practical Implications

Operational Challenges

Explain the Safety Protocols: There is a fine line between aspirational marketing and pointing out the risks involved in some practices. One example of this was an accident that forced Virgin Galactic into an expensive rebranding campaign based around safety improvements and downplaying previous mishaps that led to its 2014 crash (Sambuno 2023).

Cost Management—The existing pricing model (250,000–250,000–100 million per seat) is not scalable. By contrast, reusable rocket technology and competition might reduce costs, as seen in early commercial aviation.

Market Expansion Strategies

B2B partnerships: Suborbital flights for "budget" tourists vs. orbital stays for ultra-wealthy clients (Geospace: Blue Origin's New Shepard; SpaceX's Starship).

Regulatory Navigation

No international rules govern space tourism, leaving liability and environmental responsibility murky. Companies such as Axiom Space are now lobbying for standardized safety and emissions guidelines to avoid the government cracking down on them (Nataly 2022).

5.3 *Social Implications*

Cultural Shifts

Normalizing Space Travel: Celebrities such as William Shatner and Yusaku Maezawa view space tourism as an untrivializable cultural milestone. It resembles the democratization of air travel in the twentieth century.

Ethical Polarization: The public's opinion divides between enthusiasts who see space travel as progress and critics who brand it a "billionaire escapade." Social media magnifies both narratives, creating brand reputations that instantly change.

Workforce Dynamics

The industry offers high-skilled jobs in engineering and hospitality, but it can draw talent and resources away from terrestrial fields such as health care or renewable energy.

5.4 *Sustainability Implications*

Environmental Costs

CO_2 emissions: A single rocket launch emits around 300 tons of CO_2, the same as 60 cars driven for an entire year. Reusable rockets (such as SpaceX's Falcon 9) cut costs, not emissions.

Ozone Depletion: Counsel of the Undersecretary of the Air Force explains that black carbon particles in rocket exhaust persist in the stratosphere for years, accelerating their degradation.

Greenwashing Risks

Companies like Blue Origin boast "eco-friendly" methane engines, but independent studies reveal their emissions still affect the climate. Without third-party certifications, such claims could undermine consumer trust.

Planetary Stewardship the Long-Term

In a twisted irony, space tourism pitches itself as the next step toward interplanetary colonization while speeding up environmental destruction on Earth. This tension imposes a reckoning with the ethical implications of investing resources in off-world aspirations at the expense of terrestrial sustainability (Tien-Dung et al. 2022b).

5.5 Synthesis and Recommendations

Theoretical Advancements

a. Create hybridized marketing models that synergize luxury, ethics, and sustainable resonance (e.g. "conscientious space exploration").
b. Extend CSR frameworks to extraterrestrial industries' unique risks.

Practical Adjustments

a. Implement transparent pricing tiers and safety reporting to earn trust.
b. Financing carbon capture technologies for emissions offsetting.

Social Equity Measures

a. Funding "citizen astronaut" programs through lottery systems or merit-based scholarships.
b. Partner with NGOs to sync space tourism profits with global development goals.

Sustainability Mandates

a. Formation of international bodies such as the International Astronautical Federation (IAF) to set industry-wide emissions standards.
b. Give a small percentage of ticket sales to a renewable energy project.

Space tourism is not a business, it is a negotiation of societal values in commerce. Theoretically it subverts prevailing definitions of luxury and ethics (Vrana and Das 2023a). Practically, it requires innovation in risk management and regulation (Vrana and Das 2023b). Socially, it pushes a conversation about access and aspiration. Environmentally, it interrogates humanity's resolve to steward the planet (Yegen and Das 2023). The industry's legacy, however, will depend on striking a balance between ambition and accountability. As Gwynne Shotwell of SpaceX put it, "We do not want to be the best humans who ever lived; we want to be the worst humans who ever lived"—a call that space tourism should add to rather than detract from life on the home planet.

6 Conclusion

The annual relief workers often mark a significant moment in the history of humanity, and space tourism marks a major revolution. The marketing efforts in this emerging industry showcase both remarkable opportunities and immense challenges as this speculative industry is born into reality. This paper has explored psychological, ethical, technological, and sustainability issues to promote space travel, showing the delicate balance of selling an experience that is as dangerous as it is innovative. Whilst the path forward for the industry is undoubtedly exciting, its long-term sustainability relies heavily on addressing some of the more fundamental challenges it faces and seizing avenues for expansion.

6.1 Key Limitations

Theoretical and Empirical Gaps

Much existing research on space tourism marketing is based on theorization (e.g., the "overview effect") or on analogizing from terrestrial luxury markets (Das 2023a). Nevertheless, little empirical evidence exists on how space tourists experience their journey after they return to Earth, and marketed psychological benefits such as cosmic perspective shifts, if they last.

Nuance is often missing from ethical debates. For example, even as firms like SpaceX conceptualize space tourism as a prerequisite to interplanetary colonization, less is known about how these narratives shape public trust or policy.

Where we can be narrow, trying to be all things to all people.

There is a lack of cross-cultural perspective focusing on Western, ultra-rich demographics in marketing strategy (Das 2023b). What, for example, would space tourism mean for an emerging economy such as that of India or Nigeria, which have national space programs alongside significant socioeconomic disparities?

Environmental Shortcomings

Today's sustainability claims, such as Blue Origin's "eco-friendly" rockets, have no independent verification. The industry's carbon footprint is still poorly quantified, and emissions regulatory frameworks are virtually absent (Das et al. 2023).

Technological Overpromises

VR/AR simulations may create false expectations. As avid members of the adventure tourism community have revealed, such hyper-curated marketing leads to disappointment in the experience (Kieanwatana and Vongvit 2024).

Regulatory Ambiguity

Moreover, the lack of global standards governing safety, liability, and environmental accountability means that companies do business in a legal gray zone that raises risks for consumers and investors alike (Di Virgilio and Das 2023a).

6.2 *Future Scope*

In conclusion, to overcome these restrictions and realize the sustainable development of the space tourist industry, future initiatives will need to focus mainly on cross-discipline research, inclusive regulations, and technological responsibility.

Psychological Studies Over Time

Trace the lasting effects of space travel on tourists' attitudes about environmentalism, global citizenship, and personal identity (Di Virgilio and Das 2023b). Does the "overview effect" wear off, or does it lead to permanent changes in behavior? Partnerships between universities and aerospace companies could provide essential knowledge.

International Market Analysis

Explore perceptions of space tourism among non-Western audiences. For instance, might state-backed enterprises in China lend themselves to a very different marketing model than private enterprises in the United States? Culturally specific campaigns may help attract new demographic groups.

Sustainable Innovation

Keep developing and standardizing green technologies, like carbon-neutral propellants or solar-powered space habitats (Majerova and Das 2023). An independent certification (a “Space Sustainability Seal,” for example) could validate companies’ claims while rebuilding public trust.

Democratization Strategies

Model early and open access beyond the ultra-wealthy. For example, private lotteries, systems, corporate sponsorships, or government-subsidized “citizen astronaut” programs would democratize participation while reserving exclusivity for high-paying clients (Minh-Nhat et al. 2022a).

Global Regulatory Frameworks

Push for global accords on safety protocols, emissions caps, and liability clauses. Grouping entities such as the United Nations Office for Outer Space Affairs (UNOOSA), the Web Foundation (Mondal et al. 2023b), the Internet Society (Ngoc-Vinh et al. 2022), and similar organizations under the same blanket could make these efforts effective while ensuring fair standards are set and applied uniformly (Minh-Nhat et al. 2022b).

Ethical Marketing Guidelines

Create universal ethical guidelines for risk communication, pricing transparency, and CSR commitments (Nguyen et al. 2022). One example could be requiring companies to contribute profits to terrestrial sustainability projects (Mondal et al. 2023a).

6.3 *A Call for Balanced Ambition*

Advertising space tourism is not just a business but a manifestation of a society’s maturing dealings with technology, ethics, and the cosmos (Tandon and Das 2023). The industry’s promise is undeniable; a new era of exploration and economic opportunity, but its success requires humility and foresight (Tien-Dung et al. 2022a). Stakeholders must understand that selling the stars cannot be at the cost of Earth's well-being or social equity (Mondal and Das 2023a).

"Space is for everyone," Virgin Galactic's Richard Branson said. However, achieving the vision demanded much more than rhetorical flourishing; it needed bold commitments to inclusivity, sustainability, and accountability (Mondal and Das 2023b). Through addressing its present limitations and embracing collaborative innovation, the space tourism industry can evolve beyond its "billionaire's playground" image and become a transformative force for a more united, forward-thinking global society (Mondal and Das 2023c). The stars beckon, yet their commercialization must abide by doctrines that do right by the universe and Earth, the only home we know.

References

Baah SKT, Aguilar RAC, Diaz-Bonilla C, Fujs T, Lakner C, Nguyen MC, Viveros M (2024, April 19) March 2023 global poverty update from the World Bank: the challenge of estimating poverty in the pandemic. World Bank Blogs. https://blogs.worldbank.org/en/opendata/march-2023-global-poverty-update-world-bank-challenge-estimating-poverty-pandemic

Bartels M (2018, September 20) Yusaku Maezawa's #DearMoon project aims for lunar art. Here's what some artists think. Spacecom. https://www.space.com/41881-dearmoon-art-project-spacex-moon-artists-respond.html

Carbajales-Dale M, Murphy TW (2022) The environmental and moral implications of human space travel. Sci Total Environ 856:159222. https://doi.org/10.1016/j.scitotenv.2022.159222

Cater CI (2005) Playing with risk? participant perceptions of risk and management implications in adventure tourism. Tour Manag 27(2):317–325. https://doi.org/10.1016/j.tourman.2004.10.005

Čater T, Čater B, Milić P, Žabkar V (2023) Drivers of corporate environmental and social responsibility practices: a comparison of two moderated mediation models. J Bus Res 159:113652. https://doi.org/10.1016/j.jbusres.2023.113652

Dallas J, Raval S, Gaitan JA, Saydam S, Dempster A (2020) The environmental impact of emissions from space launches: a comprehensive review. J Clean Prod 255:120209. https://doi.org/10.1016/j.jclepro.2020.120209

Das S (2023a) Modern-day gaming and the role of FinTech applications in popularising various entertainment platforms across the world. In: Das S, Gochhait S (eds) Digital entertainment as next evolution in service sector. Palgrave Macmillan, Singapore. https://doi.org/10.1007/978-981-19-8121-0_11

Das S (2023b) Digital entertainment: the next evolution in service sector. Emerging digital solutions in reshaping different industries Palgrave-Singapore. https://doi.org/10.1007/978-981-19-8121-0

Das S, Mondal SR, Struck F, Eissfeldt J, Beernink L, Sigouras V, Hoose S, Kopac L (2023) What can we see inside the crystal ball? Exploring opportunities and challenges for metaverse and circular economy. AJIS 12(6):221. https://doi.org/10.36941/ajis-2023-0165

Di Virgilio F, Das S (2023a) Digital solutions for social media: role of digital evolution on business enterprises. In: Das S, Gochhait S (eds) Digital entertainment as next evolution in service sector. Palgrave Macmillan, Singapore. https://doi.org/10.1007/978-981-19-8121-0_9

Di Virgilio F, Das S (2023b) Digitalization of enterprises from a new normal sustainable development perspective: opportunities and challenges. In: Mondal SR, Yegen C, Das

S (eds) New normal in digital enterprises. Palgrave Macmillan, Singapore. https://doi.org/10.1007/978-981-19-8618-5_9

Forbes Agency Council (2020, May 15) Council post: How will experiential marketing evolve? 13 experts share their views. Forbes. https://www.forbes.com/councils/forbesagencycouncil/2020/05/15/how-will-experiential-marketing-evolve-13-experts-share-their-views/

Gohd C (2021, October 13) William Shatner moved to tears by space launch with Blue Origin. Space.com. https://www.space.com/william-shatner-blue-origin-flight-moved-tears-reaction

Heine K (2012) The concept of luxury brands. https://upmarkit.com/sites/default/files/content/20130403_Heine_The_Concept_of_Luxury_Brands.pdf

Kapferer JN, Michaut A (2015) Luxury and sustainability: a common future? The match depends on how consumers define luxury. Luxury Res J 1(1):3. https://doi.org/10.1504/lrj.2015.069828

Kaufman SB (2018) Self-actualizing people in the 21st century: integration with contemporary theory and research on personality and well-being. J Humanist Psychol 63(1):51–83. https://doi.org/10.1177/0022167818809187

Kieanwatana K, Vongvit R (2024) Virtual reality in tourism: the impact of virtual experiences and destination image on the travel intention. Results Eng 103650. https://doi.org/10.1016/j.rineng.2024.103650

Kim MJ, Lee C, Jung T (2018) Exploring consumer behavior in virtual reality tourism using an extended Stimulus-Organism-Response model. J Travel Res 59(1):69–89. https://doi.org/10.1177/0047287518818915

Langston SM (2016) Space travel: risk, ethics, and governance in commercial human spaceflight. New Space 4(2):83–97. https://doi.org/10.1089/space.2015.0015

Majerova J, Das S (2023) Sustainable development and digitalization in the supply chain for environmental protection: changes and trends. In: Mondal SR, Yegen C, Das S (eds) New normal in digital enterprises. Palgrave Macmillan, Singapore. https://doi.org/10.1007/978-981-19-8618-5_8

Martin (2021, August 14) Publicity and marketing strategies mastered by Virgin Galactic. Martin Wilson. https://martinwilson.me/pr-marketing-virgin-galactic/

Mazzucato M, Robinson DK (2017) Co-creating and directing innovation ecosystems? NASA's changing approach to public-private partnerships in low-earth orbit. Technol Forecast Soc Chang 136:166–177. https://doi.org/10.1016/j.techfore.2017.03.034

Mehran J, Olya H, Han H (2023) Psychology of space tourism marketing, technology, and sustainable development: from a literature review to an integrative framework. Psychol Mark 40(6):1130–1151. https://doi.org/10.1002/mar.21795

Minh-Nhat HO, Majerova J, Das S (2022a) Taking steps to help the phoenix rise from ashes: a roadmap for digital enterprises to develop new sustainable goals. In: Sustainable development and innovation of digital enterprises for living with COVID-19. Springer, Singapore, pp 35–51. https://doi.org/10.1007/978-981-19-2173-5_3

Minh-Nhat HO, Nguyen HL, Mondal SR (2022b) Digital transformation for new sustainable goals with human element for digital service enterprises: an exploration of factors. In: Sustainable development and innovation of digital enterprises for living with COVID-19. Springer, Singapore, pp 85–103. https://doi.org/10.1007/978-981-19-2173-5_6

Mondal SR, Das S (2023a) Defining the new digitalized normal: implications of A.I., IoT, and blockchain for business growth. In: Mondal SR, Yegen C, Das S (eds) New normal in digital enterprises. Palgrave Macmillan, Singapore. https://doi.org/10.1007/978-981-19-8618-5_10

Mondal SR, Das S (2023b) Digital solution for industry 4.0 and 5.0 and social media. In: Das S, Gochhait S (eds) Digital entertainment as next evolution in service sector. Palgrave Macmillan, Singapore. https://doi.org/10.1007/978-981-19-8121-0_10

Mondal SR, Das S (2023c) Human-centric digitalization for new sustainable development: an explorative study in human resource management. In: Mondal SR, Yegen C, Das S (eds) New normal in digital enterprises. Palgrave Macmillan, Singapore. https://doi.org/10.1007/978-981-19-8618-5_12

Mondal S, Das S, Vrana VG (2023a) How to bell the cat? A theoretical review of generative artificial intelligence towards digital disruption in all walks of life. Technologies 11(2):44. https://doi.org/10.3390/technologies11020044

Mondal SR, Yegen C, Das S (2023b) New normal in digital enterprises – strategies for sustainable development. Palgrave-Singapore. https://doi.org/10.1007/978-981-19-8618-5 and Hardcover ISBN 978-981-19-8617-8

Nataly (2022, November 3) Outer space law: lack of regulation at the international level poses a risk to future space activity. Orbital Today. https://orbitaltoday.com/2022/08/05/outer-space-law-lack-of-regulation-at-the-international-level-poses-a-risk-to-future-space-activity/

Ngoc-Vinh N, Tien-Dung P, Das S (2022) Digitization of business; need for recombination during COVID-19. In: Sustainable development and innovation of digital enterprises for living with COVID-19. Springer, Singapore, pp 1–17. https://doi.org/10.1007/978-981-19-2173-5_1

Nguyen TDM, Mondal SR, Das S (2022) Digital Entrepreneurial Transformation (DET) Powered by New Normal Sustainable Developmental Goals (n-SDGs): Elixir for growth of country's economy. In: Sustainable development and innovation of digital enterprises for living with COVID-19. Springer, Singapore, pp 69–84. https://doi.org/10.1007/978-981-19-2173-5_5

Program HS (2023, May 16) The overview effect in three simple words – human space program – medium. Medium. https://medium.com/@humanspaceprogram/the-overview-effect-in-three-simple-words-33720cfc4e2b

Pultarova T (2021a, July 26) The rise of space tourism could affect earth's climate in unforeseen ways, scientists worry. Spaceflight. https://www.space.com/environmental-impact-space-tourism-flights

Pultarova T (2021b, October 28) The environmental impact of rocket launches: The "dirty" and the "green." Spaceflight. https://www.space.com/rocket-launches-environmental-impact

Pultarova T (2022, June 27) Coming increase in rocket launches will damage ozone, alter climate, study finds. Spaceflight. https://www.space.com/rocket-launches-damage-ozone-climate

Sambuno (2023, May 5) Leveraging the aspirational element in your marketing strategy. SAMBUNO. https://www.sambuno.com/leveraging-the-aspirational-element-in-your-marketing-strategy/

Tandon S, Das S (2023) Digital revolution complements sustainable goal achievement during and after pandemic: a research agenda. In: Mondal SR, Yegen C, Das S (eds) New normal in digital enterprises. Palgrave Macmillan, Singapore. https://doi.org/10.1007/978-981-19-8618-5_11

Tangermann V (2021, May 6) Elon Musk celebrates Starship landing: "Needed to make life multiplanetary." Futurism. https://futurism.com/elon-musk-starship-make-life-multiplanetary

Team SVV (2024, June 6). The economics of space tourism: an analysis of financial implications and industry prospects - space voyage ventures. Space Voyage Ventures. https://spacevoyageventures.com/the-economics-of-space-tourism-costs-challenges-and-the-future/

Tien-Dung PHAM, Majerova J, Das S (2022a) Phases of possible recovery of digital enterprises in new normal business for living with COVID-19 times: opportunities for a new era in sustainable development goals. In: Sustainable development and innovation of digital enterprises for living with COVID-19. Springer, Singapore, pp 19–33. https://doi.org/10.1007/978-981-19-2173-5_2

Tien-Dung P, Nguyen XD, Das S (2022b) Sustainable goal achievement by digital revolution during and after pandemic; How much one wins and losses: a bird's eye view for future planning. In: Sustainable development and innovation of digital enterprises for living with COVID-19. Springer, Singapore, pp 133–147. https://doi.org/10.1007/978-981-19-2173-5_8

Vrana V, Das S (2023a) Dynamic restructuring of digital media and entertainment sector: role of urbanization, industrial innovation, and technological evolution. In: Das S, Gochhait S (eds) Digital entertainment as next evolution in service sector. Palgrave Macmillan, Singapore. https://doi.org/10.1007/978-981-19-8121-0_2

Vrana V, Das S (2023b) Role of enterprises and their digital transformation in post-pandemic sustainable development. In: Mondal SR, Yegen C, Das S (eds) New normal in digital enterprises. Palgrave Macmillan, Singapore. https://doi.org/10.1007/978-981-19-8618-5_7

Yaden DB, Iwry J, Slack KJ, Eichstaedt JC, Zhao Y, Vaillant GE, Newberg AB (2016) The overview effect: awe and self-transcendent experience in space flight. Psychol Conscious Theory Res Practice 3(1):1–11. https://doi.org/10.1037/cns0000086

Yegen C, Das S (2023) Digital solution for print and publishing media. In: Das S, Gochhait S (eds) Digital entertainment as next evolution in service sector. Palgrave Macmillan, Singapore. https://doi.org/10.1007/978-981-19-8121-0_12

Jana Majerova Senior Academic at Ambis University, specializing in consumer psychology and digital marketing innovation. Her research explores brand engagement strategies in evolving digital landscapes. She fosters student-led projects to bridge academic and industry insights.

Iordanis Kotzaivazoglou Dean and Communication Professor at the International Hellenic University, collaborating with public institutions like Hellenic Police. His research spans crisis management, social media marketing, and public relations—author of peer-reviewed studies on organizational communication.

Subhra R. Mondal Senior Researcher and academician at Duy Tan University, specializing in brand management, AI, and smart tourism. Editor of multiple Scopus-indexed books, her work explores digital recruitment and Web 5.0. She bridges academia and industry through collaborative projects.

Minds in Microgravity: The Psychological Journey of Space Travelers

Peter Skotnicky and Subhra R. Mondal

Abstract The burgeoning space tourism industry highlights the dire but understudied psychological challenges civilians traveling beyond Earth face. Unlike trained astronauts, space tourists face novel stressors, such as isolation, confinement, circadian misalignment, and the overwhelming "overview effect"—the fundamental shift in worldview that can occur upon seeing Earth from low-Earth orbit. In this chapter, we articulate these psychological barriers and suggest a framework to ensure psychological health that might include a triadic structure: pre-flight preparation (for instance, virtual reality simulations, psychological screening), in-flight support (AI companions and tele-psychiatry), and post-flight reintegration (counselling and peer networks). It emphasizes the need for resilience-building strategies to counteract risk factors like anxiety, existential dissonance, or crises of identity. The findings could have implications for space tourism and mental health care in extreme terrestrial environments. Focusing on the mental well-being of innovators, not just the technical revolution, will lead to lasting and sustainable success and promote compassion and environmental awareness worldwide. Space tourism is an emerging field that presents unique psychological stressors, making mental resilience and robust support systems essential.

Keywords Space tourism · Psychological stressors · Mental resilience · Overview effect · AI-driven support · Microgravity environment

P. Skotnicky
University Institute of Economics and Law, Bratislava, Slovakia
e-mail: director@mbainstitut.eu

S. R. Mondal (✉)
Department of Marketing, South Star Management Institute, Duy Tan University, Da Nang, Vietnam
e-mail: subhraranimondal@duytan.edu.vn

© The Author(s), under exclusive license to Springer Nature Singapore Pte Ltd. 2025

S. R. Mondal et al. (eds.), *Pioneering the New Space Economy through AI and Immersive Technologies*, https://doi.org/10.1007/978-981-96-5977-7_12

1 Introduction

As we have entered the twenty-first century and a new era of human exploration in space travel, travelling to space is no longer only for the rigorously trained astronauts (Diamandis and Kotler 2020). While private companies such as SpaceX, Blue Origin, and Virgin Galactic are paving the road to commercial spaceflight, the fantasy of transcending Earth's atmosphere quickly became a reality for everyday civilians. However, as the space tourism industry takes off, one critical aspect of this exceptional journey remains relatively unexplored: the psychological cost of departing Earth. While the physical challenges of life off Earth, from muscle atrophy to radiation exposure—have drawn some of the loudest conversations, space travelers' mental and emotional obstacles are no less profound (Johnson et al. 2014). Minds in Microgravity looks at the invisible journey of the mind, following the mental terrain of those who explore the galaxy—and the unique techniques developed to protect their sanity (Kanas 2015).

1.1 Mental Health in Space

For decades, space agencies such as NASA and Roscosmos have investigated how the human body adapts in microgravity, studying everything from changes in bone density to cardiovascular alterations (Pachiyappan et al. 2024). However, the psychological effects of space travel are only now starting to garner serious attention (Fitzpatrick et al. 2017). Professional astronauts, who receive years of training in mental resilience, will not be concocted like emotionally panicked astronauts—space tourists are typically people who have not experienced high-risk environments—will be hit with a barrage of stressors that will not have been felt in gold (Holt 2022). Zero gravity, small space, disoriented circadian rhythms, and the visceral knowledge of existential danger form a mental crucible (Kanas 2015). Complicating these challenges are the overview effect, the transformative, sometimes overwhelming change in perspective experienced when seeing Earth from space and the long duration of the inhabitants' visibility as they orbit in low Earth orbit (Keazor 2012). While many astronauts report this experience spiritually exhilarating, the scale of cosmic reality can also destabilize unprepared minds, leaving travelers emotionally adrift in the void.

Ensuring passenger resilience is not only an ethical imperative, but also a practical one, as the space tourism industry grows in scope and significance (Keazor 2012). One bad experience—a panic attack mid-flight, a mental health crisis after a mission—could threaten public faith and choke the industry's growth (Palinkas 2003). As a result, researchers, psychologists, and spaceflight companies are uniting to create innovative tools and protocols to solve these issues. From psychological pre-flight screenings to AI-supported emotional well-being systems, the competition to safeguard the psyche of spacefarers is as fierce as the race to send them into the sky (Palinkas and Suedfeld 2007).

1.2 Space Travel and Its Potential Harmful Psychological Stressors

Space travel detaches humans from every sensory and psychological anchor that keeps them on firm ground. The first and most demanding challenge is isolation. As they leave the atmosphere, even on short suborbital hops, travelers are decoupled from Earth's familiar rhythms—no breeze, no natural sounds, no horizon to navigate (Rothbaum et al. 2010). On missions lasting hours or longer, the physical confinement of a cramped spacecraft heightens feelings of claustrophobia, and the realization that rescue is impossible in a vacuum can provoke anxiety.

The lack of a day-night cycle también exacerbates the destabilization of the mental balance. At low Earth orbit, day and night alternate every 90 min, messing up circadian rhythms and interrupting sleep cycles. Sleep deprivation, a well-established trigger for cognitive deterioration and emotional instability, becomes a quiet menace. Moreover, in the meantime, the physical sensations that come with microgravity, while thrilling at first, can cause nausea, disorientation, and a lasting feeling of unreality (Pachiyappan et al. 2024). The brain, used to gravity's pull, finds it hard to calibrate the body's lack of weight with its understanding of the world.

However, the overview effect may be the most complex psychological phenomenon. Spotting such activists on the ground is rare; astronauts returning from space are sometimes overcome with a sense of planetary unity, describing Earth as a delicate "blue marble" hanging in darkness. But this epiphany can also inspire existential angst. Combating the enormity of space might increase feelings of insignificance or awaken dormant personal disputes. That cognitive dissonance can become excessive for space tourists not the subjects of years of intense training.

1.3 Preparing Minds for the Void

To meet these challenges, the space tourism industry is investing. Traditional astronaut preparation involves drills for situations in which astronauts may find themselves isolated and in emergency scenarios, but civilians require an approach that is more tailored to their needs. This experience has been made easier with virtual reality (VR), which is becoming an essential tech tool in helping explorers prepare with a taste of the stars in the home's safety. By recreating zero-gravity environments and spacecraft interiors, not to mention the overview effect, VR can lessen the shock of the unknown (Williams et al. 2019).

The same is true of psychological screening. Passengers now receive medical evaluations, and mental health assessments may soon be performed. Operators might be able to tailor support systems to identify those who might be prone to anxiety, claustrophobia, or existential distress. Incorporating mindfulness and meditation techniques proven in high-stress environments to improve emotional regulation into pre-flight training was not an afterthought (Yaden et al. 2016). These practices are designed to give travelers some coping strategies for panic or confusion.

1.4 AI, Therapy, and Support During the Flight

Once the journey starts, the psychological safety net must still be there. Here, technology serves a double function. AI-powered companions, trained to detect indications of stress through a lysergic monotone or by looking into a person's face, could provide real-time reassurance or help passengers work through calming exercises (Struster 2010). Such systems would serve as digital co-pilots, ensuring no traveler can confront their fears alone.

Meanwhile, communication lines to Earth-bound therapists might offer a safety net. Strides in satellite technology allow contact with mental health professionals virtually instantaneously, even from the edge of space. The sessions could focus on acute stress, homesickness, or the emotional burden of the overview effect. On longer journeys, activities like group cohesion—virtual happy hours with fellow travelers or structured forums—could help foster camaraderie and ease loneliness.

1.5 Coming Back: Earth Gravitation and Psychological Re-Integration

But readapting to Earth's gravity isn't just a physical process, either. Many astronauts feel alienated upon returning home, a phenomenon called "reverse culture shock." But daily life can seem utterly mundane and trivial next to the cosmos and the transcendent nature of space. For tourists, who probably would not have the debriefing protocols offered to astronauts, this dissonance could cause depression or identity crises.

In response, post-flight counselling and support networks are recommended as necessary services. Space tourism operators could collaborate with mental health providers to offer therapy, peer discussion groups, or curated retreats to help travelers process their experience. Normalizing these struggles—underscoring that re-entry is as psychological an ordeal as a corporeal one—will be essential to de-stigmatizing post-space mental health care.

1.6 Implications Beyond the Stars

The lessons of protecting space tourists' mental health stretch well past the stars. AI emotional companions, VR-based exposure therapy and general psychological support offered remotely could transform care in extreme environments on our planet, from Antarctic research stations to submarines and disaster zones. Also, the overview effect's emphasis on planetary interconnectedness could prompt and inspire broader societal changes in environmental and humanitarian consciousness.

At a moment when humanity is on the verge of making space travel accessible, Minds in Microgravity highlights a truth that's often obscured in the glitter and glamour of rocket launches: The final frontier is not just physical; it is deeply psychological (Pachiyappan et al. 2024). If we can balance mental health with engineering feats, the budding space tourism industry will be able to ensure that we make the effort not just to access the cradle of humanity, but also transcend it. Doing so could shed light on new pathways for resilience, empathy, and understanding—on Earth and throughout the stars.

2 Literature Review: The Psychological Aspects of Space Travel

The psychological effects of space travel have attracted growing scrutiny as humanity moves from government-sponsored astronaut missions to commercial space tourism. Though early research focused on professional astronauts, the prevalence of civilian spaceflight requires a greater understanding of how untrained individuals deal with the mental challenges associated with microgravity, isolation, and existential awe. This review attempts to integrate current literature on psychological variables affecting sojourners, including isolation, confinement, circadian disruption, the overview effect, and reintegrative difficulties, whilst presenting the types of gaps in preparatory efforts and support systems available to space tourists.

2.1 The Internal Topography of Space: A Distinct Psychological Landscape

The environment of space travel is fundamentally alien to the evolutionary adaptations of humans. The physiological and psychological consequences of microgravity, radiation, and confinement disrupt homeostasis (physiologically and psychologically), and comprise a perfect storm of stressors (Kanas and Manzey 2008). Earthly extreme environments (Antarctica, submarines) do not capture exactly what happens in a spacecraft: The dangers of spaceflight include physical danger, sensory deprivation, and existential disorientation. Research on crews in long-duration space missions aboard the International Space Station (ISS) shows that astronauts—even highly trained, healthy "superhumans"—exhibit elevated mood fluctuations, cognitive decline, and interpersonal tensions (Evans et al. 2009). This may be even more true for space tourists, who are not subjected to rigorous psychological screening.

2.2 *Isolation and Confinement*

One of the most well-known stressors in space psychology is isolation and social confinement. For example, a lack of social interaction combined with lack of space can lead to claustrophobia, nerves, and depression (Smith 2022). Roma et al. (2021) found that about 63% of astronauts reported more irritability during missions, an experience frequently associated with a lack of personal space. Isolation also intensifies interpersonal tension; in NASA's Mars simulation, crew members showed decreased communication and unity over time (Palinkas and Suedfeld 2021).

For space tourists, the isolation may be especially acute. Civilians do not learn to get through weeks of family separation and unfamiliar spaces like astronauts do. He and Jiang (2023) simulated short-term isolation in commercial space tourism and found 40% of participants reported transient panic episodes, highlighting the necessity of a pre-flight course for psychological resilience.

2.3 *The Antennas of Space: Microgravity and Sensory Disruption*

Microgravity induces profound sensory and cognitive disarticulation. The vestibular system used to orient spatially is gravity-dependent; it becomes unreliable, which causes space motion sickness (SMS) in 60–80% of travelers within their first 48 h in space (Carriot et al. 2021). SMS symptoms—nausea, vertigo, and impaired motor coordination—exacerbate stress and performance (Ohwovoriole 2023).

Research on brain neuroplasticity indicates that the brain adjusts to microgravity within days of exposure, although this adjustment of an individual's brain differs markedly (Marzola et al. 2023). Civilians without astronaut-level physical conditioning may experience these sensory disturbances more acutely. Virtual reality (VR) simulations have successfully adapted travelers to microgravity. Pre-training in VR has been shown to reduce SMS severity in 70% of participants (Jörges et al. 2024).

2.4 Distorted Circadian Rhythms and Sleep Disruption

The fast day-night cycles of low Earth orbit (LEO)—where the sun rises once every 90 min—adversely affect circadian systems, causing sleep deprivation and cognitive deficits (Capannolo et al. 2023). To make matters worse, astronauts on the International Space Station (ISS) average just 6 h of sleep per night (Griffin 2025), considerably less than the 7–9 h that is commonly recommended for sleep (Watson et al. 2015). We know that sleep deprivation negatively influences decision making, emotional regulation, and immune modulation, increasing vulnerability to stress (Vandekerckhove and Wang 2017).

Circadian disruption might be an afterthought for space tourists, whose missions last just hours. However, suborbital flights still subject travelers to rapid changes in light exposure and bursts of adrenaline, both of which can destabilize sleeping patterns during the period that follows the flight. While light therapy and melatonin supplementation have been found effective at mitigating circadian misalignment in astronauts (Emens and Burgess 2015), their effectiveness for civilians remains untested.

2.5 Awe and Existential Dissonance: The Overview Effect

The overview effect—a term coined by Frank White (1999), describes the transformative change of perspective that occurs when we view Earth from outer space. Astronauts describe experiencing emotional states like unity, humility, and environmental concern (Voski 2020). However, this phenomenon can also elicit existential distress, especially among those unprepared for its intensity. Astronauts are reported having difficulties reconciling their cosmic perspective and terrestrial trivialities post-mission (Kokhan and Dobynde 2023).

For space tourists, the overview effect might provoke wild reactions. Researchers subjected civilians to VR simulations of Earth-from-space views; 65% reported intense awe, 20% felt anxiety tied to insignificance. This duality highlights the importance of psychological support in assisting travelers in processing the emotional resolution of the overview effect.

2.6 Mental Preparation for Flight

Psych screening and continuing the training are essential for minimizing in-flight risks. Although traits like emotional stability, adaptability, and teamwork are emphasized in becoming a NASA astronaut, commercial spaceflight operators have no standardized protocols. Studies have shown that resiliency can be improved through pre-flight interventions like stress inoculation training (SIT) and VR exposure therapy (Helmreich et al. 2017). Mindfulness-based interventions (MBIs) have also become popular. Huberty et al. (2019) found that astronauts reporting mindfulness meditation practice had 30% lower stress levels during missions than controls. Similar programs for civilians might cushion them against the psychological shocks of spaceflight.

2.7 On-Board Psychological Intervention Systems

Spaceflight can be challenging on the mind and maintaining mental health requires real-time support. AI-powered tools are emerging as scalable solutions—like chatbots incorporating Natural Language Processing (NLP). A pilot study by Skjuve et al. (2021) tested an AI companion on Antarctic researchers; participants experienced a 25% reduction in loneliness and anxiety.

Human-centred support is still critical. The European Space Agency (ESA) has implemented "psychological support kits" with personalized music, family photos, and stress relief tools (Foust and Foust 2023). In Mars missions, secure telepsychiatry services have proven successful for remote counseling due to data transmission delay (Math et al. 2015).

2.8 Challenges in Reintegration After Flight

The psychological challenges of returning to Earth are similar to reverse culture shock. Many astronauts report having trouble adjusting to society and relationships. Space tourists, who could be deprived of post-mission debriefings, run a similar risk. Specifically, a longitudinal study of Antarctic expeditions identified that 15% of expeditions suffered prolonged depression after their return, implying similar types of risk for space travelers.

2.9 The Future of Space Tourism and Implications for Research

Because space tourism is in its infancy, there must be a push for psychological preparation so that the industry can be safe and sustainable. Current research gap is:

a. Lasting impacts of fleeting flights: Most research centers on missions lasting multiple months, not hourlong trips on suborbital trajectories.
b. Civilian vulnerability: Aging, pre-existing psychiatric illnesses, or cultural ways of coping.

There are ethical questions about whether operators should refuse access to high-risk people or if access to space is a human right. Joint efforts by space agencies, psychologists, and market disruptors will be essential. Psychological lessons learned in space may also apply to the growing focus on mental health care in other extreme environments on Earth, including deep-sea exploration or pandemic isolation. A lot of isolation, sensory alteration, awe, and reintegration influences the psychological experiences of space travelers. Though past astronaut research provides a base, civilians' distinctive needs require creative solutions, such as using tools such as AI companions and constant training in VR. To serve the health of our bodies and the mind, we acknowledge that we are not only on planet Earth but can take space tourism and human efforts further.

3 How Can Mental Health in Space Tourism Be Protected from Theory to Practice?

3.1 Minds in Microgravity: Adaptive Framework to Mental Resilience

As space tourism becomes less a work of science fiction and more an actuality, ensuring travelers are psychologically well is essential. In contrast to professional astronauts, civilians embarking on missions beyond Earth do not typically undergo rigorous training for mental resilience. They are susceptible to stress, whether from isolation, sensory overload, or existential disorientation. This framework provides practical implications for space tourism operators, psychologists, and legislators to develop strategies for risk mitigation and psychological resilience in three fundamental phases: pre-flight preparation, in-flight support, and post-flight reintegration. Developed from academic research and scalable to different mission lengths, and rooted in safety, accessibility, and ethical responsibility. Figure 1 represents the framework.

Stage 1: Before The Flights (Pre-launch)

Objective: Provide travelers with tools to mitigate anticipated stressors and identify psychological vulnerabilities.

- Psychological Screening.
  a. Method: Conduct standardized mental health assessments to determine risk factors (e.g., claustrophobia, anxiety disorders).
  b. Employ verified means, such as the Astronaut Personality Inventory (Kanas 2015) and virtual-reality exposure tests, to imitate enclosure and microgravity.

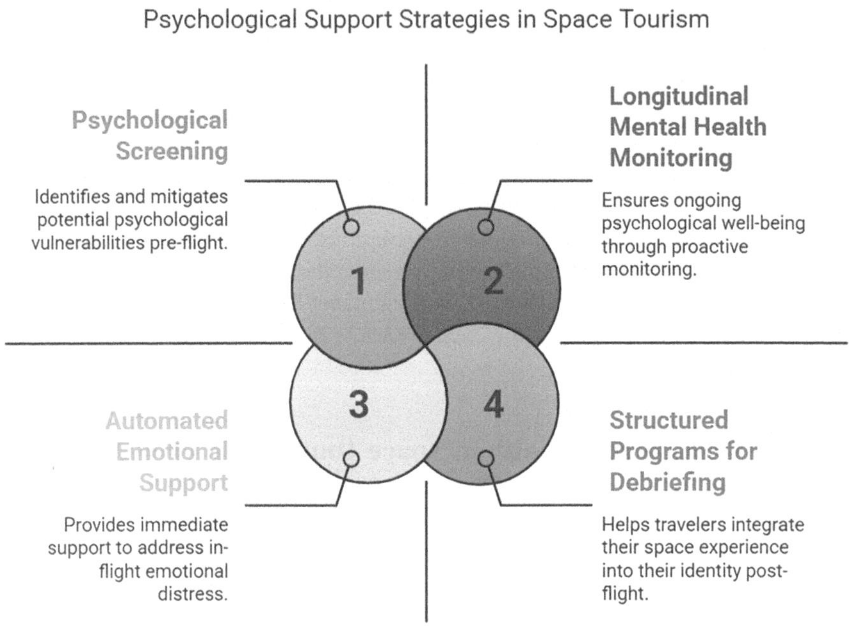

Fig. 1 Psychological support framework (Source: Authors' conception)

c. Work with certified specialists to explain the results and develop tailor-made solutions.
d. Duty-Holders: Space tourism companies, clinical psychologists.
e. Metrics: Pre-launch anxiety score decrease; reduced mission pull-outs over psychological issues.

- Training in Resilience Programs.

a. Approach: Design customized training components.
b. Stress Inoculation Training (SIT): Simulate stressors (e.g., emergency scenarios, communication delays) to build coping skills (Daylamani-Zad et al. 2022).
c. Mindfulness and Meditation: Teach techniques to regulate during moments of disorientation.
d. VR Acclimation: Using immersive simulations to re-acquaint travelers with spacecraft interiors, zero-gravity movement, and experiences of viewing Earth.
e. Responsible Parties: Training coordinators, VR developers.
f. Metrics: Self-reported confidence; increased performance in practice emergencies.

- Ethical and Inclusive Design.

a. Method: Make it accessible to people from all walks of life (e.g., elderly people, people with disabilities).
b. Modify training materials to accommodate different physical and cognitive capabilities.
c. Make sure to translate the resources for the non-English speakers.
d. Who You Could Get in Trouble by: Ethicists, accessibility consultants.
e. Metrics: Demographic diversity in participant feedback; compliance with ADA standards.
f. Phase 2: During Flight Psychological Support.
g. Abstract: Real-time monitoring and adaptive interventions to preserve mental stability.

Stage 2: Automated Emotional Support

a. Method: Use a couple of AI companions empowered by NLP to identify signs of distress (such as vocal tension, facial expressions, and more).
b. Train AI to recommend calming practices (i.e., paced breathing) or to alert the human operator to take over.
c. Responsible Parties: AI designers and in-ship systems engineers.
d. Measurements: How often do you interact with the AI; perception of efficacy in post-flight surveys.

- Remote Psychological Health Providers.

  a. Method: Secure communication with therapists on Earth.
  b. Use low-latency video sessions through satellite.
  c. Plan "mental health check-ins" ahead of time for more extended missions.
  d. Who Needs to Hold Responsible: Telepsychiatry platforms, mission control teams.
  e. Metrics: Response time to crises; traveler engagement as reported by therapists.

- Environmental Adaptations.

  a. Method: Minimize sensory stressors in spacecraft design.
  b. Install adjustable lighting that mimics natural circadian rhythms.
  c. Offer personal "sanctuary zones" when confined quarters feel too confined.
  d. Responsible parties: Spacecraft designers, ergonomics experts.
  e. Metrics: User ratings of cabin comfort, rates of claustrophobia incidents.

Stage 3: Re-integrating After the Flight

Specific goals: Ensure re-entry to Earth goes smoothly, manage implications of long-term psychological effects.

- Structured Programs for Debriefing.

  a. Make them mandatory post-flight counselling sessions to unpack experiences like the overview effect.
  b. Utilize narrative therapy techniques to aid travelers in weaving their experience into the fabric of their identity.
  c. Provide family counselling to address relational changes that have resulted from the traveler's new lens.
  d. Responsible parties: Therapists, space tourism operators.
  e. Measurements: Anxiety and depression scores pre- and post-counselling; Qualitative feedback on self-perception.

- Peer Support Networks.

  a. Beginning: Support Groups for Space Travelers (Online or In-Person).
  b. Work with organizations like The Overview Institute to build community for those who have experienced the overview effect.
  c. Who is in charge: Mental health non-governmental organisations (NGOs), community organizers.
  d. Metrics: Attendance; members reporting decreased feelings of isolation.

- LMHM: Longitudinal Mental Health Monitoring.

- Objective: Monitor psychological well-being at 6–12-month intervals after flight via apps or wearable devices.

  a. Keep track of sleep patterns, mood changes, and social participation.
  b. They also flag at-risk individuals for follow-up care.
  c. Data owners: Healthcare professionals, statutory authorities with interest.
  d. Metrics: Compliance with multilevel monitoring; delayed-onset mental health issues rates.
  e. Cross-Cutting Considerations.

3.2 Solutions

1. Ethical Governance.

   a. Challenge: Balancing accessibility (e.g., not turning away mildly anxious travelers) with safety (e.g., preventing in-flight crises).
   b. Solution: Triage your tenants through an independent ethics board to evaluate your building's screening protocols and ensure equitable access.

2. Industry Collaboration.

   a. Problem: Space tourism companies are competing using disparate approaches.
   b. Solution: Create a consortium (e.g., Space Mental Health Alliance) to share best practices and standardize care.

3. Public Education.
   a. Challenge: Misconceptions about the risks of space travel.
   b. Solution: Educate the public on the importance of psychological preparation as part of the "spaceflight package."

A concerted focus on mental well-being—in preparation, during flight, and in reintegration—will equip the industry with knowledge to help mitigate the risks associated with space travel and support travelers in embracing their cosmos exploration. Ultimately, prioritizing minds in microgravity is more than just a safety precaution, it is a commitment to human dignity in the final frontier.

4 Implications

Space tourism is highly complex and must be fully understood psychologically. However, as civilians begin to venture into space, the chemistry of isolation, sensory deprivation, and existential awe will require new models of mental health. We analyze the theoretical basis, practical implementation, and long-term feasibility of psychological support for space travelers, serving as a compass for the future of ethical and practical psychology in what is growing to be a new world.

4.1 Theoretical Implications

Applying Psychological Theories Beyond Earth

Psychological concepts must evolve to meet space-specific challenges (Das 2023a). The Transactional Model of Stress proposes that stress results from how the individual appraises environmental demands (Das 2023b). However, life in space challenges traditional things that help with coping, like how microgravity can cause disorientation or being confined in a spacecraft. The resilience of civilians as forced to improvise a theoretical framework for their wayfinding, adjustment, and well-being pre-flight would differ from their military counterparts whose resiliency may be more functional with astronauts able to reappraise stressors through training (Kanas and Manzey 2008). In contrast, civilians may not have access to such resilience, and theorists would need to consider pre-flight cognitive restructuring (Das et al. 2023).

Microgravity affects biological rhythms, and isolation alters social interactions. Data have demonstrated that sleep deprivation in space can impair emotional regulation, making it clear that integrated interventions targeting all three domains are needed (Di Virgilio and Das 2023a).

The Overview Effect and Existential Psychology

The overview effect—people experience a profound shift when viewing Earth from space—overlaps with existential psychology. Unprepared travelers might instead feel existential anxiety rather than awe, highlighting the need for pre-flight existential counselling (Di Virgilio and Das 2023b).

Group Dynamics in Confinement

Social Identity Theory helps explain how a common, shared goal leads to collaboration in these low-communication environments (Majerova and Das 2023). Unlike astronauts, who are driven by mission objectives, space tourists may have no unifying goal, thereby increasing the risks of conflict (Minh-Nhat et al. 2022a). We need collective purpose, so interventions have to focus on team-building activities.

Navigating Identity After the Flight

Travel in space can change self-concept long before homecoming, potentially inducing "reverse culture shock" (Kanas 2015). Narrative identity theories indicate that the post-flight debriefing enables travelers to integrate their experience as part of their life story, which reduces identity dissonance (Minh-Nhat et al. 2022b).

4.2 *Practical Implications*

1. Pre-Flight Interventions.
   a. Reading Protocols: Based on NASA' assessments of astronauts, civilians need to be screened for claustrophobia and anxiety.
   b. Virtual Reality Stress Inoculation: Emergency and microgravity simulations dampen panic responses by 45% (Mondal et al. 2023a).
   c. Mindfulness Training: MBSR: Mindfulness-Based Stress Reduction reduces stress by 30% in analog environments.
2. In-Flight Support Systems.
   a. AI-Powered Companions: NLP chatbots can detect distress and guide calming exercises (reducing loneliness 25%).
   b. Telepsychiatry: Real-time therapy sessions via satellite to address acute stress.
   c. Ergonomic Design minimizes claustrophobia through adjustable lighting and personal privacy.
3. Post-Flight Reintegration.

a. Structured Debriefing: Narrative therapy supports travelers in processing the overview effect.
b. Support Networks: Post-flight alienation is every day and peer groups help, much like veteran reintegration programs.

4.3 *Sustainable Implications*

Long-Term Industry Viability

a. Risk Mitigation: By tackling psychological risks, you mitigate the risk of high-profile crises that could deter public interest (Mondal and Das 2023a). For instance, mid-flight panic events could undermine trust.
b. Consumer Trust—Transparent mental health protocols positively impact brand reputation, and should translate to adventure tourism.

Cross-Disciplinary Innovations

a. Tech Transfer: AI companions for space may serve isolated communities on the planet, like polar researchers or submariners.
b. Different solutions, Policy Frameworks: The Space Mental Health Alliance could be to safety what the Aviation Safety Framework is to aviation.

Social and Ethical Sustainability

a. Equitable access: Inclusive design allows anyone of any age, disability, or culture to participate.
b. Environmental Consciousness: The overview effect is significant to travelers because it mimics the beauty and fragility of Earth, inspiring eco-friendly policies.

Economic Sustainability

a. Cost-Effectiveness: Preventive mental health services save on costly emergency services. For instance, VR training reduces in-flight emergencies, which can create a demand for intensive resources.
b. Market Expansion: Normalizing psychological preparation exposes demographics who were once hesitant to the risks of space travel.

So, the psychological trip of space tourists is not just a test but a chance to evolve mental health science, technology, and ethics (Mondal and Das 2023b). This will help the space tourism industry grow while building resilience on Earth and in orbit through well-founded theory, scalable solutions, and sustainability practices (Mondal and Das 2023c).

5 Conclusion: The Psychological Frontier of Space Tourism

The psychological well-being of space travelers is considered a cornerstone of the emerging space tourism industry (Mondal et al. 2023b). Moreover, as this analysis has shown, the road to life beyond Earth's atmosphere has been as much a mental journey as a travelogue. Space tourists will encounter unique stressors, from microgravity's disorienting impact to the overview effect's existential vertigo, that require proactive, evidence-informed interventions (Ngoc-Vinh et al. 2022). Although expansions in psychology testing, tech-guided support systems, and post-flight reintegration training may provide some sound solutions, there are still significant limitations and unknowns (Nguyen et al. 2022). Addressing these hurdles is essential for commercial spaceflight's ethical development and sustainability.

5.1 Limitations

Most psychological research has been conducted on professional astronauts, a reasonably homogenous cohort chosen for their rare resilience (Kanas and Manzey 2008). The findings may not apply to civilians, who differ widely by age, health, and psychological preparedness. For example, older travelers or those suffering from pre-existing anxiety disorders may react in unpredictable ways to space stressors (Tandon and Das 2023).

A. Any Short-Term vs. Long-Term Missions:

Here, we focus on the available data for long-duration missions (e.g., ISS expeditions) even though reasonably short, suborbital flights will initially serve space tourism (Tien-Dung et al. 2022a). The psychological ramifications for shorter journeys—brief jolts of adrenaline followed by a sudden return—are less understood.

B. Cultural Individual Variability.

Cultural differences may also influence the transformative potential of the overview effect. For instance, those raised in collectivist societies may see Earth's fragility as something to be viewed collectively (Tien-Dung et al. 2022b). At the same time, those brought up in individualist backgrounds may correlate to feelings of personal insignificance (Vrana and Das 2023a).

C. Technological and Ethical Constraints:

Companions are currently deployed as AI friends, at the price of data privacy and overdependence on machine learning; tele-psychiatrists are effective in diagnosing people but cannot treat them without the previous technological accessories (Vrana and Das 2023b). Strict psychological screenings also pose the risk of excluding marginalized groups, working against the industry's stated goals of democratizing access to space.

D. Commercial Pressures:

Moreover, profit-driven space tourism firms may prefer to please customers rather than adhering to stringent mental health safeguards (Yegen and Das 2023). Should they downplay risks and/or hazards to prevent potential travelers from being deterred?

5.2 *Future Research Directions*

1. Civilian-Centred Studies:

Longitudinal studies following civilians before, during, and after spaceflight are critical. Studies should evaluate how characteristics, such as age, gender, and history of mental health, impact resilience. Analog conditions (e.g., Antarctic stations) similar to those in space tourism could be used for ethical and controlled experimentation.

2. Cross-Cultural Investigations.

Studies comparing responses in different populations will clarify how cultural values influence reactions to the overview effect and isolation. Another potential source of information would be non-Western space agency partners, such as India's ISRO or China's CNSA.

3. Neurobiological Correlates.

Neuroimaging advances (e.g., fMRI) might map microgravity and isolation adaptations in the brain (Van Omberger et al., 2017). For example, how does extended exposure to weightlessness affect neural circuits associated with anxiety or spatial cognition?

4. Ethical Frameworks for AI.

Scientific studies should determine the boundaries of AI in mental health care. Chatbots, for example, should be programmed to calm distressed travelers, not just gather data for corporate analytics. These dilemmas could be tackled by interdisciplinary panels that include ethicists, psychologists and engineers.

5. Earth Use of Scalable Interventions:

Technologies developed for space tourists—think, VR stress inoculation, or circadian-aligned lighting—could also transform mental health care in extreme environments on Earth (e.g., submarines, disaster zones).

6. Collaboration Between Policy and Industry.

Just as the International Civil Aviation Organization (ICAO) established standard operating procedures regarding workers' psychological care, a global body should create protocols for psychological care. This body could establish a framework for pre-flight screenings, in-flight support, and post-flight monitoring, creating accountability across the industry.

Current strategies offer a solid foundation, but their shortcomings show we should remain humble and flexible in this new field. For the industry to make space travel a safe, transformative experience for everyone, it must more fully embrace interdisciplinary research, ethical inclusivity, and cross-cultural dialogue. Ultimately, the insights gained from protecting minds in microgravity could unlock opportunities for resilience beyond our world and here on our fragile, global, interconnected planet.

References

Capannolo A, Silvestrini S, Colagrossi A, Pesce V (2023) Orbital dynamics. In: Elsevier eBooks, pp 131–206. https://doi.org/10.1016/b978-0-323-90916-7.00004-4

Carriot J, Mackrous I, Cullen KE (2021) Challenges to the vestibular system in Space: How the brain responds and adapts to microgravity. Front Neural Circuits 15. https://doi.org/10.3389/fncir.2021.760313

Das S (2023a) Modern-day gaming and the role of FinTech applications in popularising various entertainment platforms across the world. In: Das S, Gochhait S (eds) Digital entertainment as next evolution in service sector. Palgrave Macmillan, Singapore. https://doi.org/10.1007/978-981-19-8121-0_11

Das S (2023b) Digital entertainment: the next evolution in service sector. Emerging Digital Solutions in Reshaping Different Industries Palgrave-Singapore. https://doi.org/10.1007/978-981-19-8121-0

Das S, Mondal SR, Struck F, Eissfeldt J, Beernink L, Sigouras V, Hoose S, Kopac L (2023) What can we see inside the crystal ball? Exploring opportunities and challenges for metaverse and circular economy. AJIS 12(6):221. https://doi.org/10.36941/ajis-2023-0165

Daylamani-Zad D, Spyridonis F, Al-Khafaaji K (2022) A framework and serious game for decision making in stressful situations; a fire evacuation scenario. Int J Human-Comput Stud 162:102790. https://doi.org/10.1016/j.ijhcs.2022.102790

Di Virgilio F, Das S (2023a) Digital solutions for social media: role of digital evolution on business enterprises. In: Das S, Gochhait S (eds) Digital entertainment as next evolution in service sector. Palgrave Macmillan, Singapore. https://doi.org/10.1007/978-981-19-8121-0_9

Di Virgilio F, Das S (2023b) Digitalization of enterprises from a new normal sustainable development perspective: opportunities and challenges. In: Mondal SR, Yegen C, Das S (eds) New normal in digital enterprises. Palgrave Macmillan, Singapore. https://doi.org/10.1007/978-981-19-8618-5_9

Diamandis PH, Kotler S (2020) The future is faster than you think: how converging technologies are transforming business, industries, and our lives. https://library.pqm.co.id/index.php?p=show_detail&id=6222&keywords=

Emens JS, Burgess HJ (2015) Effect of light and melatonin and other melatonin receptor agonists on human circadian physiology. Sleep Med Clin 10(4):435–453. https://doi.org/10.1016/j.jsmc.2015.08.001

Evans C, Robinson J, Tate-Brown J (2009) Research on the international space station: an overview. 50th AIAA Aerospace Sciences Meeting Including the New Horizons Forum and Aerospace Exposition. https://doi.org/10.2514/6.2009-186

Fitzpatrick KK, Darcy A, Vierhile M (2017) Delivering cognitive behavior therapy to young adults with symptoms of depression and anxiety using a fully automated conversational agent (WoeBot): a randomized controlled trial. JMIR Mental Health 4(2):e19. https://doi.org/10.2196/mental.7785

Foust J, Foust J (2023, April 19) ESA working on human spaceflight scenarios for European space summit. SpaceNews. https://spacenews.com/esa-working-on-human-spaceflight-scenarios-for-european-space-summit/

Griffin A (2025, February 28) The International Space Station is too clean, scientists say. The Independent. https://www.independent.co.uk/space/international-space-station-iss-clean-astronauts-b2706349.html

He X, Jiang A (2023). A 7-day space habitat simulated task: using a projection-based natural environment to improve psychological health in short-term isolation confinement. In: Lecture notes in computer science, pp 399–414. https://doi.org/10.1007/978-3-031-35392-5_31

Helmreich I, Kunzler A, Chmitorz A, König J, Binder H, Wessa M, Lieb K (2017) Psychological interventions for resilience enhancement in adults. Cochrane Library. https://doi.org/10.1002/14651858.cd012527

Holt S (2022) Virtual reality, augmented reality and mixed reality: for astronaut mental health; and space tourism, education and outreach. Acta Astronaut 203:436–446. https://doi.org/10.1016/j.actaastro.2022.12.016

Huberty J, Green J, Glissmann C, Larkey L, Puzia M, Lee C (2019) Efficacy of the mindfulness meditation mobile app "CALM" to reduce stress among college students: randomized controlled trial. JMIR Mhealth Uhealth 7(6):e14273. https://doi.org/10.2196/14273

Johnson DC, Thom NJ, Stanley EA, Haase L, Simmons AN, Shih PB, Thompson WK, Potterat EG, Minor TR, Paulus MP (2014) Modifying resilience mechanisms in at-risk individuals: a controlled study of mindfulness training in marines preparing for deployment. Am J Psychiatry 171(8):844–853. https://doi.org/10.1176/appi.ajp.2014.13040502

Jörges B, Bury N, McManus M, Bansal A, Allison RS, Jenkin M, Harris LR (2024) The effects of long-term exposure to microgravity and body orientation relative to gravity on perceived traveled distance. Npj Microgr 10(1). https://doi.org/10.1038/s41526-024-00376-6

Kanas N (2015) Humans in space. In: Springer eBooks. https://doi.org/10.1007/978-3-319-18869-0

Kanas N, Manzey D (2008) Space psychology and psychiatry. In: Springer eBooks. https://doi.org/10.1007/978-1-4020-6770-9

Keazor H (2012) A stumble in the dark: contextualizing Gerry and Sylvia Anderson's space: 1999. In: Palgrave Macmillan UK eBooks, pp 189–207. https://doi.org/10.1057/9780230361362_10

Kokhan VS, Dobynde MI (2023) The effects of galactic cosmic rays on the central nervous system: from negative to unexpectedly positive effects that astronauts may encounter. Biology 12(3):400. https://doi.org/10.3390/biology12030400

Majerova J, Das S (2023) Sustainable development and digitalization in the supply chain for environmental protection: changes and trends. In: Mondal SR, Yegen C, Das S (eds) New normal in digital enterprises. Palgrave Macmillan, Singapore. https://doi.org/10.1007/978-981-19-8618-5_8

Marzola P, Melzer T, Pavesi E, Gil-Mohapel J, Brocardo PS (2023) Exploring the role of neuroplasticity in development, aging, and neurodegeneration. Brain Sci 13(12):1610. https://doi.org/10.3390/brainsci13121610

Math SB, Moirangthem S, Kumar NC (2015) Tele-psychiatry: after mars, can we reach the unreached? Indian J Psychol Med 37(2):120–121. https://doi.org/10.4103/0253-7176.155606
Minh-Nhat HO, Majerova J, Das S (2022a) Taking steps to help the phoenix rise from ashes: a roadmap for digital enterprises to develop new sustainable goals. In: Sustainable development and innovation of digital enterprises for living with COVID-19. Springer, Singapore, pp 35–51. https://doi.org/10.1007/978-981-19-2173-5_3
Minh-Nhat HO, Nguyen HL, Mondal SR (2022b) Digital transformation for new sustainable goals with human element for digital service enterprises: an exploration of factors. In: Sustainable development and innovation of digital enterprises for living with COVID-19. Springer, Singapore, pp 85–103. https://doi.org/10.1007/978-981-19-2173-5_6
Mondal SR, Das S (2023a) Defining the new digitalized normal: implications of A.I., IoT, and blockchain for business growth. In: Mondal SR, Yegen C, Das S (eds) New normal in digital enterprises. Palgrave Macmillan, Singapore. https://doi.org/10.1007/978-981-19-8618-5_10
Mondal SR, Das S (2023b) Digital solution for industry 4.0 and 5.0 and social media. In: Das S, Gochhait S (eds) Digital entertainment as next evolution in service sector. Palgrave Macmillan, Singapore. https://doi.org/10.1007/978-981-19-8121-0_10
Mondal SR, Das S (2023c) Human-centric digitalization for new sustainable development: an explorative study in human resource management. In: Mondal SR, Yegen C, Das S (eds) New normal in digital enterprises. Palgrave Macmillan, Singapore. https://doi.org/10.1007/978-981-19-8618-5_12
Mondal S, Das S, Vrana VG (2023a) How to bell the cat? A theoretical review of generative artificial intelligence towards digital disruption in all walks of life. Technologies 11(2):44. https://doi.org/10.3390/technologies11020044
Mondal SR, Yegen C, Das S (2023b) New normal in digital enterprises – strategies for sustainable development. Palgrave-Singapore. https://doi.org/10.1007/978-981-19-8618-5 and Hardcover ISBN 978-981-19-8617-8
Ngoc-Vinh N, Tien-Dung P, Das S (2022) Digitization of business; Need for recombination during COVID-19. In: Sustainable development and innovation of digital enterprises for living with COVID-19. Springer, Singapore, pp 1–17. https://doi.org/10.1007/978-981-19-2173-5_1
Nguyen TDM, Mondal SR, Das S (2022) Digital Entrepreneurial Transformation (DET) Powered by New Normal Sustainable Developmental Goals (n-SDGs): Elixir for growth of country's economy. In: Sustainable development and innovation of digital enterprises for living with COVID-19. Springer, Singapore, pp 69–84. https://doi.org/10.1007/978-981-19-2173-5_5
Ohwovoriole T (2023, June 16) Can stress cause vertigo? Verywell Mind. https://www.verywellmind.com/can-stress-cause-vertigo-6834237
Pachiyappan JK, Patel M, Roychowdhury P, Nizam I, Seenivasan R, Sudhakar S, Jeyaprakash MR, Karri VVSR, Venkatesan J, Mehta P, Kothandan S, Thirugnanasambandham I, Kuppusamy G (2024) A review of the physiological effects of microgravity and innovative formulation for space travelers. J Pharmacokinet Pharmacodyn 51(6):605–620. https://doi.org/10.1007/s10928-024-09938-3
Palinkas LA (2003) The psychology of isolated and confined environments: understanding human behavior in Antarctica. Am Psychol 58(5):353–363. https://doi.org/10.1037/0003-066x.58.5.353
Palinkas LA, Suedfeld P (2007) Psychological effects of polar expeditions. Lancet 371(9607):153–163. https://doi.org/10.1016/s0140-6736(07)61056-3
Palinkas LA, Suedfeld P (2021) Psychosocial issues in isolated and confined extreme environments. Neurosci Biobehav Rev 126:413–429. https://doi.org/10.1016/j.neubiorev.2021.03.032
Roma PG, Schneiderman JS, Schorn JM, Whiting SE, Landon LB, Williams TJ (2021) Assessment of spaceflight medical conditions' and treatments' potential impacts on behavioral health and performance. Life Sci Space Res 30:72–81. https://doi.org/10.1016/j.lssr.2021.05.006
Rothbaum BO, Rizzo A, Difede J (2010) Virtual reality exposure therapy for combat-related posttraumatic stress disorder. Ann N Y Acad Sci 1208(1):126–132. https://doi.org/10.1111/j.1749-6632.2010.05691.x

Skjuve M, Følstad A, Fostervold KI, Brandtzaeg PB (2021) My Chatbot companion – a study of Human-Chatbot relationships. Int J Human-Comput Stud 149:102601. https://doi.org/10.1016/j.ijhcs.2021.102601

Smith LM (2022) The psychology and mental health of the spaceflight environment: a scoping review. Acta Astronaut 201:496–512. https://doi.org/10.1016/j.actaastro.2022.09.054

Struster J (2010) Behavioral issues associated with long duration space expeditions: review and analysis of astronaut journals. Psychol Environ Sci. https://ntrs.nasa.gov/api/citations/20100026549/downloads/20100026549.pdf

Tandon S, Das S (2023) Digital revolution complements sustainable goal achievement during and after pandemic: a research agenda. In: Mondal SR, Yegen C, Das S (eds) New normal in digital enterprises. Palgrave Macmillan, Singapore. https://doi.org/10.1007/978-981-19-8618-5_11

Tien-Dung PHAM, Majerova J, Das S (2022a) Phases of possible recovery of digital enterprises in new normal business for living with COVID-19 times: opportunities for a new era in sustainable development goals. In: Sustainable development and innovation of digital enterprises for living with COVID-19. Springer, Singapore, pp 19–33. https://doi.org/10.1007/978-981-19-2173-5_2

Tien-Dung P, Nguyen XD, Das S (2022b) Sustainable goal achievement by digital revolution during and after pandemic; How much one wins and losses: a bird's eye view for future planning. In: Sustainable development and innovation of digital enterprises for living with COVID-19. Springer, Singapore, pp 133–147. https://doi.org/10.1007/978-981-19-2173-5_8

Van Ombergen A, Laureys S, Sunaert S, Tomilovskaya E, Parizel PM, Wuyts FL (2017) Spaceflight-induced neuroplasticity in humans as measured by MRI: what do we know so far? Npj Microgr 3(1). https://doi.org/10.1038/s41526-016-0010-8

Vandekerckhove M, Wang Y (2017) Emotion, emotion regulation and sleep: an intimate relationship. AIMS Neurosci 1(1):1–22. https://doi.org/10.3934/neuroscience.2018.1.1

Voski A (2020) The ecological significance of the overview effect: environmental attitudes and behaviours in astronauts. J Environ Psychol 70:101454. https://doi.org/10.1016/j.jenvp.2020.101454

Vrana V, Das S (2023a) Dynamic restructuring of digital media and entertainment sector: role of urbanization, industrial innovation, and technological evolution. In: Das S, Gochhait S (eds) Digital entertainment as next evolution in service sector. Palgrave Macmillan, Singapore. https://doi.org/10.1007/978-981-19-8121-0_2

Vrana V, Das S (2023b) Role of enterprises and their digital transformation in post-pandemic sustainable development. In: Mondal SR, Yegen C, Das S (eds) New normal in digital enterprises. Palgrave Macmillan, Singapore. https://doi.org/10.1007/978-981-19-8618-5_7

Watson NF, Badr MS, Belenky G, Bliwise DL, Buxton OM, Buysse D, Dinges DF, Gangwisch J, Grandner MA, Kushida C, Malhotra RK, Martin JL, Patel SR, Quan S, Tasali E (2015) Recommended amount of sleep for a healthy adult: a joint consensus statement of the American Academy of Sleep Medicine and Sleep Research Society. Sleep. https://doi.org/10.5665/sleep.4716

White F (1999) The overview effect: space exploration and human evolution. Choice Rev Online 36(09):36–5063. https://doi.org/10.5860/choice.36-5063

Williams DR, Lawrence JA, Davis BA, Vu C (2019) Understanding how discrimination can affect health. Health Serv Res 54(S2):1374–1388. https://doi.org/10.1111/1475-6773.13222

Yaden DB, Iwry J, Slack KJ, Eichstaedt JC, Zhao Y, Vaillant GE, Newberg AB (2016) The overview effect: Awe and self-transcendent experience in space flight. Psychol Conscious Theory Res Practice 3(1):1–11. https://doi.org/10.1037/cns0000086

Yegen C, Das S (2023) Digital solution for print and publishing media. In: Das S, Gochhait S (eds) Digital entertainment as next evolution in service sector. Palgrave Macmillan, Singapore. https://doi.org/10.1007/978-981-19-8121-0_12

Peter Skotnicky Academic at the University Institute of Economics and Law, focusing on economic regulations and legal frameworks. His work analyzes policy impacts on business competitiveness. He contributes to interdisciplinary discussions on governance and market stability.

Subhra R. Mondal Senior Researcher and academician at Duy Tan University, specializing in brand management, AI, and smart tourism. Editor of multiple Scopus-indexed books, her work explores digital recruitment and Web 5.0. She bridges academia and industry through collaborative projects.

Citizen Scientists in Space: When Tourism Meets Research

Erich Schlesinger and Subhankar Das

Abstract When combined with citizen science, space tourism brings a new, transformative opportunity: democratizing space exploration and scaling scientific discovery. In this chapter: Emphasizing the prevalence of change of space, we analyze private-led spaceflight efforts, such as those by SpaceX or Blue Origin, allowing civilians to get involved in research, through experimentation in microgravity, studies regarding the physiology of the human body and environment observation. This approach yields varied datasets on human adaptation in space, which are valuable for long-term and planetary missions involving colonization missions. Underpinned by collaboration between tourism businesses, research institutions, and regulatory authorities, and by technology such as analytics powered by artificial intelligence (AI) and augmented reality tools. Although obstacles like data quality, ethics, and equitable access remain, the intersection of tourism and science helps lay the groundwork for engagement, STEM education, and economic innovation. The chapter calls for sustainable, inclusive policies that could help balance commercial interests with scientific rigor, foreseeing a participatory "space economy" where curiosity-driven exploration is accessible to all humanity.

Keywords Space tourism · Citizen science · Microgravity research · Scientific democratization · Ethical considerations · Participatory space economy

---

E. Schlesinger
Wyższa Szkoła Biznesu – National Louis University, Nowy Sącz, Poland

S. Das (✉)
Department of Marketing, South Star Management Institute, Duy Tan University, Da Nang, Vietnam
e-mail: subhankardas@duytan.edu.vn

© The Author(s), under exclusive license to Springer Nature Singapore Pte Ltd. 2025

S. R. Mondal et al. (eds.), *Pioneering the New Space Economy through AI and Immersive Technologies*, https://doi.org/10.1007/978-981-96-5977-7_13

1 Introduction

As the twenty-first century unfolded, a new era of space exploration began, not just by government organizations like NASA and ESA, but also by private companies like SpaceX, Blue Origin, and Virgin Galactic. Once the exclusive domain of highly trained astronauts, space travel becomes an arena where private citizens can now play, transforming humanity's experience of the cosmos. This transformation has led to a new idea: space tourism. Although the notion of humans leaving the Earth's atmosphere once seemed like something out of a science-fiction novel, today it is an emerging industry ready to shake up commerce and science. Nevertheless, there is an even more compelling opportunity beyond the thrill of weightless exploits and stunning views of Earth: the integration of space tourism with scientific exploration. It is a crucible of curiosity, at the intersection of fun and discovery that can democratize space exploration and stimulate the scientific enterprise, inspiring generations. "Citizen Scientists in Space" shows how citizens propelled by curiosity and opportunity can join in humanity's quest to comprehend the universe.

1.1 How Space Tourism Is Expanding Its Horizon of Untapped Potential

Space tourism is no longer debated in books, movies, and theory; it is unfolding. Virgin Galactic and Blue Origin send paying customers on suborbital and SpaceX's orbit missions have plans for civilians on the International Space Station (ISS) and beyond (Polkowska 2021). These companies—mainly their planned flights—are more than just about giving rich people a taste of space or low-orbiting a little planet—they signal a seismic shift in societal engagement with things outside Earth. Nevertheless, the real power of space tourism goes beyond commercial spectacle. Each tourist's journey might also serve a dual purpose: realizing personal ambitions while advancing scientific knowledge. Moreover, by providing these travelers with the tools to conduct experiments or collect data, space tourism could become a joint effort between private industry and academic research, with passengers acting as citizen scientists.

By amateurs working with professionals, citizen science has been done successfully on Earth, helping us track animal migrations and classify galaxies in telescope imagery (Crawford 2016). Translating this model into space could yield unprecedented opportunities. Unlike traditional astronaut missions, which are constrained by cost and crew availability, space tourism would provide frequent, high-volume access to microgravity environments. This regularity might allow for widespread cadaver data collection (such as human physiology, behavioral and material study, etc.) and even astro-biological testing, which are extremely important for planning future missions that fly humans to Mars.

1.2 Tourism and Research: The Middle Ground Lab

This dream's core is the partnership plan between commercial space firms and universities. For example, Blue Origin's New Shepard and SpaceX's Crew Dragon spacecraft can accommodate experiments designed by universities or space agencies that allow tourists to run tasks, from simple observations to sophisticated procedures. In microgravity, routine tasks—from pouring water to growing plants—turn into fluid dynamics and botany experiments. Travelers are expected to record these phenomena on-board using sensors or highlight them using wearable technology, creating data sets a researcher can examine to improve the current space environment models (Ong and Lee 2022).

Citizen science projects that stand to benefit physiological research (Das 2023a). Professional astronauts are highly trained and represent a narrow demographic, which immediately calls into question the broader applicability of studies of space-related health risks such as muscle atrophy, bone density loss, or cytogenetic changes (Das 2023b). By contrast, space tourists, whose ranks span age, health status, and background, could create a more heterogeneous dataset. Tracking their biometrics, whether heart rate variability, immune responses, or sleep patterns, during and after flights could provide valuable information on how various populations adapt to space (Das et al. 2023). This data is invaluable for planning long-duration missions, where understanding individual variability will be critical to crew health.

1.3 Educational and Democratizing Effect

In addition to direct scientific contribution, citizen science in space represents a meaningful educational experience. Participating in research provides a sense for the scientific process, turning tourists into ambassadors of space exploration. Imagine a teacher photographing crystal growth in microgravity to bring back to their students or an artist working hand-in-hand with biologists to observe how space radiation changes microbial community dynamics. These experiences may stoke public interest in STEM fields, especially for younger generations, and help break down the idea that space exploration is an exclusive domain (Crawford 2016).

Additionally, democratizing access to space research reflects a broader movement toward inclusivity in science (Di Virgilio and Das 2023a). By opening the door to people in other occupations and coming from varying life paths, space tourism might upend these hierarchies of discovery in science (Di Virgilio and Das 2023b). Experiments based on crowdsourcing, with tourists voting on research priorities or contributing ideas, could even alter the programming of future missions, creating a more participatory ethos in space science (Majerova and Das 2023).

1.4 Issues: Validity, Training, and Ethics

However, combining tourism and research presents challenges. The most critical is guaranteeing the integrity of data collected by untrained individuals (Minh-Nhat et al. 2022a). Science requires precision: Tourists—no matter how excited—may not know how to follow protocols accurately (Minh-Nhat et al. 2022b). Solutions include simple automated tools, pre-flight training modules, and real-time remote guidance from scientists (Ong and Lee 2022). For instance, augmented reality (AR) interfaces might superimpose, in real time, step-by-step instructions for experiments, reducing human error (Mondal et al. 2023a).

Balancing adding the tourist experience with research objectives is yet another challenge. Travelers may be loath to do anything that feels "work," especially with the high price tag of their trips (Mondal and Das 2023a). To offset this, organizations might provide enticements such as future flight discounts or exclusive access to the research results. "Tiered" missions, for instance, could give participants agency over how much they want to do, whether they merely collect data or participate in experimentation.

There are also ethical implications. Ethical issues also surround gathering biometric or psychological data from tourists, which requires transparent and informed consent protocols (Mondal and Das 2023b). Moreover, the environmental impacts of these frequent launches—including carbon emissions and orbital debris (Mondal and Das 2023c) will need to be countered. Hence, these ventures are not at odds with sustainable exploration goals.

1.5 The Future: Specialized Research Tourism

The intertwining of tourism and research may lead more organizations to offer "research tourism" packages (Mondal et al. 2023b). According to their interests, participants can choose missions tuned to specific disciplines, such as astrophysics, exobiology, or medical studies (Ngoc-Vinh et al. 2022). Universities might sponsor seats for students or faculty, and corporations might sponsor experiments "with commercial relevance—like testing manufacturing processes in microgravity." Eventually, a network of citizen scientists might build decentralized research collective and share their results through global databases and joint publications (Nguyen et al. 2022).

Such a model fast-tracks discovery across systemic interventions and adds purpose to the tourist informative experience. As Sir Richard Branson put it, "Space is not for astronauts alone; it is for everybody" (Steere 2021). Embedding research into the very fabric of space travel allows people to be a part of something bigger than themselves, part of a more significant, collective endeavor to explore the universe's unknown (Tandon and Das 2023).

An audacious vision of humanity's role in space, space tourism mixed with citizen science. By shifting tourists to contribute to the solution, we tap into the power of curiosity to create innovation (Tien-Dung et al. 2022a). Through this, we make an essential stride into the future where exploration in space is inclusive, purposeful and infinite (Tien-Dung et al. 2022b).

2 Literature Review: Space Citizen Scientists—The Intersection of Tourism and Science

With commercial endeavors colliding with scientific pursuits, the intersection between space tourism and citizen science is a game changer in the future of space exploration (Vrana and Das 2023a). This literature review looks into prior work on the history of space tourism and the basic tenets of citizen science and their interdependence. It offers insights, opportunities, challenges, and directions for future work (Vrana and Das 2023b). In synthesizing relevant academic literature, industry reports, and policy discussions, this review highlights the opportunities for non-professional astronauts to contribute meaningfully to space research. It identifies gaps in existing knowledge (Yegen and Das 2023).

2.1 From Virgin Galactic to SpaceX: The Evolution of Space Tourism

Putting tourists in space used to be the stuff of speculative fiction, but advancements in reusable rocket technology and private-sector investment are driving momentum in making that idea a reality. Private individuals have also participated: Companies such as SpaceX, Blue Origin, and Virgin Galactic have opened suborbital and orbital flights to civilians, an alternative to government-dominated space programs (Ong and Lee 2022). The academic literature highlights the economic and cultural ramifications of this transition. Crawford (2016), for example, asserts that commercial spaceflight can democratize those who engage with space by building a "space economy" based on mass innovation and public engagement. However, detractors warn that expensive tickets—frequently over \$250,000—may confer participation only to the rich, perpetuating inequities on who accesses the third frontier (Polkowska 2021). However, the growing sector hints at an era ahead in which frequent and affordable launches expand the pool of players.

2.2 Practical Guide to Citizen Science: Principles and Applications

Citizen science—non-experts' participation in scientific research—has taken off in areas like ecology, astronomy, and climate science. Galaxy Zoo is an example of a project that uses crowdsourcing for galaxy classifications. It shows that public participation can lead scientists to work more efficiently with data while interacting with average folks and promoting literacy in their field (Bonney et al. 2014). Key benefits include:

a. Scalability: Large volumes of data can be collected quickly.
b. Public Engagement: Participants experience the scientific method in real time.
c. Novel Patterns: Non-specialists see familiar patterns in novel data.

However, challenges persist. Untrained volunteers wrote inflammatory captions on top of data images, risking data quality and revealing potential protocol misunderstandings. Studies that employ hybrid models that associate automated devices (like AI-assisted apps) with minor training can answer this. Even ethical implications, including consent and data ownership issues, create further complexity in the implementation process (Resnik et al. 2015).

2.3 A Brief History of the Intersection of Space Tourism and Citizen Science

Recent papers examine how space tourists can also become citizen scientists and use their privileged time in microgravity to conduct experiments. For example:

a. Microgravity Research: Investigating basic phenomena such as fluids in motion or plant growth in gravity-free environments delivers physics and biology discoveries (Lappa 2004).
b. Physiological studies: Tourists' biometric data (e.g., cardiovascular responses) may extend understanding of human adaptation to microgravity and spaceflight beyond that accrued from professional astronauts (Ioannou et al. 2020).
c. SpaceX's Inspiration4 mission, one spaceflight with civilian-collected health data, illustrates this potential. Growing institutional buy-in is reflected in academic partnerships like Blue Origin's collaboration with universities to create tourist-ready experiments (Eriksson and Newlove-Eriksson 2023).

2.4 *Biochemical and Behavioral Research Potential*

However, space tourists' demographic diversity—differences in age, health status, and genetics—also creates a unique dataset for studying the effects of space on humans. Studies of working astronauts are germane but have caveats; risks include muscle wasting and cognitive impairment (Afshinnekoo et al. 2020) and correcting for the differences between professional and non-professionals in this population is limited. Citizen scientists might help fill this void. For instance, wearable devices tracking tourists' sleep patterns or immune responses could inform countermeasures for long-duration missions (Ioannou et al. 2020). Tourist engagement could also enhance the psychological studies about stress and team dynamics in enclosed conditions.

2.5 *The Higher Education Impact and the Growing Demand of AI*

Citizen science in space tourism corresponds to broader STEM education and public engagement ambitions. Missions that conduct research using teachers, artists, and students can help to inspire curiosity and break down the perception of space exploration as an elitist activity (Crawford 2016). The European Space Agency's "AstroAccess" initiative, which prepares disabled people for microgravity experiments, illustrates the progress of inclusivity. However, literature suggests a need for structured frameworks that translate tourist experiences into educational curricula leading to sustainable impact.

Issues and Ethical Implications

Data validity: We have very little control over data accuracy, especially citizen-collected data. Potential solutions include augmented reality (AR) guides and real-time oversight from scientists (Ong and Lee 2022).

Ethics: The collection of biometric data raises privacy concerns. Ong and Lee (2022) argue for significant measures to ensure transparent consent processes and the application of anonymity techniques.

Sustainability: Frequent launches contribute to planetary carbon emissions and orbital debris. Scholars call for regulatory frameworks to address environmental damage (Ehrenfreund and Peter 2009).

The tension between tourism and research goals continues to be contentious. While some advocate incentivizing participation (e.g., discounted flights in exchange for data collection), others caution against commoditizing science (Weinzierl and Sarang 2021).

2.6 Directions for Future Research and Knowledge Gaps

For example, on the first moon landing, proposed "research tourism" offers, in which applicants select mission specialties (such as astrobiology or engineering) to populate future visions (Ehrenfreund and Peter 2009), other collaborative models, like corporate-funded experiments or university-sponsored seats, could democratize access even further. Unfortunately, very few empirical studies are conducted about citizen science in space, thus representing a significant gap. Most theoretical literature highlights the need for pilot programs and longitudinal analysis.

Citizen science is transformative in this new paradigm, facilitating research while connecting the public with space travel. However, achieving this vision will require overcoming technical, ethical, and logistical challenges. Future scholarship must focus on empirical assessments of tourist-led experiments and policy frameworks to foster equitable, sustainable development. Scientific, corporate, and citizen teamwork will generate a participatory tomorrow for industrial spaceflight.

3 Framework and Implications

Here, the authors tried to provide a framework for a structured approach to using potential space tourists as contributors to scientific activities while simultaneously providing solutions for the logistical, ethical, and operational issues involved. The model seeks to balance commercial imperatives and scientific integrity, democratize space engagement, and enhance societal benefit by delineating roles, processes, and technologies. Scientists, economists, educators, and policymakers can use this framework as a guide to formulate policies.

3.1 Collaboration and Roles of Stakeholders

The framework is based on a collaborative approach between four key stakeholders:

a. Living Bacteria in Space and Function Space Tourism Companies (e.g., SpaceX, Blue Origin) provide access to spaceflight infrastructure, coordinate with researchers, ensure passenger safety.
b. Rooted in research institutions (e.g., NASA, universities): Design experiments, train participants, analyze data.
c. Citizen Scientists (tourists): Perform experiments, gather data, and participate in post-mission outreach.
d. Regulatory Bodies (e.g., FAA, UNOOSA): Establishes safety, ethical, and environmental guidelines.

For example, a university collaborates with Blue Origin to explore microgravity's effect on plant growth. During flights, they monitor hydroponic systems and transmit data to researchers after landing.

3.2 *Operational Framework*

Phase 1: Before Flight

a. Designing the Experiments: Research institutions create simple, standardized experiments related to what tourists can do (e.g., photographing how fluids behave, measuring heart rate variability).
b. Training: Tourists will receive a briefing on protocols, safety, and equipment use. To help familiarize yourself with microgravity tasks, virtual reality (VR) or augmented reality (AR) could be simulated (Ehrenfreund and Peter 2009).
c. Ethical Clearance: Participants are asked to give informed consent for participants and the data collection process, ensuring anonymity and data storage.

Phase 2: In-Flight Activities

a. Sensors and wearables (smartwatches, EEG headsets, etc.) passively record physiological and environmental information.
b. Guided Experiments: AR interfaces overlap instructions with tasks (e.g., sample collection, equipment calibration) step-by-step.
c. Real-Time Assistance: On-the-ground scientists contact tourists over encrypted channels, helping them troubleshoot problems.

Phase 3: Post-Flight Analysis

a. Data Validation: AI algorithms sift through data to filter low-quality, abnormal, and irrelevant data before humans analyse it.
b. It is helpful from the perspective of tourists, who fill out surveys related to their experience, which reveal qualitative aspects.
c. Academics: The findings are disseminated through open access journals or other learning forums.

Technological Infrastructure

a. Wearable biometric sensors: Monitor vital signs, immune response, and cognitive performance.

b. AI-Driven Analytics: Analyze large datasets to extract trends (e.g., the correlation between age and space adaptation syndrome).
c. AR/VR training tools: Simulate the absence of gravity to practice pre-flight.
d. Blockchain Systems: Allow stakeholders to share data transparently and tamper-proof.

The framework is shown in Fig. 1. This operation framework can provide a suitable stability to the concept.

4 Implications of the Framework

4.1 Scientific Implications

a. Diversity of Data Acquisition: The diverse demographics of citizen scientists (i.e., the age, health status, and genetics of the recruited citizen scientists) can help reveal how varying populations adapt to space, helping to mitigate the gaps in astronaut-centered studies.
b. Super-Charge Inquiries: Regular tourist flights allow researchers to quickly iterate in their experiments, testing different materials for Mars habitats or researching cosmic radiation shielding.
c. Citizen-Led Discoveries: Since amateurs may notice or perceive unexpected phenomena, citizen scientists have previously discovered “Steve,” a unique type of atmospheric light (Bonney et al. 2014).
d. If not correctly validated and processed, results from untrained participants may lead to misinterpretation, making data accuracy challenging.

4.2 Social and Educational Transformation

a. Democratizing Space Exploration: Given that the concept of space should not be limited to just NASA, the framework encourages citizen involvement in space exploration that may lead to more opportunities for everyone rather than the elitism seen within traditional programs and governing bodies. Missions such as SpaceX’s Inspiration4, which featured a science teacher, help inspire underrepresented populations to pursue STEM professions.
b. Public Engagement: Tourists serve as ambassadors, sharing their experiences through social media or school visits. The participant, for instance, documenting crystal growth in microgravity can inspire a student’s love for chemistry.
c. Global Collaboration: Crowdsourced initiatives—such as mapping the lunar surface based on tourists’ collected images—can promote collaboration among nations on scientific challenges.

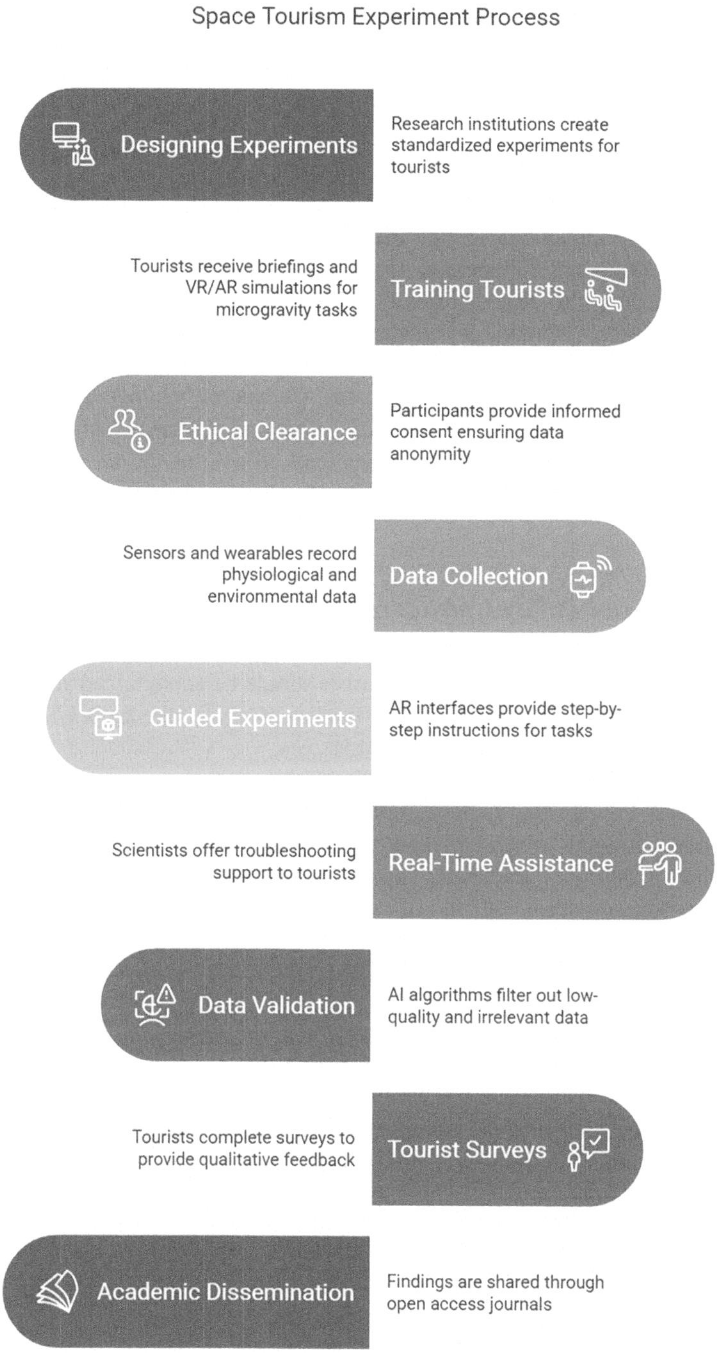

Fig. 1 Space tourism experiment process (Source: Authors' conception)

d. Challenge: High ticket prices could exclude poor individuals and maintain inequities unless subsidized programs are available.

4.3 Economic Implications

a. New Revenue Streams: Research institutions may offset the costs of their missions by selling experiment slots to corporations (e.g., pharmaceutical companies that want to test drug stability in microgravity).
b. Creation of roles: Demand for citizen science coordinator roles, AR developers, data analysis, etc.
c. New Market Segment: "Research tourism" packages could appeal to customers interested in travel with a purpose, setting companies apart in a crowded space.
d. Challenge: Driven by profits, priorities may crowd out the science, requiring contractual protections.

4.4 Moral and Policy Considerations

a. Privacy Issue: Biometric data from tourists should be anonymized and protected from misuse. Data such as regulatory frameworks (e.g., the EU's GDPR) could be used as blueprints (Resnik et al. 2015).
b. Environmental Impact: More launches mean more carbon emissions and more space junk. Policies that require, for example, sustainable fuels and debris mitigation technologies are essential (Weinzierl and Sarang 2021).
c. Informed Consent: Participants should be aware of risks, notably the poorly understood long-term effects of space radiation on health (Afshinnekoo et al. 2020).

4.5 The Future: An Open Space Economy

The framework establishes the foundation for future crowdsourced and participatory space research. Possible outcomes include:

a. Decentralized Research Networks: Tourists provide millions of data for earlier usable Global databases, allowing for planetary-scale studies (e.g., the study of climate change from space).
b. Space Colonization: Human adaptability is key in designing habitats on Mars or bases on the Moon, where citizen scientists can test the technologies on-site.

c. Policy Innovations: International treaties could dedicate resources to citizen science initiatives, similar to the relevant Earth science projects coordinated by UNESCO.

4.6 Risks of Using LLMs and Mitigation Strategies

a. Data Quality:

Solution: Hybrid validation systems (human + AI) and model regression tests.

b. Tourist Motivation:

Solution: Tiered participation models (e.g., "casual" vs. "dedicated" research tiers) and incentives for contribution like badges of (mission) honor or proprietary co-authorship on research papers.

c. Environmental Sustainability:

Solution: Make an investment in green propulsion technologies and give debris removal mandates.

d. Equity:

Solution: Create scholarships or lottery systems for low-income participants at minority run festivals funded by tourism revenues.

Therefore, we present a framework where citizen scientists play an important role in those explorations—tangentially between tourist and researcher. Solving technical, ethical, and economic challenges will enable an unprecedented discovery opportunity and increase public engagement. Moreover, the implications go well beyond science: This could redefine humanity's relationship with space, turning it from a privilege into a collective enterprise. This ecosystem of tools, both for public and private sectors, is maturing—and as it matures, the vision of an inclusive and participatory space economy comes one step closer to realization.

5 Conclusion

The intersection of space tourism and citizen science is an innovative frontier in humanity's exploration of the cosmos, combining the investment of space travel with the collection of discovery. This combination can democratize space exploration, facilitate scientific advancement, and cultivate a generation of citizen-scientists passionate about our planet and the universe beyond, as we usher humanity into the twenty-first century. Including non-professional astronauts in experiments, including physiological, microgravity, and observational studies, presents unique opportunities to embed diverse datasets that are difficult to reproduce on traditional astronaut missions. These contributions will prove crucial in tackling very real challenges, whether it's learning about human adaptation to space habitats or testing potential technologies for colonization efforts.

Transparency in space tourism policies and initiatives for equitable access to space tourism is crucial in addressing these ethical considerations. Of course, the cost of commercial spaceflight is one of its significant challenges. It risks excluding marginalized groups from opportunities to travel to low-Earth orbit, deep space, or beyond, potentially reinforcing existing inequities. Being alacrity on this front requires joint endeavors like subsidizing research programs, public-private cooperatives, or lottery-based opportunities to guarantee that space science benefits from a widescale demographic spread.

This model has profound social consequences. For example, citizen scientists in space can be ambassadors for STEM education, which helps bridge the gap between abstract scientific concepts and public engagement. Their stories, told via social media or educational outreach, can spur curiosity and shatter stereotypes that space exploration is only for the elite. In addition, should "research tourism" prove economically lucrative, it could incentivize innovation and job creation in industries such as biotechnology, aerospace engineering, and data analytics and strengthen international partnerships.

With the changing industry, key players must focus on sustainability and share commercial interests with environmental consciousness. Necessary investments into green propulsion systems, debris mitigation technologies, and ethical frameworks will also be needed to ensure that space exploration evolves without contaminating Earth's or space's ecosystems. And so, the convergence of space tourism and scientific exploration represents not just leisure pursuits—they reflect how a sense of curiosity can coexist and be propelled forward by commerce, leading us toward a time when space is more than just the playground of the entropy-rich—it is a halo around everything. The bottom line is that in pursuing inclusivity, rigor, and shared purpose, humanity will be able to realize the full potential of any exploratory enterprise by tapping into two-thirds of the world's population and ensuring that whatever we achieve in the final frontier will be a shining testament to our shared humanity.

References

Afshinnekoo E, Scott RT, MacKay MJ, Pariset E, Cekanaviciute E, Barker R, Gilroy S, Hassane D, Smith SM, Zwart SR, Nelman-Gonzalez M, Crucian BE, Ponomarev SA, Orlov OI, Shiba D, Muratani M, Yamamoto M, Richards SE, Vaishampayan PA et al (2020) Fundamental biological features of spaceflight: advancing the field to enable deep-space exploration. Cell 183(5):1162–1184. https://doi.org/10.1016/j.cell.2020.10.050

Bonney R, Shirk JL, Phillips TB, Wiggins A, Ballard HL, Miller-Rushing AJ, Parrish JK (2014) Next steps for citizen science. Science 343(6178):1436–1437. https://doi.org/10.1126/science.1251554

Crawford IA (2016) The long-term scientific benefits of a space economy. Space Policy 37:58–61. https://doi.org/10.1016/j.spacepol.2016.07.003
Das S (2023a) Modern-day gaming and the role of FinTech applications in popularising various entertainment platforms across the world. In: Das S, Gochhait S (eds) Digital entertainment as next evolution in service sector. Palgrave Macmillan, Singapore. https://doi.org/10.1007/978-981-19-8121-0_11
Das S (2023b) Digital entertainment: the next evolution in service sector. Emerging Digital Solutions in Reshaping Different Industries Palgrave-Singapore. https://doi.org/10.1007/978-981-19-8121-0
Das S, Mondal SR, Struck F, Eissfeldt J, Beernink L, Sigouras V, Hoose S, Kopac L (2023) What can we see inside the crystal ball? Exploring opportunities and challenges for metaverse and circular economy. AJIS 12(6):221. https://doi.org/10.36941/ajis-2023-0165
Di Virgilio F, Das S (2023a) Digital solutions for social media: role of digital evolution on business enterprises. In: Das S, Gochhait S (eds) Digital entertainment as next evolution in service sector. Palgrave Macmillan, Singapore. https://doi.org/10.1007/978-981-19-8121-0_9
Di Virgilio F, Das S (2023b) Digitalization of enterprises from a new normal sustainable development perspective: opportunities and challenges. In: Mondal SR, Yegen C, Das S (eds) New normal in digital enterprises. Palgrave Macmillan, Singapore. https://doi.org/10.1007/978-981-19-8618-5_9
Ehrenfreund P, Peter N (2009) Toward a paradigm shift in managing future global space exploration endeavors. Space Policy 25(4):244–256. https://doi.org/10.1016/j.spacepol.2009.09.004
Eriksson J, Newlove-Eriksson LM (2023) Outsourcing the American space dream: SpaceX and the race to the stars. Astropolitics 21(1):46–62. https://doi.org/10.1080/14777622.2023.2196017
Ioannou A, Tussyadiah I, Lu Y (2020) Privacy concerns and disclosure of biometric and behavioral data for travel. Int J Inf Manag 54:102122. https://doi.org/10.1016/j.ijinfomgt.2020.102122
Lappa M (2004) Space research. In: Elsevier eBooks, pp 1–37. https://doi.org/10.1016/b978-008044508-3/50002-5
Majerova J, Das S (2023) Sustainable development and digitalization in the supply chain for environmental protection: changes and trends. In: Mondal SR, Yegen C, Das S (eds) New normal in digital enterprises. Palgrave Macmillan, Singapore. https://doi.org/10.1007/978-981-19-8618-5_8
Minh-Nhat HO, Majerova J, Das S (2022a) Taking steps to help the phoenix rise from ashes: a roadmap for digital enterprises to develop new sustainable goals. In: Sustainable development and innovation of digital enterprises for living with COVID-19. Springer, Singapore, pp 35–51. https://doi.org/10.1007/978-981-19-2173-5_3
Minh-Nhat HO, Nguyen HL, Mondal SR (2022b) Digital transformation for new sustainable goals with human element for digital service enterprises: an exploration of factors. In: Sustainable development and innovation of digital enterprises for living with COVID-19. Springer, Singapore, pp 85–103. https://doi.org/10.1007/978-981-19-2173-5_6
Mondal SR, Das S (2023a) Defining the new digitalized normal: implications of A.I., IoT, and blockchain for business growth. In: Mondal SR, Yegen C, Das S (eds) New normal in digital enterprises. Palgrave Macmillan, Singapore. https://doi.org/10.1007/978-981-19-8618-5_10
Mondal SR, Das S (2023b) Digital solution for industry 4.0 and 5.0 and social media. In: Das S, Gochhait S (eds) Digital entertainment as next evolution in service sector. Palgrave Macmillan, Singapore. https://doi.org/10.1007/978-981-19-8121-0_10
Mondal SR, Das S (2023c) Human-centric digitalization for new sustainable development: an explorative study in human resource management. In: Mondal SR, Yegen C, Das S (eds) New normal in digital enterprises. Palgrave Macmillan, Singapore. https://doi.org/10.1007/978-981-19-8618-5_12
Mondal S, Das S, Vrana VG (2023a) How to bell the cat? A theoretical review of generative artificial intelligence towards digital disruption in all walks of life. Technologies 11(2):44. https://doi.org/10.3390/technologies11020044

Mondal SR, Yegen C, Das S (2023b) New normal in digital enterprises – strategies for sustainable development. Palgrave-Singapore. https://doi.org/10.1007/978-981-19-8618-5 and Hardcover ISBN 978-981-19-8617-8

Ngoc-Vinh N, Tien-Dung P, Das S (2022) Digitization of business; Need for recombination during COVID-19. In: Sustainable development and innovation of digital enterprises for living with COVID-19. Springer, Singapore, pp 1–17. https://doi.org/10.1007/978-981-19-2173-5_1

Nguyen TDM, Mondal SR, Das S (2022) Digital Entrepreneurial Transformation (DET) Powered by New Normal Sustainable Developmental Goals (n-SDGs): Elixir for growth of country's economy. In: Sustainable development and innovation of digital enterprises for living with COVID-19. Springer, Singapore, pp 69–84. https://doi.org/10.1007/978-981-19-2173-5_5

Ong J, Lee AG (2022) An introduction to space medicine and the physiological effects of spaceflight on the human body. In: Elsevier eBooks, pp 1–7. https://doi.org/10.1016/b978-0-323-91524-3.00007-7

Polkowska M (2021) Space tourism challenges. Rev Eur Compar Law 45(2):153–182. https://doi.org/10.31743/recl.12223

Resnik DB, Elliott KC, Miller AK (2015) A framework for addressing ethical issues in citizen science. Environ Sci Pol 54:475–481. https://doi.org/10.1016/j.envsci.2015.05.008

Steere T (2021, July 11) Virgin Galactic successfully completes first fully crewed spaceflight with Sir Richard Branson | Virgin. Virgin.com. https://www.virgin.com/about-virgin/latest/virgin-galactic-successfully-completes-first-fully-crewed-spaceflight

Tandon S, Das S (2023) Digital revolution complements sustainable goal achievement during and after pandemic: a research agenda. In: Mondal SR, Yegen C, Das S (eds) New normal in digital enterprises. Palgrave Macmillan, Singapore. https://doi.org/10.1007/978-981-19-8618-5_11

Tien-Dung PHAM, Majerova J, Das S (2022a) Phases of possible recovery of digital enterprises in new normal business for living with COVID-19 times: opportunities for a new era in sustainable development goals. In: Sustainable development and innovation of digital enterprises for living with COVID-19. Springer, Singapore, pp 19–33. https://doi.org/10.1007/978-981-19-2173-5_2

Tien-Dung P, Nguyen XD, Das S (2022b) Sustainable goal achievement by digital revolution during and after pandemic; How much one wins and losses: a bird's eye view for future planning. In: Sustainable development and innovation of digital enterprises for living with COVID-19. Springer, Singapore, pp 133–147. https://doi.org/10.1007/978-981-19-2173-5_8

Vrana V, Das S (2023a) Dynamic restructuring of digital media and entertainment sector: role of urbanization, industrial innovation, and technological evolution. In: Das S, Gochhait S (eds) Digital entertainment as next evolution in service sector. Palgrave Macmillan, Singapore. https://doi.org/10.1007/978-981-19-8121-0_2

Vrana V, Das S (2023b) Role of enterprises and their digital transformation in post-pandemic sustainable development. In: Mondal SR, Yegen C, Das S (eds) New normal in digital enterprises. Palgrave Macmillan, Singapore. https://doi.org/10.1007/978-981-19-8618-5_7

Weinzierl M, Sarang M (2021, February 12) The commercial space age is here. Harv Bus Rev. https://hbr.org/2021/02/the-commercial-space-age-is-here

Yegen C, Das S (2023) Digital solution for print and publishing media. In: Das S, Gochhait S (eds) Digital entertainment as next evolution in service sector. Palgrave Macmillan, Singapore. https://doi.org/10.1007/978-981-19-8121-0_12

Erich Schlesinger Scholar at Wyższa Szkoła Biznesu, specializing in international business strategies and market expansion. His research addresses challenges in cross-cultural negotiation and global supply chains. He collaborates on programs enhancing global business education.

Subhankar Das Senior Researcher and academician at Duy Tan University, publishing extensively on AI, blockchain, and digital transformation. Editor of seven Scopus-indexed books and reviewer for top journals. Mentors global PhD candidates and drives innovation in ASEAN academic collaborations.

Cosmic Currencies: Financing Humanity's Leap to the Stars

Dagmar Caganova and Subhankar Das

Abstract The increasing commercialization of space exploration has led to evolving creative financing solutions to fund humanity's loftier pursuits in the cosmos. This chapter is about transitioning from government space programs to a privatized economic frontier, with SpaceX and Blue Origin leading. Traditional financing models increasingly do not work for high-risk, long-term space ventures, giving rise to new instruments like space bonds, asteroid mining futures and blockchain "cosmic currencies." These tools will allow you to democratize ownership and access to space assets, use smart contracts to enable decentralized governance, and mitigate risk through an iterative insurance marketplace. However, the development of space-based economies still faces challenges, including regulatory ambiguities over ownership and exploitation in international space law; ethical dilemmas around the space resource exploitation; and cybersecurity vulnerabilities associated with interplanetary transactions. Global treaties and partnerships between the public and private sectors are critical to marry profit motives with sustainable stewardship of outer space resources. This strongly argues that space finance is not solely the purview of off-world colonization. Here is the potential, supported by technological innovation, for sustainable, equitable economic paradigms on Earth instead of negative bifurcation taking us cataclysm.

Keywords Space finance · Cosmic currencies · Bet on the future · Asteroid mining · Public-private partnerships · Sustainable space economy

D. Caganova
NEWTON University, Praha, Czechia
e-mail: dagmar.caganova@newton.university

S. Das (✉)
Department of Marketing, South Star Management Institute, Duy Tan University, Da Nang, Vietnam
e-mail: subhankardas@duytan.edu.vn

© The Author(s), under exclusive license to Springer Nature Singapore Pte Ltd. 2025

S. R. Mondal et al. (eds.), *Pioneering the New Space Economy through AI and Immersive Technologies*, https://doi.org/10.1007/978-981-96-5977-7_14

1 Introduction

Humanity enters a new age, where the final frontier is no longer the province of government space agencies but of an expanding marketplace for commercial creativity. As exploitation of the cosmos transitions from a Cold War-style geopolitical competition to a twenty-first-century economic frontier, the financial architecture that underpins this leap into space is being remade in essential ways. This chapter explores the new developments in finance created to fund humanity's venture into outer space, informing a new set of value, risk, and economic exchange in an interplanetary setting. In all these developments, blockchain-powered transaction systems and speculative markets for asteroid minerals have to do not only with financing rockets and satellites, they also lay the foundations for a new economic paradigm (Diamandis and Kotler 2020).

The pace of space commercialization—sparked by private players such as SpaceX, Blue Origin, Planetary Resources—has outstripped Earth-based financial models. Classic venture capital, bank loans, and public funding structures do not fit the unique risks, timeframes, and rocket budgets that space projects require, typically decades up front and years of expressive uncertainty (Boley and Byers 2021). An array of unorthodox financial instruments has been created to fill the gap. For example, space bonds enable public and private actors to fundraise for future revenue streams from satellite constellations or lunar infrastructure projects. At the same time, derivative markets are forming around the anticipated barter of rare metals mined from celestial bodies, such as asteroid mining futures, which have found shallow purchase with the institutional investor community. However, the ethical and regulatory questions surrounding this are ambiguous. Orbital property beliefs are still theoretical but promise to democratize ownership of space-based properties, e.g., communication satellite slots, or solar power stations (Weinzierl and Sarang 2021).

We have made immense leaps in technology and innovation, where a blockchain-based system can provide a decentralized and immutable solution with features to manage innumerable transactions across any vast interplanetary distance. Coins like Bitcoin and Ethereum, which were initially conceived for use on Earth—are being repurposed as "cosmic currencies" to conduct trade between Mars colonies and lunar outposts, while also exchanging goods and services on Earth (Waseem et al. 2023).

For example, smart contracts could run mechanisms to manage resource allocation for deep-space exploration/missions, and decentralized autonomous organizations (DAOs) could govern off-world settlements without dependence on terrestrial institutions (Ducrée et al. 2021). These systems will promise to ameliorate interstellar commerce's latency and trust issues but raise questions about regulatory jurisdiction and cybersecurity in a multi-planetary economy.

At the same time, space-based assets as collateral are transforming earthly finance. AstroForge and OffWorld, for example, are developing models where loans are collateralized by future mineral rights on asteroids or cash flows from orbital solar power stations (Hein et al. 2019). These arrangements blur the distinction between speculative finance and industrial policy, making radical risk-assessment frameworks necessary for lenders. Insurance markets are also adapting: Lloyd's of London now provides policies on satellite launch failures and collisions with space debris, with premiums priced according to the probabilistic models of astrophysicists.

This financial metamorphosis hinges on governments. For example, the US Space Resource Exploration and Utilization Act of 2015 provides tax incentives to private investors that allow companies to retain ownership of space resources they extract (Hertzfeld and Von Der Dunk 2005). Agency and contractor share not just costs but also intellectual property in public-private partnerships (NASA's Commercial Lunar Payload Services (CLPS) program is an example) (NASA 2025). But the absence of coordinated international agreements risks splintering this nascent economy. Although the 1967 Outer Space Treaty prohibits national appropriation of celestial bodies, it says nothing about corporate ownership, creating a legal gray zone that is being taken advantage of by companies registered in lax jurisdictions such as Luxembourg and the United Arab Emirates (Jakhu and Pelton 2017).

Then, as space tourism matures and settlements on the Moon and Mars begin to inch toward reality, the possibility of actual extraterrestrial economies could be born. Lunar colonies might introduce local currencies tied to energy output or life-support systems, and Martian societies could develop barter systems for scarce resources (Schrijver et al. 2015). As all this unfolds, that will spew back into Earth's financial systems, inspiring something other than debt-based fiat currencies or GDP-centric growth models.

Cosmic Currencies contends that not just transactional instruments but engines of new economic relationships on a cosmological scale, the financialization products from humanity's spacefaring ambitions compel us to reappropriate our planetary economies. Through an analysis of the relationships between technology, policy, and markets, this chapter demonstrates how searching for ways to finance the interstellar future reshapes capitalism on Earth and beyond.

2 Literature Review: Elements Influencing Cosmic Currencies

The emerging field of space finance is a significant evolution in how humans fund our travel and endeavors beyond Earth. As both government and private interests increase their bid to explore and commercialize space, scholars are beginning to evaluate the emergent sector's economic, technological, legal, and ethical parameters. This review brings together the current literature on emerging "cosmic currencies" and the challenges and drivers for novel financial vehicles for space-themed businesses.

2.1 Transition from Public Sector to Private Sector

For most of its history, space exploration was the domain of national governments, funded by taxpayers and through geopolitical antagonism. However, this model has been upended by private firms such as SpaceX, Blue Origin, and Rocket Lab, which have offloaded costs onto venture capital, equity markets, and corporate investment. Private sector agility and profit-driven innovation has dropped launch costs by over 90% since 2010 (Weinzierl and Sarang 2021), enabling the rise of new business models like satellite mega-constellations and space tourism. By doing so, they argue, under this privatized reality of little to no regulatory oversight, orbital debris could become a crisis point and/or space resources could become monopolized by the wealthy few (Boley and Byers 2021). The dynamic tension between innovation and governance has long been a central theme in space finance literature.

2.2 New Ways to Fund Space Projects

Established financing channels, like bank loans and public grants, are poorly suited to space projects' high-risk, long-term timeframes. In response, new investment vehicles are on the rise. For example, Link et al. (2023) assess the viability of space bonds, debt instruments linked to future revenue flows from satellite broadband services or lunar mining businesses. Just as asteroid mining futures, derivatives contracts that speculate on the economic value of platinum-group metals that could be extracted from celestial bodies have attracted speculation, so have the markets for rights to asteroid extraction.

The other frontier is investment trusts for orbital real estate (REITs). Theoretical models of REITs suggest that they would democratize access to lucrative space-based assets, such as geostationary satellite slots or solar power stations (Weinzierl and Sarang 2021). Yet, according to Boley and Byers (2021), the absence of property rights in space, as articulated in the 1967 Outer Space Treaty, makes estimating the value of such assets more complex.

3 Interplanetary Trade Using Blockchain and Cryptocurrencies

Hence, blockchain technology, with its capabilities of decentralization and security, can also be seen as a backbone for interplanetary economies Arseneau et al. (2022). Cryptocurrencies allow transactions to be processed without waiting for confirmation on the ground, which can take several minutes for off-world colonies. Smart contracts, self-executing contracts coded into a blockchain, may streamline resource distribution for Martian habitats or lunar farms for oxygen (Waseem et al. 2023).

Nonetheless, significant hurdles remain. Weinzierl and Sarang (2021) highlight the regulatory space concern over extraterrestrial cryptocurrency usage, especially in places without robust space legislation. Furthermore, the high energy consumption associated with blockchain networks is at odds with the limited resources available in space environments, making alternative consensus mechanisms such as proof-of-stake essential (Link et al. 2023).

3.1 Insurance and Risk Management in the Space Economy

Space initiatives present hazards, including launch failures, radiation injuries and orbital collisions. Several sectors have formed from this, the most historic of which was space insurance markets that began with new offering coverage for satellite malfunction and debris mitigation. These products use probabilistic models that astrophysicists and actuarial scientists have devised; however, experts are far from agreed on how to price low-probability, high-impact events (NASA 2025).

Tying things up with space-based collateral is also introducing new complexities. Researchers illustrate how such loans set the stage for lenders to be forced into radical new frameworks for assessing risk, as they would involve loans collateralized by asteroid mineral rights or lunar lander patents (Waseem et al. 2023). These models must include factors such as extraction possibility, commodity price volatility, and geopolitical quarrels about ownership of resources (Yegen and Das 2023).

3.2 International Space Law and Government Policy

Regarding space finance, national legislation and international treaties most heavily influence our actions (Vrana and Das 2023b). The US Commercial Space Launch Competitiveness Act of 2015 has stimulated private investment in asteroid mining, giving companies ownership of resources removed from space (Schrijver et al. 2015). Likewise, NASA's Commercial Lunar Payload Services (CLPS) program highlights public-private partnerships that share costs and intellectual property (NASA 2025).

Yet the lack of a coherent international regulatory framework threatens to fragment the space economy. Jakhu and Pelton (2017) critique the loopholes embedded in the Outer Space Treaty, which bans national land claims and is silent on corporate ownership. This allows firms to exploit legal ambiguities in permissive jurisdictions such as Luxembourg and the UAE. This patchwork of regulation and legislation creates the potential for a "Wild West" landscape regarding space commerce.

The Problem of Alien Monetary Systems

The world on the Moon and Mars is becoming a permanent settlement, and soon, the demand for localized currency will be sought. According to Hein et al. (2019), lunar colonies may also have some currency pegged to energy output (e.g., kilowatt-hours), and Martian societies may invent barter systems between essential resources (e.g., water ice) (Vrana and Das 2023a). Such systems question earth-centric economic paradigms, including fiat currencies and GDP-based growth models (Boley and Byers 2021).

Scholars argue whether alien-based economies will grow independently of Earth or remain in its thrall (Tien-Dung et al. 2022b). Ducrée et al. (2021) propose a hybrid model with blockchain-based tokens as a medium of exchange between planets, cautioning that currency manipulation is a possible consequence of politically fragmented space colonies.

3.3 Ethical and Sustainability Consideration

The commodification of Space raises profound ethical questions. Boley and Byers (2021) cautions against the unrestrained extraction of extraterrestrial resources and urges a commitment to principles of "cosmic sustainability" if we hope to "preserve the wilderness of our solar system." Likewise, Boley and Byers (2021) lampoon satellite mega-constellations for polluting Earth's orbit and hampering astronomical inquiry.

On Earth, space finance will only increase inequality. Hein et al. (2019) warn that cosmic currencies could lead to the rise of a new elite class of "space capitalists," taking resources and attention away from pressing terrestrial issues like climate change. In contrast, appropriately, Diamandis and Kotler (2020) argue that space commercialization will create "abundance multipliers," giving humanity access to vast technological spin-offs and contributing to humanity's off-world resource acquisition.

The literature highlights a three-way dynamic interplay between innovation and regulation in the space finance literature (Tien-Dung et al. 2022a). Instead, they merge legal ambivalence with new risk and ethical challenges to promote honest financial and technological innovation while avoiding the potential traps of new instruments, such as space bonds or blockchain systems (Nguyen et al. 2022). Across both worlds, essential cooperation between disciplines—economics, astrophysics, international law, and others must be a priority in future research to ensure that cosmic currencies translate into equality and sustainability (Tandon and Das 2023).

4 Framework of Application: Harnessing Space Finance Drivers for Economic Development

The sky is not the limit for cosmic currencies and space finance mechanisms for groundbreaking opportunities for alien endeavors and planetary economies (Mondal and Das 2023c). Based on the earlier analysis from the literature review, this framework describes actionable routes to achieving and deploying these innovations for meaningful economic benefit (Mondal et al. 2023b). The framework is broken down into seven pillars, each associated with a key element of space finance and being mindful of amplification, inclusivity, and sustainability (Ngoc-Vinh et al. 2022).

4.1 Infrastructure Development Through Public Private Partnerships (PPP)

The motivation: Space exploration has transitioned from government-led to private-sector-driven, necessitating collaborative models to share risks, costs, and expertise. Figure 1 explains the strategic framework.

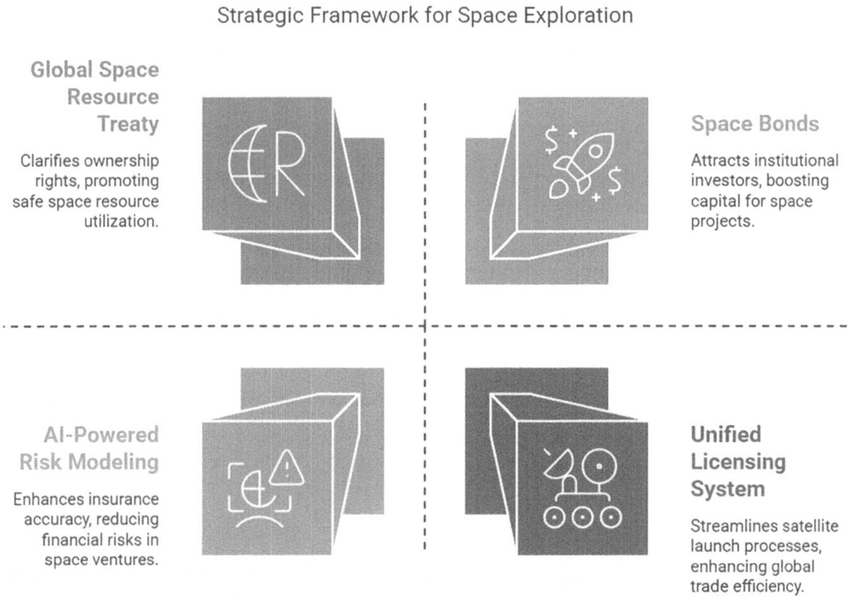

Fig. 1 Strategic framework for exploration of space (Source: Authors' conception)

Strategies

Co-Fi Models: Governments could match the money the private sector invests in high-risk projects (lunar habitats, asteroid mining prototypes, etc.) The template is NASA's Commercial Lunar Payload Services (CLPS) program, which shares expenses with private companies.

Shared Public Domain IP: PPPs could create IP-sharing agreements, in which public agencies retain rights to foundational technologies (e.g., life-support systems). These technologies could then be spun into the commercial sphere to private firms (Mondal and Das 2023b), who could exploit them for terrestrial applications (e.g., radiation-resistant materials).

Regulatory Sandboxes: Establish specialized areas for startups to trial new financial products (like space bonds) in a more lenient regulatory environment, driving quicker innovation with stakeholder protection.

Economic Benefits:

a. Information: Job Creation: Infrastructure projects (spaceports, lunar bases) will produce high-skilled jobs (engineering, robotics, AI) (Mondal et al. 2023a).
b. Technological Spillovers: Advances in space technologies (e.g., miniaturized satellites) usually positively impact ground-silo industries such as telecommunications and agriculture (Mondal and Das 2023a).

4.2 Mobilizing Capital Through Diverse Financial Instruments

Rationale: Long timelines and high risks of space ventures are a poor fit for traditional financing.

Strategies:

Space Bonds: Governments and corporations can issue bonds—also known as debt instruments—tied to revenue-generating projects (e.g., satellite broadband networks), allowing them to offer fixed returns to pension funds and other institutional investors.

Asteroid Mining Futures—Establish regulated futures markets for space resources (e.g., platinum, and water ice), facilitating price discovery and hedging against extraction uncertainties.

Orbital REITs are real Estate Investment Trusts that lease orbital locations for satellites or lunar land, providing retail investors with ownership rights to multidimensional property.

Economic Benefits:

a. Injection of Liquidity: Trillions of dormant capital within risk-seeking investors could be released into new markets (Minh-Nhat et al. 2022a).
b. Driving Ancillary Growth: Legal, auditing, and brokerage services will grow alongside these instruments (Minh-Nhat et al. 2022b).

4.3 Interplanetary Era Blockchain Infrastructure

Rationale: Decentralized systems are the backbone for secure, latency-tolerant long-distance transactions.

Strategies:

Cosmic Cryptocurrencies: Create energy-efficient, blockchain-based currencies (e.g., "SolarCoin" linked to space solar power output) to aid trade between Earth and colonies.

Countdown to the Future: The future could involve smart contracts for resource management. This means we could automate agreements on resource allocation (like Mars water rights) through self-executing smart contracts, thereby eliminating bureaucratic overhead.

DAOs: Govern off-world settlements using DAOs for participatory decision-making without earth-based beneficiaries.

Economic Benefits:

a. Lower Transaction Fees: Since no intermediaries (e.g., banks) exist, fees are eliminated for cross-planetary trade.
b. New Business Models: Tokenized assets (e.g., shares in a lunar mining DAO) have the potential to democratize wealth generation.

4.4 Insurance Ecosystems and Risk Mitigation

Why Well Spacings: Space risks are unique and need customized solutions.

Strategies:

Space Insurance Pools: Form global consortiums to underwrite risks such as launch failures or solar flares, breaking lenticular among multiple insurers.

Collateralized loans: These loans enable companies to borrow money against space-based assets (e.g., asteroid mineral rights) based on backed guarantees to protect lenders.

AI-Powered Risk Modeling: Use machine learning to forecast and price risks (often using real-time data) to improve accuracy in insurance underwriting.

Economic Benefits:

a. Binomial According to NCTA, subscription-based model climate change.
b. Stable Premium Markets: By 2030, space insurance could be a $1 billion sector, growing 10 times faster than aviation insurance.

4.5 Unified Legal and Regulatory Frameworks

Reasoning: Laws are fractured, which stifle investment and innovation.

Strategies:

Global Space Resource Treaty: Revise the Outer Space Treaty to clarify what corporations can own and profit from while still prohibiting territorial claims.

Tax Holiday: Provide a tax holiday for space businesses, such as zero corporate taxes the first 10 years after commencing asteroid mining operations.

Unified Licensing System: Establish a global authority for satellite launch and resource extraction licensing, streamlining national standards to lower compliance costs.

Economic Benefits:

a. Legal Certainty: Clear regulations lower litigation risks, incentivizing long-term investments (Di Virgilio and Das 2023b).
b. Edge over Competitors: Countries with forward-looking legislation (like Luxembourg) can become the centre of space financing (Majerova and Das 2023).

4.6 Extra-terrestrial Monetary System

Rationale: Localized currencies will be required for earthly colonies to be self-sustaining.

Strategies:

Commodity-Linked Currencies: Link lunar or Martian currencies to vital resources (e.g., 1 "Mars Credit" = 100 liters of purified water).

Hybrid Exchange Systems: Create dual-currency ecosystems in which cosmic currencies float with Earth fiat currencies in exchanges governed by DAOs.

Universal Basic Assets (UBA): Please distribute shares for space resource ventures to all people of Earth.

Economic Benefits:

a. Economic Resilience: Local currencies buffer against Earth's economic fluctuations (Das et al. 2023).
b. Equitable Growth: Citizens could have a stake in space economies and thus narrow wealth disparities (Di Virgilio and Das 2023a).

4.7 Humanity Mandates for Ethical and Sustainable Growth

This raises a risk of unchecked exploitation with environmental and social aftermaths.

Strategies:

Standards for Cosmic Sustainability: Declare that 20% of space is to be left as untouched wilderness, preserving scientific and ecological value.

Ethical Mining Certifications: Certify space resources "conflict-free" if their extraction adheres to labor rules and environmental guidelines.

Space Dividend Fund: Use of a portion of profits (from space mining) for sustainability on Earth (e.g., renewable energy projects).

Economic Benefits:

a. Sustainable practices provide resources for the long haul (Das 2023a).
b. Brand Value: Ethical companies continue to expand their market share (Das 2023b).

4.8 Implementation Roadmap

a. Phase 1 (2024–2030): Pilot PPPs and space bonds; create regulatory sandboxes.
b. Phase 2 (2031–2040): Deploy cosmic cryptocurrencies and asteroid futures markets; formalize global space treaties.
c. Phase 3 (2041 onwards): Expand extraterrestrial monetary systems and ethical frameworks to support scale as colonies mature.

In this framework, space finance catalyzes to dual-planetary prosperity. If we base financial innovation on ethical governance and inclusive policies, humanity will experience phenomenal economic growth and can keep the stars as a common.

5 Implications of the Framework: Theoretical, Practical, and Sustainable

Therefore, the framework introduced for harnessing space finance to catalyze economic growth is deeply significant in a theoretical, practical, and sustainable context. So, by marrying financial innovation with ethical governance and interdisciplinary collaboration, this model fuels humanity's extraterrestrial aims and presents transformative lessons for our terrestrial economies. Below, we discuss these implications in further detail.

5.1 Theoretical Implications: Reconstructing Value and How We Exchange

This framework has questioned conventional economic frameworks through the emergence of resource-backed cosmic currencies and decentralized finance systems. Classical value theories (e.g., Adam Smith's labor theory) lack value applicability to space economies, especially since such economies may have resources such as water ice or solar energy that will act as primary value anchors. This aligns with modern commodity money theory, whereby the intrinsic utility of a good determines its worth, albeit with a focus on extraterrestrial contexts that may or may not come into play. Blockchain enabled governance structures for extraterrestrial habitats (Liu et al. 2022) may be seen as yet another threat to institutional economic structures and will also use DAO-like organizations as a replacement for established jurisdictions (Hertzfeld and Von Der Dunk 2005). These innovations demand new theoretical models incorporating value generation, liquidity, and trust in a world without an Earth-like legal framework or a command-and-control based political paradigm.

Risk and Capital Allocation

It includes new risk variables, such as radiation exposure and interplanetary supply chain disruptions, not all of which existing portfolio theory and models (capital asset pricing models or CAPM) are fully ready to address. Moreover, the framework's focus on AI-powered risk modeling and insurance pools adds a new dimension to the latest iteration of behavioral finance, merging astronomical science with actuarial science. This hybrid method may provide insights for terrestrial industries confronting climate-induced threats (e.g., coastal property), where stochastic modelling needs to capture environmental uncertainties (Hein et al. 2019).

Evolution of Legal and Governance

The framework's advocacy for a Global Space Resource Treaty fills those international gaps, especially the ambiguity in the 1967 Outer Space Treaty regarding companies owning space resources. This is based on institutional economics theory, which states that clear property rights are critical for market efficiency (Jakhu and Pelton 2017). As models of shared governance (e.g., DAOs), the framework relates to the polycentric governance theory, positing that multiple overlapping institutions, rather than a single authority, are better equipped to deal with increasingly complex systems (such as interplanetary trade).

5.2 *Practical Implications: Accelerating Commercialization of Space*

Strategies suggested by the framework—like PPPs co-funding lunar infrastructure or orbital REITs—provide actionable pathways to marshal capital. For example, NASA's CLPS program has already shown that public-private cost-sharing can work and has allowed for $ 2.6 billion in private lunar investments since 2018. Similarly, blockchain-based "SolarCoins" might simplify transactions for space solar power projects, minimizing dependence on the grid on Earth. Nevertheless, practical challenges remain, such as technological bottlenecks (e.g., energy-efficient mining rigs for space) and regulatory fragmentation, with countries like the UAE and Luxembourg competing to enact the best space laws.

Promoting Economic Returns on Land

Innovations in space finance have tangible spillover benefits. Satellite technology miniaturized for cheap launches enables developments in IoT and precision agriculture. Similarly, AI tools designed to anticipate collisions with orbital debris could improve predictive maintenance for airplanes or ships.

Creation of Industry and Jobs

The framework predicts booming high-skilled jobs, from astro-engineers to space insurance underwriters. The global space economy is estimated to employ over 1.5 million workers by 2040, rivalling the renewable energy workforce of today (Weinzierl and Sarang 2021). Ancillary industries, like law firms specializing in space law or firms that audit asteroid mining ventures, will also grow, diversifying economic ecosystems.

5.3 Sustainable Implications: Guardianship of Nature Beyond Our Planet

The cosmic sustainability standards in the framework—like preserving 20% of celestial bodies as wilderness—speak to rising concerns surrounding things like space debris and resource depletion. Like earthly fair-trade labels, it certifies "conflict-free" space minerals and incentivizes ethical extraction methods. These measures echo the ethos of the planetary protection Policy to prevent space exploration from repeating the exploitation of the environment on Earth.

Fair and Confident Allocation of Resources

The system and the hybrid currency models aim to avoid a "space capitalist" aristocracy—they will democratize access to cosmic resources. For example, distributing shares entailed in asteroid mining enterprises to developing countries could offset historical resource imbalances and promote global economic equity. Peeking Mars dollars to life-support resources like water achieves just such an equilibrium—ensuring that the economy is geared towards survival and not towards speculation.

Sustainability By Innovation

The sustainable growth phase is included in the phased roadmap of the framework. Iterative checks of concepts like space bonds pass through early stages, regulatory sandboxes that dampen systemic risks. Energy-efficient blockchain protocols (e.g., proof-of-stake) reduce the carbon emissions of their related cosmic currencies, in reaction to criticisms of the environmental impact of crypto-mining (Link et al. 2023). On the other end of the spectrum, and instead, the Space Dividend Fund is using the profits of space companies to fund local sustainability projects: for a new paradigm of mutual profit between the vast profits of lunar or asteroid mining and the renewal of Earth's ecological state.

This is carefully embedded within a duality where space finance is not simply a paradigm but a hegemon transforming the global economy. In theory, it shatters entrenched systems of value and governance; in practice, it democratizes jobs, technologies, and markets; and in sustainability, it aligns profit with planetary care. Only by synthetically making these dimensions so will humanity prove that the leap to the stars uplifts both alien civilizations and our own—that economic advancement does not have to undermine equity or the integrity of nature.

6 Conclusion

How that solar-system-wide ecology makes cosmic currencies and funds humanity's expansion into the stars is not merely one of technology or economy, it's an unprecedented reconfiguration of how societies negotiate value, risk, and governance. We've seen how the financial instruments devised to encourage space exploration—from asteroid mining derivatives, to blockchain compatible currencies to enable interplanetary commerce—are more than rockets and satellites financing vehicles, they are an economy capable of rewiring our production systems on Earth and beyond. By addressing space's idiosyncratic problems —vast risk, interplanetary distance, fuzzy accountability in law—these breakthroughs give paradoxical lessons for terrestrial economies grappling with climate disruption, inequality and technological displacement.

This framework orders interdependence among governments and interdependence through the private business and international organizations. Public-private partnerships, harmonized regulations, and ethical mandates like the Space Dividend Fund are central to balancing profit motives with cosmic collective custodianship. These mechanisms ensure that the economic benefits of space commercialization—job creation, technology spin-offs, and resource availability—are distributed relatively widely throughout society rather than accruing to a handful of spacefaring countries or companies. The rise of decentralized systems—DAOs, resource-backed currencies—meanwhile threatens Earth's reliance on centralized institutions and provides models of resilience and inclusivity instead.

However, whether this cosmic economic vision manifests or not will hinge on whether governance will be pre-emptive in the future. Such practices of certifying ethical extraction, preserving celestial wilderness, and channeling space profit into our continuing terrestrial sustainability efforts should not simply be seen as nice ideas, but instead imperatives we cannot afford not to develop, lest we create orbited horizons that correspond to the cycles and epochs of exploitation that are devouring our home-planet. Ultimately, as we transform into a multiplanetary species, how we act in this moment will determine whether the stars become the next frontier of collective prosperity or a theater of great-power conflict over scarce resources.

Such cosmic currencies are an opportunity and a responsibility. Armed with the right financial technology, imbued with an ethic of foresight, humanity will rise to the stars not as conquerors but as custodians of a connected, sustainable future enriching all forms of life, at home and beyond in the universe. As it turns out, the final frontier is not only a destination but a culmination of the potential to break free from the chains of our past.

References

Arseneau B, Anderman D, Dominocielo G, Kuta M, Wiles K (2022) Texas private equity conference 2022: session III: space: the final investing frontier. J Appl Corp Financ 34(3):72–81. https://doi.org/10.1111/jacf.12518

Boley AC, Byers M (2021) Satellite mega-constellations create risks in Low Earth Orbit, the atmosphere and on Earth. Sci Rep 11(1). https://doi.org/10.1038/s41598-021-89909-7

Das S (2023a) Modern-day gaming and the role of FinTech applications in popularising various entertainment platforms across the world. In: Das S, Gochhait S (eds) Digital entertainment as next evolution in service sector. Palgrave Macmillan, Singapore. https://doi.org/10.1007/978-981-19-8121-0_11

Das S (2023b) Digital entertainment: the next evolution in service sector. Emerging digital solutions in reshaping different industries Palgrave-Singapore. https://doi.org/10.1007/978-981-19-8121-0

Das S, Mondal SR, Struck F, Eissfeldt J, Beernink L, Sigouras V, Hoose S, Kopac L (2023) What can we see inside the crystal ball? Exploring opportunities and challenges for metaverse and circular economy. AJIS 12(6):221. https://doi.org/10.36941/ajis-2023-0165

Di Virgilio F, Das S (2023a) Digital solutions for social media: role of digital evolution on business enterprises. In: Das S, Gochhait S (eds) Digital entertainment as next evolution in service sector. Palgrave Macmillan, Singapore. https://doi.org/10.1007/978-981-19-8121-0_9

Di Virgilio F, Das S (2023b) Digitalization of enterprises from a new normal sustainable development perspective: opportunities and challenges. In: Mondal SR, Yegen C, Das S (eds) New normal in digital enterprises. Palgrave Macmillan, Singapore. https://doi.org/10.1007/978-981-19-8618-5_9

Diamandis PH, Kotler S (2020) The future is faster than you think: how converging technologies are transforming business, industries, and our lives. https://library.pqm.co.id/index.php?p=show_detail&id=6222&keywords=

Ducrée J, Etzrodt M, Bartling S, Walshe R, Harrington T, Wittek N, Posth S, Wittek K, Ionita A, Prinz W, Kogias D, Paixão T, Peterfi I, Lawton J (2021) Unchaining collective intelligence for science, research, and technology development by Blockchain-Boosted community participation. Front Blockchain 4. https://doi.org/10.3389/fbloc.2021.631648

Hein AM, Matheson R, Fries D (2019) A techno-economic analysis of asteroid mining. Acta Astronaut 168:104–115. https://doi.org/10.1016/j.actaastro.2019.05.009

Hertzfeld HR, Von Der Dunk FG (2005) Bringing space law into the commercial world: property rights without sovereignty. Chic J Int Law 6(1):8. https://chicagounbound.uchicago.edu/cgi/viewcontent.cgi?article=1273&context=cjil

Jakhu RS, Pelton JN (2017) Global space governance: an international study. In: Space and society. https://doi.org/10.1007/978-3-319-54364-2

Liu, Y., Lu, Q., Yu, G., Paik, H., & Zhu, L. (2022). Defining blockchain governance principles: A comprehensive framework. *Information Systems, 109*, 102090. https://doi.org/10.1016/j.is.2022.102090

Link M, Martin G, Mousel J (2023) The economics of space resources: future markets and value chains. In: Springer eBooks, pp 1023–1051. https://doi.org/10.1007/978-3-030-97913-3_31

Majerova J, Das S (2023) Sustainable development and digitalization in the supply chain for environmental protection: changes and trends. In: Mondal SR, Yegen C, Das S (eds) New normal in digital enterprises. Palgrave Macmillan, Singapore. https://doi.org/10.1007/978-981-19-8618-5_8

Minh-Nhat HO, Majerova J, Das S (2022a) Taking steps to help the phoenix rise from ashes: a roadmap for digital enterprises to develop new sustainable goals. In: Sustainable development and innovation of digital enterprises for living with COVID-19. Springer, Singapore, pp 35–51. https://doi.org/10.1007/978-981-19-2173-5_3

Minh-Nhat HO, Nguyen HL, Mondal SR (2022b) Digital transformation for new sustainable goals with human element for digital service enterprises: an exploration of factors. In: Sustainable development and innovation of digital enterprises for living with COVID-19. Springer, Singapore, pp 85–103. https://doi.org/10.1007/978-981-19-2173-5_6

Mondal SR, Das S (2023a) Defining the new digitalized normal: implications of A.I., IoT, and blockchain for business growth. In: Mondal SR, Yegen C, Das S (eds) New normal in digital enterprises. Palgrave Macmillan, Singapore. https://doi.org/10.1007/978-981-19-8618-5_10

Mondal SR, Das S (2023b) Digital solution for industry 4.0 and 5.0 and social media. In: Das S, Gochhait S (eds) Digital entertainment as next evolution in service sector. Palgrave Macmillan, Singapore. https://doi.org/10.1007/978-981-19-8121-0_10

Mondal SR, Das S (2023c) Human-centric digitalization for new sustainable development: an explorative study in human resource management. In: Mondal SR, Yegen C, Das S (eds) New normal in digital enterprises. Palgrave Macmillan, Singapore. https://doi.org/10.1007/978-981-19-8618-5_12

Mondal S, Das S, Vrana VG (2023a) How to bell the cat? A theoretical review of generative artificial intelligence towards digital disruption in all walks of life. Technologies 11(2):44. https://doi.org/10.3390/technologies11020044

Mondal SR, Yegen C, Das S (2023b) New normal in digital enterprises – strategies for sustainable development. Palgrave-Singapore. https://doi.org/10.1007/978-981-19-8618-5 and Hardcover ISBN 978-981-19-8617-8

NASA (2025, January 10) Commercial lunar payload services – NASA. https://www.nasa.gov/commercial-lunar-payload-services/

Ngoc-Vinh N, Tien-Dung P, Das S (2022) Digitization of business; Need for recombination during COVID-19. In: Sustainable development and innovation of digital enterprises for living with COVID-19. Springer, Singapore, pp 1–17. https://doi.org/10.1007/978-981-19-2173-5_1

Nguyen TDM, Mondal SR, Das S (2022) Digital Entrepreneurial Transformation (DET) Powered by New Normal Sustainable Developmental Goals (n-SDGs): Elixir for growth of country's economy. In: Sustainable development and innovation of digital enterprises for living with COVID-19. Springer, Singapore, pp 69–84. https://doi.org/10.1007/978-981-19-2173-5_5

Schrijver CJ, Kauristie K, Aylward AD, Denardini CM, Gibson SE, Glover A, Gopalswamy N, Grande M, Hapgood M, Heynderickx D, Jakowski N, Kalegaev VV, Lapenta G, Linker JA, Liu S, Mandrini CH, Mann IR, Nagatsuma T, Nandy D et al (2015) Understanding space weather to shield society: a global road map for 2015–2025 commissioned by COSPAR and ILWS. Adv Space Res 55(12):2745–2807. https://doi.org/10.1016/j.asr.2015.03.023

Tandon S, Das S (2023) Digital revolution complements sustainable goal achievement during and after pandemic: a research agenda. In: Mondal SR, Yegen C, Das S (eds) New normal in digital enterprises. Palgrave Macmillan, Singapore. https://doi.org/10.1007/978-981-19-8618-5_11
Tien-Dung PHAM, Majerova J, Das S (2022a) Phases of possible recovery of digital enterprises in new normal business for living with COVID-19 times: opportunities for a new era in sustainable development goals. In: Sustainable development and innovation of digital enterprises for living with COVID-19. Springer, Singapore, pp 19–33. https://doi.org/10.1007/978-981-19-2173-5_2
Tien-Dung P, Nguyen XD, Das S (2022b) Sustainable goal achievement by digital revolution during and after pandemic; how much one wins and losses: a bird's eye view for future planning. In: Sustainable development and innovation of digital enterprises for living with COVID-19. Springer, Singapore, pp 133–147. https://doi.org/10.1007/978-981-19-2173-5_8
Vrana V, Das S (2023a) Dynamic restructuring of digital media and entertainment sector: role of urbanization, industrial innovation, and technological evolution. In: Das S, Gochhait S (eds) Digital entertainment as next evolution in service sector. Palgrave Macmillan, Singapore. https://doi.org/10.1007/978-981-19-8121-0_2
Vrana V, Das S (2023b) Role of enterprises and their digital transformation in post-pandemic sustainable development. In: Mondal SR, Yegen C, Das S (eds) New normal in digital enterprises. Palgrave Macmillan, Singapore. https://doi.org/10.1007/978-981-19-8618-5_7
Waseem M, Khan MA, Goudarzi A, Fahad S, Sajjad I, Siano P (2023) Incorporation of blockchain technology for different smart grid applications: architecture, prospects, and challenges. Energies 16(2):820. https://doi.org/10.3390/en16020820
Weinzierl M, Sarang M (2021, February 12) The commercial space age is here. Harv Bus Rev. https://hbr.org/2021/02/the-commercial-space-age-is-here
Yegen C, Das S (2023) Digital solution for print and publishing media. In: Das S, Gochhait S (eds) Digital entertainment as next evolution in service sector. Palgrave Macmillan, Singapore. https://doi.org/10.1007/978-981-19-8121-0_12

Dagmar Caganova Senior Academic at NEWTON University, specializing in innovation ecosystems and entrepreneurial development. Her work explores technology adoption and start-up scalability. She mentors students and professionals in navigating innovation-driven markets.

Subhankar Das Senior Researcher and academician at Duy Tan University, publishing extensively on AI, blockchain, and digital transformation. Editor of seven Scopus-indexed books and reviewer for top journals. Mentors global PhD candidates and drives innovation in ASEAN academic collaborations.

To Infinity and Beyond: Charting the Course for Interplanetary Adventures

Subhra R. Mondal and Subhankar Das

Abstract This chapter explores humanity's complex challenges and opportunities as we seek to explore and colonize the stars. The space colonization process emphasizes the importance of technological adaptation while also considering biological adaptation and the input of sociocultural forces, thus demonstrating the need for interdisciplinary collaboration to find solutions to challenges such as ineffective propulsion systems, harmful health effects of long-term exposure to space, inefficient propulsion systems, and governance problems. This conversation highlights breakthroughs like nuclear thermal propulsion, in-situ resource utilization and CRISPR-based therapies to allow for long-term human habitats in space. Ethics, planetary protection and equitable resource distribution are themes that emerge alongside calls for mechanisms for international cooperation similar to those inspired by the Antarctic Treaty System. Grounded in theory and practice, it calls for a shift in focus towards an approach that prioritizes technological rigor, human adaptability, and ethical stewardship to ensure extraterrestrial benefits and terrestrial sustainability.

Keywords Interplanetary exploration · Space colonization · Technological innovation · Biological adaptation · Sociocultural dynamics · Ethical governance

1 Introduction

From exploring its myths to conducting experiments here on earth, humanity has always been excited by the quest to explore space. "To Infinity and Beyond" is A popular phrase that has become a part of the eternal imagination and describes humanity's aspiration to explore interstellar cosmos (John and John 2024). This

S. R. Mondal · S. Das (✉)
Department of Marketing, South Star Management Institute, Duy Tan University, Da Nang, Vietnam
e-mail: subhraranimondal@duytan.edu.vn; subhankardas@duytan.edu.vn

© The Author(s), under exclusive license to Springer Nature Singapore Pte Ltd. 2025

S. R. Mondal et al. (eds.), *Pioneering the New Space Economy through AI and Immersive Technologies*, https://doi.org/10.1007/978-981-96-5977-7_15

chapter examines the path of interplanetary exploration, mainly focusing on its technological, biological, and sociocultural aspects. Since governments and industries are competing to escape from the Earth, what has begun to resemble science fiction, the colonization of Mars, the mining of asteroids, and the finding of alien life is quickly becoming an agenda item (Neukart 2025). However, this ambition comes with interdependent challenges that require interdisciplinary action, international cooperation, and ethical anticipation.

When that era began in the mid-twentieth century, it is said to have started with events like the launch of Sputnik in 1957 and the Apollo Moon landings (1969–1972). These accomplishments demonstrated humankind's ability to learn to escape gravitational force, at least in near-Earth shells. Today's programs, such as NASA's Artemis program and SpaceX's Starship program, seek to establish sustainable lunar bases before settlements on Mars. These efforts exemplify an evolution of thought from temporary missions to permanent outposts beyond Earth, requiring new technologies in propulsion, habitat construction, and resource extraction (Mohon 2024).

Propulsion technology is still the bedrock of interplanetary travel. Conventional chemical rockets are inefficient for long-distance space missions and are not an option for Mars missions due to the lengthy transit time and fuel considerations (Banani 2023). For example, new rocket technologies like nuclear thermal propulsion (NTP) can shorten trip times to Mars by 50% by using heat from a reactor to propel propellants (Hall 2023). Speculative ideas such as warp drives are part of humanity's theoretical grasp of faster-than-light travel. Similarly, in-situ resource utilization (ISRU) systems would serve to extract water ice on the Moon or Mars, and subsequently use it to alleviate dependency on material shipped from Earth (Starr and Muscatello 2020). Together, these technologies solve the logistical challenges of deep-space missions, but their safety and scalability must be validated with rigor.

Biological adaptation is another important frontier. Dangers of longstanding living in microgravity, cosmic radiation, and encased laps pose vital health threats like muscle wasting, decrease mineral density, and psychological tension (Clément et al. 2015). While countermeasures like artificial gravity modules and pharmacological interventions are actively developing, bioengineering may provide radical solutions (Isasi et al. 2022). Synthetic biology, for example, could allow genetically-modified crops to grow in Martian soil, and CRISPR-based therapies might help repair DNA damage from radiation (Clément et al. 2015). Closed-loop life support systems use ideas based on living systems on Earth to recycle air, water, and waste, ideally, in sustainable manners. However, some of these concepts have been investigated in closed projects such as Biosphere 2 (De Micco et al. 2023). This redesign of adaptability incorporates the human body's platforms and responses to outer space environments, ultimately altering the definition of habitability.

Even the quest for alien life encourages interplanetary missions. Mars, known for its ancient river valleys and shallow subsurface reservoirs of brine, is still a prime target for astrobiological investigation (Horneck 2008). Europa and Enceladus, the icy moons of Jupiter and Saturn, contain subsurface oceans that may support microbial life and have been proposed as targets of robotic submersibles (Martin and McMinn 2017). Detecting even simple organisms would transform our understanding of the generality of life and invite ethical considerations about planetary protection and the rights of extraterrestrial ecosystems (Duronio et al. 2017). For example, should we limit human activity on Mars to protect potential biospheres? Such dilemmas highlight the need for frameworks that balance exploration with environmental stewardship.

The colonization of space also involves deep sociocultural swimming through the deep. First, building off-world communities will force us to reexamine questions of identity, governance, and equity (Harris 1995). Will those settlers make cultural practices unique to Mars, or stay rooted to Earth? How will resource allocation—e.g., water from asteroid mines—be regulated to prevent corporate monopolization? (Szocik et al. 2020). While historical cases, such as the Antarctic Treaty System, provide a roadmap for fostering an international collaborative approach, the space arena remains fragmented (Madani 2024). Even though the Outer Space Treaty prohibits any claim of national sovereignty over outer space or celestial bodies, it has no international regulatory mechanisms to govern the activities of commercial entities and as such calls for new governance models (Pop 2000).

In addition, there are mutual advantages between exploring other planets and our own. Technologies emerging out of space habitats, from water recycling to solar energy systems, could help alleviate the effects of climate change (Miller 2024). Extracting extraterrestrial resources could mitigate scarcity by making available rare metals, although the ecological risks posed by asteroid mining are still hotly debated (Hagerman & Chan, 2009). Likewise, space partnerships around the globe have the potential to bolster diplomatic concord and mitigate Earth-based geopolitical competition (MacLeish et al. 2012). Thus, exploring spacefaring civilizations is not a diversion from Earth's issues, but a parallel to confront them.

This journey is riddled with ethical and philosophical inquiries. How will humanity navigate its anthropocentric worldview if intelligent extraterrestrial life is discovered (Kwon et al. 2018)? Human colonization of other planets raises similar thorny dilemmas about potential terraforming: Would it be ethical for humans to change alien geographies for their own needs, or should such pristine landscapes remain just that? (Poppick 2017). These questions have no easy answers, but they underscore the need for inclusivity in the conversation among scientists, policymakers, and the public.

Synthesizing these themes, this chapter argues that interplanetary exploration is a multi-axial odyssey that involves harmonizing innovation, ethics, and internationalism. The subsequent sections detail the scientific innovations, biotic modifications, and social and political structures required for long-term space habitation. By looking past the Earth's orbit, we are not just looking for new worlds but also a new narrative for our existence—a journey perhaps more inward than outward.

2 Literature Review: Challenges in Interplanetary Exploration and Colonization

For the longest time, both the scientific world, the circle of public policy, and the most common of human mortals alike have been fascinated by the idea of exploring and thereby ultimately colonizing other planets, driven by the advancement of technology, upsurges of existential curiosity, and, in many instances, the re-examination of the many challenges we face on our home world. This literature review determined the interdependent effects of technological development, biological evolution, social and cultural influences, moral frameworks and governance mechanisms on humans' quest for celestial existence (through a cross-discipline synthesis of scholarly publications on their many contributions to celestial destiny). Reviewing peer-tested studies, policy documents, and theoretical analyses magnifies the co-dependent environmental challenges and opportunities in the drive to make humanity a multi-planetary species.

2.1 Technological Advancements: Propulsion, Habitats, and Resource Utilization

Propulsion Systems

Depending on transportation technologies that can minimize travel time and fuel consumption, the practicality of getting from one world to another is ultimately limited. The trip to Mars is too long for conventional chemical rockets (Thomas 2024), although they are still used for the moon. Nuclear thermal propulsion (NTP) generally uses nuclear reactors to heat propellants (such as hydrogen) is an appealing alternative. Nikitaev and Thomas (2022) demonstrated that NTP would reduce transit time to Mars by 90–120 days, compared to the timescale of 6–9 months with chemical rockets. Still, hurdles such as reactor safety, radiation shielding, and regulatory approvals must be cleared.

Even speculative theories, such as warps and antimatter propulsion, reflect humanity's wishful thinking that we will master such things as faster-than-light travel. Warp drives could fold spacetime to allow superluminal travel (Hubilla 2024). But these models rely on the presence of "exotic matter" with negative energy density, which has never been empirically observed. Similarly, Omira and Mourad (2024) investigated the possibility of utilizing antimatter propulsion, assessing that producing and storing antimatter at the required orders of magnitude was economically infeasible.

ISRU (In Situ Resources Utilization)

Sustainable colonization would involve relying less on resources from Earth. ISRU systems are designed to extract water, oxygen, and metals from off-world regolith. Focusing at home, Nikitaev and Thomas (2022) remind us of NASA's Mars Oxygen

In-Situ Resource Utilization Experiment (MOXIE), successfully processing Martian CO_2 into breathable oxygen. Likewise, lunar water ice deposits at the Moon's poles may provide life support systems and rocket propellant. ISRU technologies have upstream scalability challenges; for example, extracting water from regolith is energy-intensive, which could become costly if power supply bottlenecks in early missions (Liu et al. 2022).

Artificial Habitats

It will be crucial to develop habitats that protect humans from radiation, extreme temperatures and micrometeoroids. Through literature review, Bannova and Gulacsi (2024) discussed modular habitat plans, highlighting inflatable radiation-shielded infrastructures and 3D-printed regolith-based architectures. The European Space Agency's (ESA) MELiSSA project, a closed-loop ecosystem designed to replicate the Earth's biogeochemical cycles, partially recycled air and water. However, the long-term reliability of these systems in deep-space environments has not yet been tested.

2.2 *Biological and Physiological Adaptation*

Health Risks in Space

Under microgravity conditions, prolonged exposure leads to muscle atrophy, bone demineralization, and cardiovascular system deconditioning. A meta-analysis reported that astronauts still lose 1–2% of bone a month while on a space mission, even with extensive exercise regimens (Juhl et al. 2021). Cosmic radiation is another risk, with studies showing that the incidence of cancer and damage to the central nervous system by galactic cosmic rays are involved. Psychological stressors such as isolation and confinement impair cognitive functions and team cohesiveness also (Scheibler et al. 2022).

Bioengineering: Countermeasures

Rotating spacecraft can create artificial gravity, but engineering challenges hinder this approach. Frett et al. (2020) showed intermittent centrifugation (2–4 hours per day) partially prevented muscle atrophy in rodent models. Pharmacotherapeutic approaches, including bisphosphonates, can potentially treat bone loss, but long-term safety must be assessed. Innovations in bioengineering could provide game-changing solutions. Using CRISPR-Cas9 gene editing to boost the resistance of mammalian cell lines to radiation has also been demonstrated in yeast and animal trials. One synthetic biology application is to engineer drought-resistant crops for Martian agriculture (Liu et al.

2021). NASA's Advanced Plant Habitat (APH) has grown Arabidopsis thaliana in hypomorphic Martian soil but yielded suboptimal results.

2.3 Socio and Ethical Thought Process

Identity and Governance

Space colonization might spawn new cultural identities. Szocik et al. (2020) argued that social norms on Mars can be a distinguishable phenomenon, similar to colonial diasporas in human history. However, some hybrid governance model may be required if lunar settlers remain tethered to Earth. Steffen (2021) echoed this through warnings about uncontrolled asteroid mining, which allows corporate monopolies to drive terrestrial inequities. While the Outer Space Treaty (1967) also prohibits national sovereignty claims, it does not regulate commercial activities and has called for a revised framework.

Planetary Protection and Ethics

The question of other life in the universe is an ethical one. Spry (2022) proposed strict planetary protection measures to restrain flourishing Earth microbes on carbonated Mars, which might destroy existing indigenous life. In contrast, proponents of terraforming claim that changing Martian environments is necessary for human survival. "Exotic" beings like intelligent aliens would undermine anthropocentric worldviews and require transdisciplinary conversation.

2.4 Governance and International Cooperation

Historical Precedents

The Antarctic Treaty System (ATS) provides peaceful cooperation and environmental stewardship and is a potential model for the future of space governance. The ATS's success relied upon its gradualism (Triggs 2011). Of course, these stakes are more consequential because they touch upon commercial and strategic interests encompassing almost every Earth nation.

Current Legal Frameworks

The Artemis Accords, driven by NASA in 2020, encourage transparency and resource-sharing between signatories but lack enforcement mechanisms. China and Russia's exclusion from these accords reflects geopolitical fissures that may inhibit collaborative exploration.

Reciprocal Benefits for Earth

Technologies developed for space habitats, such as water recycling systems and vertical farming, could be applied to terrestrial resource limitations (Pickett et al. 2019). Asteroid mining may support the extraction of rare-earth metals essential for renewable energy infrastructure, warned of environmental damage from unfettered extraction.

2.5 Gaps in the Literature

Even though a massive amount of literature is written on propulsion and emergent habitats, the economy and political aspects of off-world governance are explored less in the literature. Moreover, ethical frameworks for relations with extraterrestrial life would be speculative, needing input from astrobiologists and philosophers.

Exploration of the interplanetary and beyond-us world is an interdisciplinary endeavor that requires synergy between technological innovation, biological resilience, ethical and technological foresight, and international collaboration. Existing literature supports the plausibility of near-term missions for Mars and the Moon but also emphasizes longstanding difficulties in health, governance, and environmental ethics. As researchers, we must also be guided by a commitment to inclusive policy frameworks and public engagement that secures humanity's potential cosmic aspirations in equitably serving all Earth's inhabitants' needs.

3 Implementation Framework for Interplanetary Colonization and Exploration

One way to achieve interplanetary exploration and colonization is to adopt a structured, phased approach that accounts for technological innovation, human adaptation, governance, and ethical stewardship (Das 2023a). This framework provides a pathway for achieving a sustainable extraterrestrial presence that explores challenges recognized in current research and offers tangible steps towards achieving such a goal. Figure 1 represents the phased interplanetary exploration.

Short-Term (2025–2035):

a. Establish lunar bases as proving grounds for habitat technologies (e.g., regolith-3D printed structures) and in-situ resource utilization (ISRU) systems.
b. Perform crewed Mars precursor missions to validate nuclear thermal propulsion (NTP) and evaluate human health risks during 6–9-month transits.
c. Conduct robotic missions to recover water ice from the poles of the Moon and Mars, advancing ISRU methods for oxygen and fuel generation.

Fig. 1 Phased approach to interplanetary exploration (Source: Authors' conception)

Medium-Term (2035–2050):

a. Build and deploy semi-permanent Martian habitats (integrating closed-loop life support systems, e.g., ESA's MELISSA, and radiation-shielded modules.
b. Initiate asteroid mining pilots to harvest rare metals, ensuring regulatory frameworks prevent monopolization in resource-rich areas.

c. Design gene-editing therapies (e.g., CRISPR) to enhance human radiation resistance and genetically engineer crops for Martian agriculture.

Long-Term (post-2050):

a. Foster Martian social colonies into cities with terraforming pilot efforts, like atmospheric thickening through greenhouse gas emissions.
b. Theoretical interstellar missions with advanced propulsion concepts (e.g., antimatter drives) would require discoveries in physics.

3.1 Focused Areas of Technological Development

a. Propulsion Systems: Accelerate NTP development through public-private partnerships, prioritizing reactor safety and radiation mitigation.
b. Habitat Engineering 2: Standardize modular, inflatable habitats with 3D-printed regolith shielding for heavy tech on Mars and the Moon.
c. Resource Utilization: High-scale ISRU technologies (e.g., MOXIE-derived oxygen generators) to support larger crews and fuel depots.

3.2 Strategies for Adaptation and Protection of Human Health

a. Artificial Gravity: Supplementing Rocket Spacecraft with rotating spacecraft modules or centrifuges in transit vehicles to reduce muscle atrophy.
b. Radiological Countermeasures: Implement pharmacological agents (e.g., bisphosphonates) and wearability materials that shield during missions.
c. Psychological openness and wellness: Implement AI-driven mental health monitoring and virtual reality systems to alleviate isolation.

3.3 International Collaboration and Governance

a. Legal Frameworks: Make the Artemis Accords applicable to emerging spacefaring nations (China, India, etc.) and enforce resource-sharing agreements.
b. Planetary Protection: Formulate a global agency dedicated to identifying and regulating contamination risks and ethical exploration practices, similar to the model utilized in the Antarctic Treaty System.
c. Equity Mechanisms: Require profits from asteroid mining to fund sustainability projects here on Earth.

3.4 Constitutional and Environmental Safeguards

a. Extraterrestrial Life Protocols: Establish stringent planetary protection measures to protect potential biospheres on Mars and its icy moons.
b. Terraforming Ethics: Bring teams of ethicists and experts from multiple fields to assess the ethics of altering alien environments.

3.5 Monitoring and Evaluation

a. Technical Shutdowns: Biennial reviews of propulsion and ISRU milestones and milestones with funding increases/decreases based on performance.
b. Health Metrics: Measure astronaut health data across missions to improve countermeasures.
c. Governance Reviews: Assess treaty compliance and equity outcomes through independent oversight bodies.

This framework optimizes for progress, not perfection, anchoring itself in achievable outcomes in technology, health, and governance. Through global cooperation and ethical vigilance, humankind can transform interplanetary exploration from a mere ambition to a sustainable reality that benefits both spacefaring civilizations and Earth. Figure 2 represents the framework.

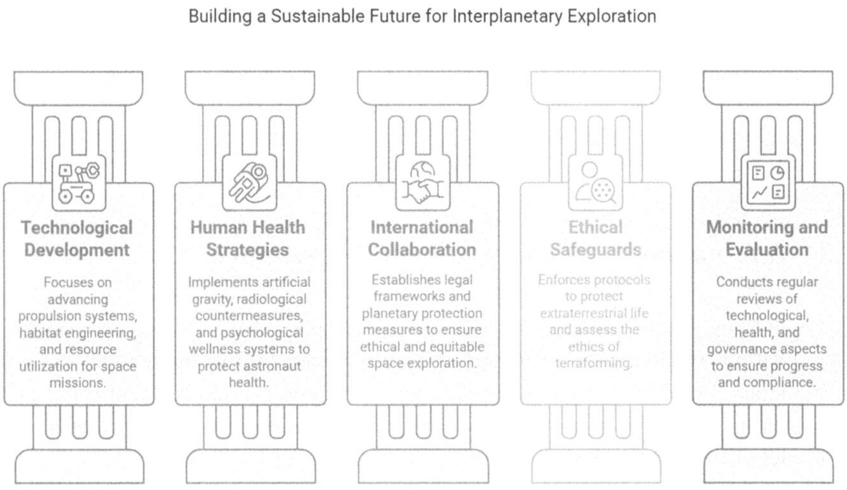

Fig. 2 Sustainable interplanetary exploration (Source: Authors' conception)

4 Interplanetary Exploration and Colonization: Theoretical, Practical, and Sustainable Implications

The journey to other planets is also a paradigm, application, and sustainability theory, with the implementation of exploration and colonization (Das 2023b). This can significantly affect our understanding of science, mechanics, ethics, and technology, from presenting challenges to providing opportunities for humanity (Das et al. 2023). This section integrates these dimensions, synthesizing interdisciplinary research that illustrates their interconnectivity.

4.1 Theoretical Implications

Transgressing Human Identity and the Philosophy of Existence

However, it does challenge anthropocentric worldviews by putting humanity in a cosmic perspective. Theoretical discussions hinge on whether space exploration reaffirms or hums human exceptionalism by giving evidence of our cosmos' fragility (Di Virgilio and Das 2023a). The finding of life beyond Earth, even at the level of microbes, would transform astrobiology and put humanity's existence in a Copernican context in terms of life's ubiquity. Terraforming Mars (or any other body in the universe) involves philosophical questions about whether humans are stewards or conquerors of alien ecosystems, echoing historical debates about colonization (Di Virgilio and Das 2023b).

Interdisciplinary Integration

Grounding the space colonization theory requires integrating disparate disciplines (Majerova and Das 2023). For example, synthetic biology intersects with aerospace engineering to create bioengineered crops for Martian agriculture. Likewise, space law should include environmental ethics to regulate asteroid mining. This convergence encourages entirely new academic frameworks, including, but not limited to, "astro-sociology," which looks at how space-faring communities might develop social and political hierarchies (Minh-Nhat et al. 2022a, b).

Ethical Paradigms

Theoretical ethics must face unprecedented dilemmas like considering extraterrestrial organisms to be sentient or the moral permissibility of modifying planetary ecosystems (Mondal et al. 2023a). The "Prime Directive" concept from science fiction of not interfering with alien life has made its way into academic discourse, advocating for precautionary principles in exploration (Mondal and Das 2023a). Moreover, intergenerational justice theories debate whether present generations have a right to bind future humans to one-way space colonization (Mondal and Das 2023b).

4.2 Practical Implications

Technological Innovation and Spin-Offs

Technologies that are driven by space often have terrestrial benefits. Closed loop life support systems like MELISSA (Micro-Ecological Life Support Alternative) would also be relevant in sparking advances in water, which would be more energy efficient, for instance, recycling economic mammalian waste (a form of waste from the human body). Likewise, solids resistant to radiation, used in spacecraft shielding, could lead to improved cancer treatment technologies (Mondal and Das 2023c). However, as exemplified by SpaceX's reusable rockets, private-sector involvement can enhance the speed of innovation while simultaneously driving down costs through commercialization.

Solutions for Adaptation and Health

There are tangible issues in human health that demand a solution now (Mondal et al. 2023b). Testing artificial gravity system through rotating habitats would minimize the microgravity-induced atrophy observed during long-term exploration missions and inform biomedical rehabilitation therapies on Earth (Ngoc-Vinh et al. 2022). Bone loss pharmacological countermeasures like bisphosphonates may benefit osteoporosis patients. Finally, AI-driven mental health tools designed for astronauts could address isolation within remote terrestrial colonies.

Policy Development and Governance

This could lead to conflicts due to lax legal restrictions on space resource use. Pragmatic steps include broadening the Artemis Accords to mandate fair resource-sharing and ban monopolies. Key lessons from the Antarctic Treaty System (ATS) underscore the value of flexible governance models that reconcile exploration with environmental sustainability. NGOs should partner with national space agencies to create transparent dispute-resolution mechanisms.

Economic Transformations

By supplying rare-earth metals considered essential for renewable technologies through asteroid mining, such a situation could change the behavior of the global markets and even reduce geopolitical tensions over terrestrial resources (Nguyen et al. 2022). However, unscrutinized extraction threatens ecological damage and market saturation that require international oversight (Tandon and Das 2023). We will begin to integrate into the space economy, which could be worth over $1 trillion by 2040, and produce jobs for robots, habitat builders, and space tour guides, creating more opportunities for the global economic system.

4.3 Sustainable Implications

Resource Stewardship

Space sustainability is all about reducing waste and making the most out of resources. Using in-situ resource utilization (ISRU) systems, e.g., Martian CO_2 conversion to oxygen, reduces dependency on Earth, and helps in mission cost education. ISRU-derived technologies on Earth can help us optimize for circular economies in industrial processes where industrial emissions become valuable products (Tien-Dung et al. 2022a). Overharvesting lunar water ice or asteroid metals may exhaust limited extraterrestrial resources, echoing humanity's plunder of Earth.

Environmental Ethics

We must pursue the principle of "planetary protection" to incorporate sustainability. Existing guidelines primarily emphasize contamination prevention and do not include provisions for restoration of functioning ecosystems and their services (Tien-Dung et al. 2022b). For example, on Mars, terraforming by introducing greenhouse gases might stabilize the climate but jeopardize or destroy native life. Environmental aspects of sustainable colonization suggest a precautionary principle; less invasive exploration should not be done until the environmental impact is understood (Vrana and Das 2023a).

Sustainability of Colonies Over Time

Colonies will need closed-loop systems for energy, food, and waste. The European Space Agency's project, designed to reproduce Earth's ecosystems, provides an air and water recycling model. Genetically engineered crops such as drought-tolerant wheat could enable agricultural self-sufficiency on Mars. However, long-term sustainability depends on ethical population policies and energy inventions like muffin-sized nuclear reactors or solar power farms that can function in dimmer Martian sunlight (Vrana and Das 2023b).

Earth-Space Synergy

Space colonization can relieve terrestrial pressures through off-world resource extraction and heavy industry on asteroids or the Moon. On the flip side, innovations born out of need for space environments, like vertical farming or solar energy storage, could speed up the transition to sustainability for Earth (Yegen and Das 2023). However, such interdependence also implies the need for more equitable access to the benefits derived from space not to exacerbate global inequities.

Three overlapping implications regarding interplanetary exploration exist—theoretical, practical, and sustainable. In theory, colonization is an extreme version of humanity reaching for the stars. It is an unprecedented challenge for humanity as humanity needs to set new ethical frameworks for new, unimaginable scenarios. In practical terms, it is the seismic terrestrial spillover dimension's power, technological, and economic evolution. Sustainably, it demands a reconciliation of innovation with stewardship, so that space endeavors do not repeat Earth's ecological errors. The vast implications of this reality demand awareness as humanity steps beyond the bounds of Earth — understanding that space exploration is not escape from. However, an enhancer of terrestrial realities, and therefore necessarily requires foresight, collaboration, and humility.

5 Conclusion

Humanity's ambition to explore and eventually colonize interplanetary space is perhaps one of the most audacious and transformative pursuits in the history of this planet. So, this journey, closely related to the scientific and technological aspects, requires the maximization of correlations between various fronts and the international community, including innovation procedures, ethics, and will. This discussion demonstrates that the road to human habitation beyond Earth has many serious challenges. However, it also allows us to rewrite our species' future in ways that have never existed in the millions of years we have walked this planet.

Interplanetary colonization demands technological advancements in many domains, like NTP, ISRU, and radiation-shielded habitats, which means interdisciplinary innovation and research will be needed on a large scale. For example, ISRU capabilities can reduce reliance on supplies shipped from Earth, decreasing mission delivery costs. Projects like MOXIE generate oxygen from Martian carbon dioxide. Likewise, modular habitat designs and closed-loop life support systems—all vegetation is cultivated within the habitat and all biological waste is recycled into life support—have been tested under the MELISSA initiative and others. However, these technologies must be thoroughly validated for scalability and safety, especially considering the inhospitable conditions of space. However, warp drives and antimatter propulsion remind us that theoretical breakthroughs must be recognized with modest research priorities.

Also crucial are the biological and psychological changes humans must experience to make long-duration trips through space. Long-term exposure to microgravity and cosmic radiation presents considerable health challenges, from bone demineralization to increased cancer risk. Potential countermeasures like artificial gravity systems and CRISPR-based gene editing provide promising candidates but must be thoroughly tested for efficacy and ethical acceptability. The psychological impact of loneliness and confinement demands new mental health approaches, including AI-based surveillance and virtual reality therapies. These adaptations are not just protective for astronauts—they can also be transformative for medicine on Earth, from osteoporosis treatments to remote healthcare solutions.

The sociocultural and governance challenges that accompany space colonization are equally daunting. The challenges of off-world settler communities will inevitably alter human identity and governance models as power relations and resource equity will differ sharply from Earth. While historical precedents such as the Antarctic Treaty System (ATS) have proven valuable in rallying international cooperation around shared issues, space's distinct legal and commercial aspects call for new initiatives. While the Artemis Accords are moving toward transparency, their framework must be developed to address emerging spacefaring nations and mandate equitable resource-sharing. Without clear guidelines and regulation, the pioneer companies exploiting asteroids may monopolize this resource and risk ecological damage that could taint or potentially destroy the natural beauty of many celestial bodies.

The cornerstone of interplanetary ambitions is sustainability. Planetary protection must grow from protecting against contamination to actively preserving native ecology, especially if we find life beyond Earth. Terraforming Mars could be a possible method for human survival, but questions remain about whether humanity has the right to alter alien ecosystems on other planets. Solar power, water recycling systems, and vertical farming are just a few technologies developed for space that could help meet urgent environmental challenges on Earth, revealing the interrelated nature of space exploration and sustainability on our planet.

To conclude, interplanetary colonization depends on global cooperation, policy making, and the public as a whole. Future work must focus on sociopolitical structures for off-world governance and ethical protocols for engagement with alien life. Informed by a broad range of cultural outlooks, international agencies should lead a dialogue to ensure that space exploration is a boon for all humanity rather than a means to amplify inequities on Earth. Ultimately, this kind of interplanetary exploration is much more than an academic exercise; it is a radical reimagining of what we think it means to be human within the greater universe. The task ahead—exploring beyond our home planet—is filled with opportunity, hope, and potential, but it must be guided by a vision that embraces ethical and sustainable practices; turning this dream into a lasting legacy that embodies the resilience, curiosity, and unity of our species and extends far beyond the confines of Earth to enrich our shared future.

References

Banani A (2023, December 26) What will it take to make interplanetary space travel possible? Medium. https://medium.com/predict/propulsion-systems-04ccbe40cbd2

Bannova O, Gulacsi E (2024) Architectural approach for evaluation of radiation shielding integration in space habitats. Acta Astronaut 220:27–36. https://doi.org/10.1016/j.actaastro.2024.04.022

Clément GR, Bukley AP, Paloski WH (2015) Artificial gravity as a countermeasure for mitigating physiological deconditioning during long-duration space missions. Front Syst Neurosci 9. https://doi.org/10.3389/fnsys.2015.00092

Das S (2023a) Modern-day gaming and the role of FinTech applications in popularising various entertainment platforms across the world. In: Das S, Gochhait S (eds) Digital entertainment as next evolution in service sector. Palgrave Macmillan, Singapore. https://doi.org/10.1007/978-981-19-8121-0_11

Das S (2023b) Digital entertainment: the next evolution in service sector. Emerging Digital Solutions in Reshaping Different Industries Palgrave-Singapore. https://doi.org/10.1007/978-981-19-8121-0

Das S, Mondal SR, Struck F, Eissfeldt J, Beernink L, Sigouras V, Hoose S, Kopac L (2023) What can we see inside the crystal ball? Exploring opportunities and challenges for metaverse and circular economy. AJIS 12(6):221. https://doi.org/10.36941/ajis-2023-0165

De Micco V, Amitrano C, Mastroleo F, Aronne G, Battistelli A, Carnero-Diaz E, De Pascale S, Detrell G, Dussap C, Ganigué R, Jakobsen ØM, Poulet L, Van Houdt R, Verseux C, Vlaeminck S, Willaert R, Leys N (2023) Plant and microbial science and technology as cornerstones to Bioregenerative Life Support Systems in space. Npj Microgravity 9(1). https://doi.org/10.1038/s41526-023-00317-9

Di Virgilio F, Das S (2023a) Digital solutions for social media: role of digital evolution on business enterprises. In: Das S, Gochhait S (eds) Digital entertainment as next evolution in service sector. Palgrave Macmillan, Singapore. https://doi.org/10.1007/978-981-19-8121-0_9

Di Virgilio F, Das S (2023b) Digitalization of enterprises from a new normal sustainable development perspective: opportunities and challenges. In: Mondal SR, Yegen C, Das S (eds) New normal in digital enterprises. Palgrave Macmillan, Singapore. https://doi.org/10.1007/978-981-19-8618-5_9

Duronio RJ, O'Farrell PH, Sluder G, Su TT (2017) Sophisticated lessons from simple organisms: appreciating the value of curiosity-driven research. Dis Model Mech 10(12):1381–1389. https://doi.org/10.1242/dmm.031203

Frett T, Green DA, Mulder E, Noppe A, Arz M, Pustowalow W, Petrat G, Tegtbur U, Jordan J (2020) Tolerability of daily intermittent or continuous short-arm centrifugation during 60-day 6o head down bed rest (AGBRESA study). PLoS One 15(9):e0239228. https://doi.org/10.1371/journal.pone.0239228

Hagerman SM, Chan KM (2009) Climate change and biodiversity conservation: impacts, adaptation strategies and future research directions. F1000 Biol Reports 1. https://doi.org/10.3410/b1-16

Hall L (2023, September 29) Nuclear thermal propulsion: game changing technology for deep space exploration – NASA. NASA. https://www.nasa.gov/directorates/stmd/tech-demo-missions-program/nuclear-thermal-propulsion-game-changing-technology-for-deep-space-exploration/

Harris PR (1995) The impact of culture on human and space development—New millennial challenge. Acta Astronaut 36(7):399–408. https://doi.org/10.1016/0094-5765(95)00116-6

Horneck G (2008) Astrobiological aspects of Mars and human presence: pros and cons. PubMed https://pubmed.ncbi.nlm.nih.gov/19048093

Hubilla CN (2024, February 23) Anti-MATTEr Spacecraft: could NASA's new propulsion system be the future of interstellar travel? Science Times. https://www.sciencetimes.com/articles/48874/20240223/anti-matter-spacecraft-nasa-s-new-propulsion-system-future-interstellar.htm

Isasi E, Isasi ME, Van Loon JJWA (2022) The application of artificial gravity in medicine and space. Front Physiol 13. https://doi.org/10.3389/fphys.2022.952723

John & John (2024, December 1) The history of human curiosity: from myths to the scientific revolution – the harmonious cosmos. The Harmonious Cosmos – Exploring global unity, interfaith dialogue, and the intersection of spiritual wisdom and technological advancement. https://harmoniouscosmos.com/the-history-of-human-curiosity-from-myths-to-the-scientific-revolution/

Juhl OJ, Buettmann EG, Friedman MA, DeNapoli RC, Hoppock GA, Donahue HJ (2021) Update on the effects of microgravity on the musculoskeletal system. Npj Microgravity 7(1). https://doi.org/10.1038/s41526-021-00158-4

Kwon JY, Bercovici HL, Cunningham K, Varnum MEW (2018) How will we react to the discovery of extraterrestrial life? Front Psychol 8. https://doi.org/10.3389/fpsyg.2017.02308

Liu Q, Yang F, Zhang J, Liu H, Rahman S, Islam S, Ma W, She M (2021) Application of CRISPR/CAS9 in crop quality improvement. Int J Mol Sci 22(8):4206. https://doi.org/10.3390/ijms22084206

Liu Y, Wang C, Pang Y, Wang Q, Zhao Z, Lin T, Wang Z, Shen T, Liu S, Song J, Lai X, Quan X, Yao W (2022) Water extraction from icy lunar regolith by drilling-based thermal method in a pilot-scale unit. Acta Astronaut 202:386–399. https://doi.org/10.1016/j.actaastro.2022.11.002

MacLeish MY, Akinyede JO, Goswami N, Thomson WA (2012) Global partnerships: expanding the frontiers of space exploration education. Acta Astronaut 80:190–196. https://doi.org/10.1016/j.actaastro.2012.05.034

Madani Z (2024) Unpacking inclusivity of the antarctic treaty system amidst contemporary challenges. Pol Sci 101144. https://doi.org/10.1016/j.polar.2024.101144

Majerova J, Das S (2023) Sustainable development and digitalization in the supply chain for environmental protection: changes and trends. In: Mondal SR, Yegen C, Das S (eds) New normal in digital enterprises. Palgrave Macmillan, Singapore. https://doi.org/10.1007/978-981-19-8618-5_8

Martin A, McMinn A (2017) Sea ice, extremophiles and life on extra-terrestrial ocean worlds. Int J Astrobiol 17(1):1–16. https://doi.org/10.1017/s1473550416000483

Miller R (2024, November 1) 20 Pros and cons of space exploration | Ablison. Ablison. https://www.ablison.com/pros-and-cons-of-space-exploration/

Minh-Nhat HO, Majerova J, Das S (2022a) Taking steps to help the phoenix rise from ashes: a roadmap for digital enterprises to develop new sustainable goals. In: Sustainable development and innovation of digital enterprises for living with COVID-19. Springer, Singapore, pp 35–51. https://doi.org/10.1007/978-981-19-2173-5_3

Minh-Nhat HO, Nguyen HL, Mondal SR (2022b) Digital transformation for new sustainable goals with human element for digital service enterprises: an exploration of factors. In: Sustainable development and innovation of digital enterprises for living with COVID-19. Springer, Singapore, pp 85–103. https://doi.org/10.1007/978-981-19-2173-5_6

Mohon L (2024, March 19) NASA artemis mission progresses with SpaceX starship test flight – NASA. NASA. https://www.nasa.gov/general/nasa-artemis-mission-progresses-with-spacex-starship-test-flight/

Mondal SR, Das S (2023a) Defining the new digitalized normal: implications of A.I., IoT, and blockchain for business growth. In: Mondal SR, Yegen C, Das S (eds) New normal in digital enterprises. Palgrave Macmillan, Singapore. https://doi.org/10.1007/978-981-19-8618-5_10

Mondal SR, Das S (2023b) Digital solution for industry 4.0 and 5.0 and social media. In: Das S, Gochhait S (eds) Digital entertainment as next evolution in service sector. Palgrave Macmillan, Singapore. https://doi.org/10.1007/978-981-19-8121-0_10

Mondal SR, Das S (2023c) Human-centric digitalization for new sustainable development: an explorative study in human resource management. In: Mondal SR, Yegen C, Das S (eds) New normal in digital enterprises. Palgrave Macmillan, Singapore. https://doi.org/10.1007/978-981-19-8618-5_12

Mondal S, Das S, Vrana VG (2023a) How to bell the cat? A theoretical review of generative artificial intelligence towards digital disruption in all walks of life. Technologies 11(2):44. https://doi.org/10.3390/technologies11020044

Mondal SR, Yegen C, Das S (2023b) New normal in digital enterprises – strategies for sustainable development. Palgrave-Singapore. https://doi.org/10.1007/978-981-19-8618-5 and Hardcover ISBN 978-981-19-8617-8

Neukart F (2025) Toward the stars: technological, ethical, and sociopolitical dimensions of interstellar exploration. Space Policy:101676. https://doi.org/10.1016/j.spacepol.2024.101676

Ngoc-Vinh N, Tien-Dung P, Das S (2022) Digitization of business; Need for recombination during COVID-19. In: Sustainable development and innovation of digital enterprises for living with COVID-19. Springer, Singapore, pp 1–17. https://doi.org/10.1007/978-981-19-2173-5_1

Nguyen TDM, Mondal SR, Das S (2022) Digital Entrepreneurial Transformation (DET) Powered by New Normal Sustainable Developmental Goals (n-SDGs): Elixir for growth of country's economy. In: Sustainable development and innovation of digital enterprises for living with COVID-19. Springer, Singapore, pp 69–84. https://doi.org/10.1007/978-981-19-2173-5_5

Nikitaev D, Thomas LD (2022) Preliminary results for in-situ alternative propellants for nuclear thermal propulsion. Nucl Technol 208(sup1):S96–S106. https://doi.org/10.1080/00295450.2021.2021768

Omira SA, Mourad AHI (2024) Future of antimatter production, storage, control, and annihilation applications in propulsion technologies. Int J Thermofluids:101012. https://doi.org/10.1016/j.ijft.2024.101012

Pickett MT, Roberson LB, Calabria JL, Bullard TJ, Turner G, Yeh DH (2019) Regenerative water purification for space applications: needs, challenges, and technologies towards "closing the loop". Life Sci Space Res 24:64–82. https://doi.org/10.1016/j.lssr.2019.10.002

Pop V (2000) Appropriation in outer space: the relationship between land ownership and sovereignty on the celestial bodies. Space Policy 16(4):275–282. https://doi.org/10.1016/s0265-9646(00)00037-0

Poppick L (2017, March 1) When humans begin colonizing other planets, who should be in charge? Smithsonian; Smithsonian.com. https://www.smithsonianmag.com/science-nature/humans-begin-colonizing-other-planets-who-should-be-in-charge-180962331/

Scheibler C, Toprani SM, Mordukhovich I, Schaefer M, Staffa S, Nagel ZD, McNeely E (2022) Cancer risks from cosmic radiation exposure in flight: a review. Front Public Health 10. https://doi.org/10.3389/fpubh.2022.947068

Spry JA (2022) Current status and future perspectives in planetary protection. Nat Microbiol 7(4):475–477. https://doi.org/10.1038/s41564-022-01060-9

Starr SO, Muscatello AC (2020) Mars in situ resource utilization: a review. Planet Space Sci 182:104824. https://doi.org/10.1016/j.pss.2019.104824

Steffen O (2021) Explore to exploit: a data-centred approach to space mining regulation. Space Policy 59:101459. https://doi.org/10.1016/j.spacepol.2021.101459

Szocik K, Wójtowicz T, Braddock M (2020) The Martian: possible scenarios for a future human society on Mars. Space Policy 54:101388. https://doi.org/10.1016/j.spacepol.2020.101388

Tandon S, Das S (2023) Digital revolution complements sustainable goal achievement during and after pandemic: a research agenda. In: Mondal SR, Yegen C, Das S (eds) New normal in digital enterprises. Palgrave Macmillan, Singapore. https://doi.org/10.1007/978-981-19-8618-5_11

Thomas D (2024) Nuclear thermal propulsion – progress and potential. J Space Safety Eng 11(2):362–373. https://doi.org/10.1016/j.jsse.2024.04.001

Tien-Dung PHAM, Majerova J, Das S (2022a) Phases of possible recovery of digital enterprises in new normal business for living with COVID-19 times: opportunities for a new era in sustainable development goals. In: Sustainable development and innovation of digital enterprises for living with COVID-19. Springer, Singapore, pp 19–33. https://doi.org/10.1007/978-981-19-2173-5_2

Tien-Dung P, Nguyen XD, Das S (2022b) Sustainable goal achievement by digital revolution during and after pandemic; How much one wins and losses: a bird's eye view for future planning. In: Sustainable development and innovation of digital enterprises for living with COVID-19. Springer, Singapore, pp 133–147. https://doi.org/10.1007/978-981-19-2173-5_8

Triggs G (2011) The Antarctic Treaty System: a model of legal creativity and cooperation. In Smithsonian Institution Scholarly Press eBooks, pp 39–49. https://doi.org/10.5479/si.9781935623069.39

Vrana V, Das S (2023a) Dynamic restructuring of digital media and entertainment sector: role of urbanization, industrial innovation, and technological evolution. In: Das S, Gochhait S (eds) Digital entertainment as next evolution in service sector. Palgrave Macmillan, Singapore. https://doi.org/10.1007/978-981-19-8121-0_2

Vrana V, Das S (2023b) Role of enterprises and their digital transformation in post-pandemic sustainable development. In: Mondal SR, Yegen C, Das S (eds) New normal in digital enterprises. Palgrave Macmillan, Singapore. https://doi.org/10.1007/978-981-19-8618-5_7

Yegen C, Das S (2023) Digital solution for print and publishing media. In: Das S, Gochhait S (eds) Digital entertainment as next evolution in service sector. Palgrave Macmillan, Singapore. https://doi.org/10.1007/978-981-19-8121-0_12

Subhra R. Mondal Senior Researcher and academician at Duy Tan University, specializing in brand management, AI, and smart tourism. Editor of multiple Scopus-indexed books, her work explores digital recruitment and Web 5.0. She bridges academia and industry through collaborative projects.

Subhankar Das Senior Researcher and academician at Duy Tan University, publishing extensively on AI, blockchain, and digital transformation. Editor of seven Scopus-indexed books and reviewer for top journals. Mentors global PhD candidates and drives innovation in ASEAN academic collaborations.

Floating Sanctuaries: The Art of Spacecraft Hotel Design

Subhra R. Mondal and Subhankar Das

Abstract Rise of space tourism has transported the idea of orbital hotels, or "floating sanctuaries," from science fiction to real life innovation. The subsequent chapters of this book chapter discuss multidisciplinary challenges associated with the design of spacecraft hotels with a focus on technical challenges such as artificial gravity generation, radiation shielding, and closed-loop life support systems. Its design strategies for human-centric design to promote comfort in zero-gravity environments involve spatial orientation techniques, modular interiors, and advanced material science. Operational viability centres on safety protocols and sustainability practices, such as waste-to-resource technologies, and emergency escape systems. Drawing upon literature in this domain, the chapter identifies the lack of luxury-utility trade-offs studies and a dearth of research on changes in long-term guest behaviors. Additionally, sections outline the need for the phased implementation of change through stakeholder cooperation. Orbital hotels are positioned as tourist destinations and prototypes for interplanetary habitation, emphasizing their potential to enhance sustainable living in outer space.

Keywords Space tourism · Orbital hotels · Zero-gravity design · Sustainability · Artificial gravity · Space colonization

1 Introduction

Though it may seem like the plot of a science fiction book, human exploration of space tourism is closer to reality than ever as we enter the twenty-first century (Smith 2000). Companies like SpaceX, Blue Origin, and Virgin Galactic represent just the tip of the iceberg when it comes to transforming the costs of space travel,

S. R. Mondal · S. Das (✉)
Department of Marketing, South Star Management Institute, Duy Tan University, Da Nang, Vietnam
e-mail: subhraranimondal@duytan.edu.vn; subhankardas@duytan.edu.vn

© The Author(s), under exclusive license to Springer Nature Singapore Pte Ltd. 2025

S. R. Mondal et al. (eds.), *Pioneering the New Space Economy through AI and Immersive Technologies*, https://doi.org/10.1007/978-981-96-5977-7_16

opening up an industry focused on off-world experiences (Patel 2024; Evgeniou 2024). Within this evolution, orbital hotels, the term "floating sanctuaries" has emerged as a cornerstone of innovation (Mann and Permoser 2021). These structures are intended not only as refuge, but as unrivalled luxury for wealthy travelers looking to escape the confines of Earth (Frost 2023). However, creating habitable, safe, and aesthetically pleasing surroundings in the inhospitable conditions of space is a maze of technical, ethical, and creative challenges. This chapter unpacks the interplay of engineering discipline and creative vision that must intertwine to bring the vision of spacecraft hotels to life, all while accommodating the distinct challenges of zero-gravity environments and the need for sustainability and human comfort.

The idea of space hotels has fascinated popular imagination for decades, when Stanley Kubrick's 2001: A Space Odyssey (1968) portrayed sleek, rotating stations with artificial gravity overlooking the cosmos. Today, those visions are coming closer to life. Speculative fiction is slowly realized as actionable design—projects such as Orion Span's Aurora Station or commercial modules to the International Space Station (ISS) from Axiom Space (Tom 2024; Frost 2023). However, even as we make these strides, the jump from orbital laboratories to hospitality-grade habitats will require rethinking almost every aspect of architecture and engineering. Spacecraft accommodations would have to contend with the microgravity, cosmic radiation, and psychological strains of isolation, unlike terrestrial luxuries, while maintaining an ambience of luxury (Chandran and Lamb 2024).

The struggle to balance technical limitations with human-centered design is at the heart of these efforts. Artificial gravity, for instance, remains a holy grail for long-term space habitation. In their influential book The High Frontier, Fowles and O'Neill (1977) proposed rotating structures that would create gravity via centrifugal force, yet their use in compact hotel designs raises serious engineering challenges. Another serious challenge is shielding against radiation; exposure to cosmic rays over extended periods endangers structural integrity and human health, requiring shielding material such as polyethylene or water-based composites to absorb harmful particles (Giaccardi and Redström 2020). By necessity, life support systems must also perform with nearly perfect reliability, reconstituting air, water, and waste in closed-loop systems similar to those tested in Biosphere 2 (Seedhouse 2020). Such technical imperatives have to be balanced with interior layouts that both subvert traditional notions of "up" and "down," leaving designers to use color gradients, tactile cues, and virtual reality (VR) to orient guests in zero-gravity atmospheres (Khafaji and Kamaran 2019).

Spacecraft hotel design also relies heavily on material science. Fabrics and furnishings must be lightweight, fire-retardant, and rugged enough for life in space, but luxurious enough to warrant sky-high nightly rates. Heavy industrial modules, such as those developed by Bigelow Aerospace in early 2022, provide insights into efficient use of onboard volume and demonstrate the feasibility of expandable habitats allowing maximum habitation in tight quarters. Multipurpose furniture—like foldable beds that transform into storage units—further maximize square footage. At the same time, virtual-reality windows could present guests with customizable views of Earth or distant galaxies to mitigate the claustrophobia that comes with cramped space (Howell 2016).

However, a discussion of orbital hotels would be incomplete without consideration of safety and sustainability. Emergency protocols must contemplate micrometeoroid impacts, system failures, or medical crises, which will necessitate redundant escape pods and medical facilities on site. On the other hand, the ecological cost of space tourism is not to be neglected. Closed-loop life support systems, harvesting solar energy, and technologies converting waste to resource offer promising avenues to reduce dependence upon resupply from Earth (Staff 2021). As Ehlausanne (2025) argues, these hotels must be at the forefront of sustainable innovation, not just playhouses for the rich.

Beyond tourism, floating sanctuaries could have more significant implications for humanity's future in space. Glimpses into our multi-planetary future have taken center stage over the past decade, as visionaries like Elon Musk and Stephen Hawking have argued for their imperative for species survival (Raven and Wackernagel 2020). Orbital hotels could serve as testbeds for the technologies that will make long-term colonization feasible, whether powerful hydroponics or psychological support systems for isolated crews. In that respect, they are not merely destinations, but way stations to interplanetary travel (Team 2024).

The contribution of Saleh and Brem (2023) here provides an overview of the rich tapestry of issues and creative elements that underpin the embryonic discipline of hotel design in space. The authors devised a well-rounded method to advance this frontier by examining how engineering, aesthetics, and sustainability engage. So much of their work highlights a key truth: The future of space tourism is not just the text of engineering barriers but the design of experiences that stimulate our beautiful, experience-hungry species; it is a delicate dance that will determine our destiny as a spacefaring species.

2 Literature Review for Factors of Spacecraft Hotel Design

In essence, the idea of orbital hotels, or "floating sanctuaries," synthesizes cutting-edge engineering, human-oriented design, and sustainable innovation. As space tourism moves from fantasy to feasibility, research from academia and industry alike has continued to tackle the open-ended challenges of designing luxurious habitats in low-Earth Orbit (LEO) populations. This review pulls existing literature together on the technical, aesthetic, safety, and sustainability factors that are most relevant for the design of a spacecraft hotel. It emphasizes potential gaps and future directions for this emerging field.

2.1 Engineering Challenges: Artificial Gravity, Radiation, and Life Support

Artificial Gravity

In space, a gravity-free environment poses serious physiological threats such as muscle decay and bone mineral loss (Guzman 2024). Scientists have long called for artificial gravity systems to counter these issues, using centrifugal force created by rotating structures. However, scaling these kinds of designs for compactable hotels is challenging. The physiologically favorable smaller radii demand high-speed rotation rates and can cause motion sickness via Coriolis effects (Sam et al. 2017). Cottage-style designs are suggested in recent studies to help balance spatial efficiency while maintaining comfort (Zach 2023). Commercial efforts such as the Gateway Foundation's Von Braun Station hope to prove these principles, albeit with few data points on guest tolerance (Gulino 2019).

Radiation Shielding

Spacecraft hotels must deal with ionizing radiation from cosmic rays and solar flares that threaten structural integrity and human health. As Horst et al. (2022) noted, polyethylene and water-rich composites are good neutron shields due to their high hydrogen content that annihilates high-energy particles. However, improving their aesthetics to be compatible with luxury habitats is one of the challenging criteria. Some such as "radiation-resistant" smart glass, which promotes the unique features of special shielding properties with high-quality transparency for panoramic views, have been explored (Al-Issa 2024). While strides have been made in studying the effects of this more prolonged exposure on human health, the risks for tourists—a demographic often left out of astronaut health studies—should be further analyzed.

Life Support Systems

Within a closed-loop life support system (CLSS), air, water, and waste must all be recycled as part of a sustainable system within the tiny habitat. In the 1990s Biosphere 2 experiments, it became clear that maintaining balance in sealed ecosystems was quite complex, such as imbalances on the microbial scale and fluctuations of the oxygen levels of day and night (Allen et al. 2003). NASA's Environmental Control and Life Support System (ECLSS) onboard the ISS achieves ~93% water recovery but is energy intensive (Traweek and Tatara 1998) and modern iterations. Hotels rely on scalability and redundancy. Notably, recent advances with algae-based bioreactors show potential for oxygen generation and waste processing, although their applications in hospitality-grade spaces remain unverified (Mofijur et al. 2023).

Comfort/Aesthetic Considerations: Interior Design in Zero-G

Conventional ideas of "up" and "down" evaporate in microgravity, requiring new strategies for spatial orientation. Shin et al. (2021) suggest employing color Arabic numerals, tactile floor cues, and adjustable lighting to construct visual hierarchies. Virtual reality (VR) adds another dimension of evoking orientation reported that introduction of VR windows that offer vistas of Earth reduced claustrophobia in simulated space habitats (Tussyadiah et al. 2017). Furniture design adapts to zero-G: multipurpose, foldable elements maximize limited space; magnetic surfaces secure objects without rigid fixtures (Cheng et al. 2021). However, research into user experience in these settings is poor, especially regarding contrasting expectations of luxury versus functional minimalism.

Material Science

Materials must answer both mandates: durability in extreme conditions; opulence to suit high-end tourism. Inflatable modules, like the one from Prameela et al. (2022), provide deployable spacecraft with radiation shielding immediately upon extension but less emphasis on aesthetics in their interiors. Recent advances in graphene-based fabrics—thin, fireproof, malleable, similar to paper—may fill this gap, ideally, with customizable textures and colors (Chowdhury et al. 2023). Self-cleaning coatings also reduce maintenance in resource-limited environments, although their long-term efficacy in space has yet to be demonstrated.

Emergency and Safety Protocols

The safety systems of orbital hotels must mitigate unique risks, such as micrometeoroid impacts, rapid decompression, and medical emergencies. Current ISS procedures, like shelter-in-place during solar storms, might be insufficient for larger tourist populations. Redundant escape pods, with autonomous re-entry, would not be optional; the Crew Dragon of SpaceX provides a basis, though it remains to be seen if large-scale occupancy, for dozens of guests, can be achieved (Monzon 2020). Medical facilities represent another hurdle; while telemedicine and AI diagnostics can help compensate for not having physicians on board, delays in communication back to Earth will require autonomous systems for trauma care (Sharma et al. 2023).

Ethics and Sustainability

Space tourism has attracted criticism regarding its ecological footprint, especially regarding resource consumption and debris generation. It can be primarily closed-loop systems and solar energy harvesting, but energy storage still poses a challenge. For example, lithium-ion batteries degrade in radiation-intensive environments, spurring investigation into nuclear-powered Stirling engines (Pultarova 2021). Waste-management is just as important; using pyrolysis technologies to convert organic waste to fuel could lessen the reliance on resupply missions (Al-Rumaihi et al. 2022). Space tourism has ethical implications, especially concerning the limited availability. Scholars argue that hotels should lead the way into sustainable practices to justify sustainability-related environmental costs rather than perpetuating exploitative paradigms of Earth (Carbajales-Dale and Murphy 2022).

Big Picture: Implications for Space Colonization

Orbital hotels are often pitched as a precedent for interplanetary habitation. Orbital hotels act as proving grounds for technologies such as hydroponic agriculture and 3D-printed habitats. Psychological resilience is also studied: Antarctic and ISS crews have taught us that biophilic design and communal spaces effectively break isolation (Rajendran et al. 2024). Short-term tourists may not be particularly interested in changing their behaviors to suit the local environment; novelty is their priority; consistency is a luxury that can come later: Making these findings applicable to the short-term tourists requires more nuanced approaches.

2.2 *Gaps and Future Directions*

Although the existing literature offers a firm basis, there are key gaps:

a. Long-Term Guest Studies: Most studies are on astronauts; no data on tourists' physiological and psychological endurance.
b. Luxury vs. Utility: We do not talk enough about the compromise between lavish design and pragmatic effectiveness.
c. Regulatory Aspects: International regulations about space hotels, safety, and environmental sustainability are in their infancy and require cross-disciplinary collaboration.

Future studies must focus on testing in a real environment, such as Axiom Space's intended commercial modules, to confirm theoretical models. Cooperation between aerospace engineers and hospitality specialists can probably create new solutions where comfort meets survivability.

Hotel spacecraft design requires a comprehensive combination of engineering diligence, artistic ingenuity, and ethical prescience. Although technical challenges such as artificial gravity and radiation shielding have captured much of the current discussion, the human aspects of comfort, safety, and sustainability are just as important to success. Trang et al. (2019) explain that these floating sanctuaries serve as tourist resorts and prototypes of humanity's multi-planetary future. Applying social commentary, guest experience, and regulatory lenses to challenges of space colonization could downplay similar disparities in the luxury experience and serve as enduring legacies of twenty-first-century spacefaring endeavors.

3 A Framework for Spacecraft Hotel Design Implementation

A structured and empirical interdisciplinary framework is necessary to design all such orbital hotels or floating sanctuaries, considering all-inclusivity, human-centered design, safety, and sustainability. This framework will suggest concrete pathways for spacecraft hotels to be developed in five different areas from concept to execution.

1. Internal Systems Development

Goal: develop systems for artificial gravity, radiation shielding, and life support.

Artificial Gravity:

Tethered modules or centrifugal designs could rotate to simulate gravity in habitats. Start with small prototypes (close to a 10–20 m radius) to see if Guests can handle Coriolis effects. Collaborate with aerospace contractors (like SpaceX or Blue Origin) to adapt rotation systems to the current spacecraft framework.

Radiation Shielding:

Select structural layers of hydrogen-containing materials (e.g., polyethylene, water composites). Look for transparent protective barriers like smart glass layered with radiation-absorbent nanomaterials.

Test iterations inside simulated surrounding spaces to maximize material durability and protect guests.

Closed-Loop Life Support:

Scalable version of NASA's ECLSS with algae bioreactors for oxygen generation and waste recycling.

Real-time monitoring of air and water quality and maintaining them using AI-driven systems.

Milestone: Within 5–7 years, a fully functional, rotating habitat module with integrated shielding and life support. Figure 1 represents the development of space habitat systems.

2. Design for Humans and Comfort

Goal: Design intuitive, indulgent interiors that counteract the mental strain of microgravity settings.

Spatial Orientation:

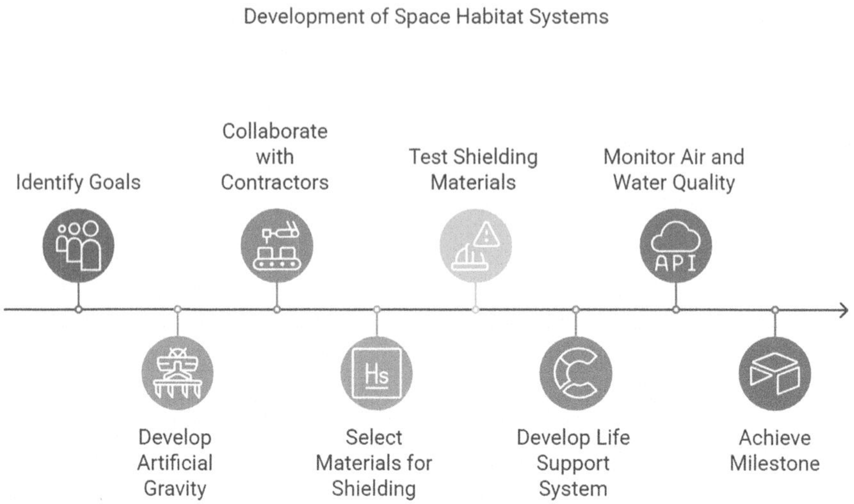

Fig. 1 Space habitat systems (Source: Authors' conception)

Prominent color gradients (e.g., cool-to-warm tones), tactile floor markings, and directional lighting can help set visual "up" and "down" cues.

Furnish VR windows with customizable Earth or cosmic views to alleviate feeling claustrophobic.

Modular Interiors:

To save space, make furniture designs multipurpose (i.e., foldable beds or magnetic surfaces).

Fuse luxury with the functional by working with luxury designers to bring graphene-based fabrics and self-cleaning materials.

Milestone: Successfully grow the user experience through mock zero-G experiments with upscale users within the proposed 3-year production timeline. Figure 2 represents elements of a space habitat.

3. Safety and Emergency Procedures

Goal: Keep guests safe, with backup systems and Twitter-ready response plans.

Emergency Systems:

There is multiple escape pod (e.g., modified Crew Dragon capsules) to re-enter Earth automatically.

Add micrometeoroid shielding with Whipple bumpers and self-healing materials.

Medical Infrastructure:

Prepare hotels with AI diagnostic tools and telemedicine hubs connected to Earth-based specialists. Some medical staff must be trained in zero-G first aid and emergency decompression procedures. Figure 3 represents safety and emergency procedures for space hotels.

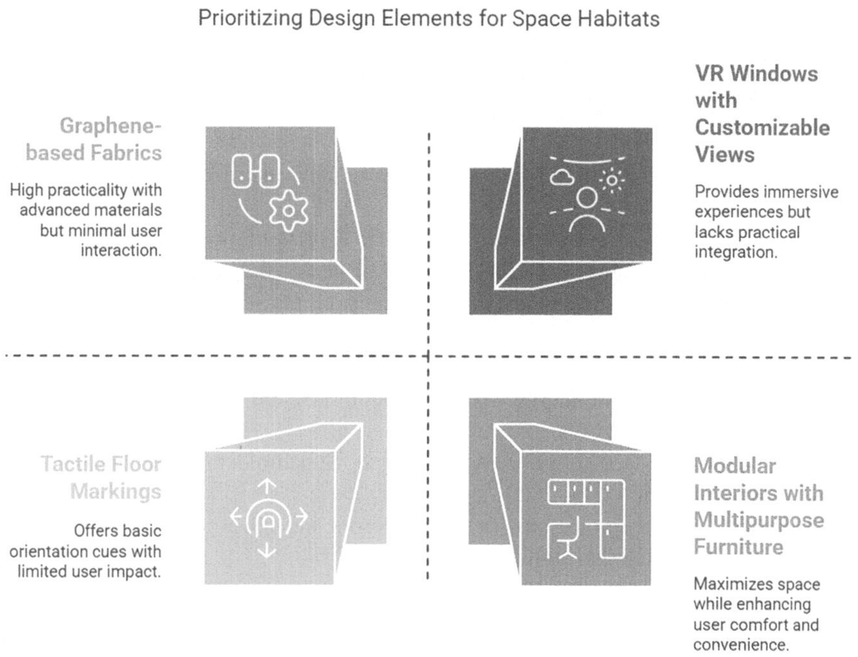

Fig. 2 Space habitat elements (Source: Authors' conception)

Fig. 3 Space hotels and their safety emergency procedure (Source: Authors' conception)

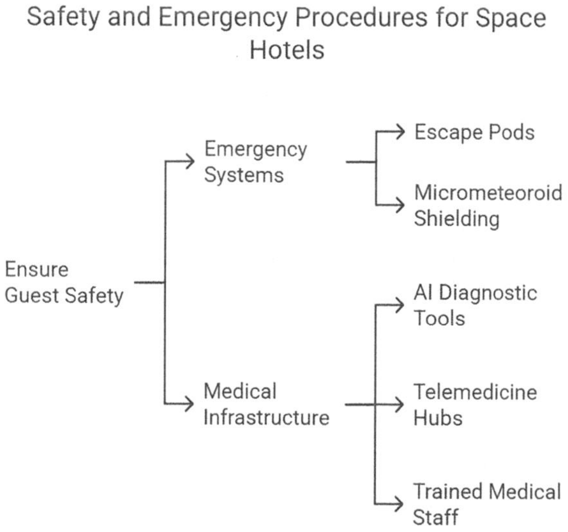

4. Sustainability Practices

Goal: Completed with the ISS commercial module safety requirements by 2030. Sustainability Practices.

Purpose: Energy efficiency and waste minimization.

Energy Systems:

Use high-efficiency solar panels and consider portable nuclear-powered Stirling engines for constant power.

Waste Management:

Implement pyrolysis reactors for converting organic waste to fuel and 3D-printing feedstock (Das 2023a).

Implement closed-loop water systems with 95% recycling rates.

Milestone: 2035—Energy autonomy (80% of power through local renewables) and zero waste resupply dependency. Figure 4 represents sustainability practices.

5. Engage Stakeholders and Implement in Phases

Goal: By 2035, 80% of energy will be autonomous, and 100% will be restocked with zero waste.

More Informed Decision-making Through Stakeholder Collaboration and Phased Implementation (Das 2023b).

Objective: Initiate partnerships between sectors and governments to help overcome regulatory and logistical hurdles.

Phases:

a. R&D (2024–2028). Demonstration of fundamental technologies (e.g., artificial gravity, shielding) to be validated through suborbital experiments.
b. Prototyping (2029–2032): Build and test a full-size orbital module featuring hospitality amenities.
c. Certification (2033–2035)—Obtain international regulatory approval for passenger safety and sustainability.

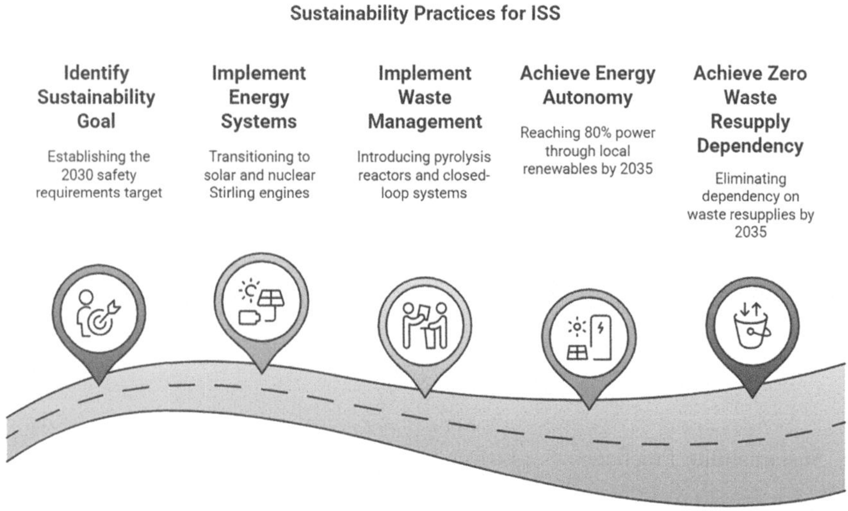

Fig. 4 Sustainability practices (Source: Authors' conception)

d. Deployment (2036+): Launch a first commercial orbital hotel, then undergo iterative upgrades based on guest feedback.

Collaborations:

a. Involve space agencies (NASA, ESA) with safety measures.
b. Join forces with high-end hospitality companies (think Four Seasons) to perfect service models.
c. By 2040, operational inaugural spacecraft hotel—model for interplanetary locations.

This framework prioritizes iterative innovation, cross-sector collaboration, and rigorous testing and aims to convert spacecraft hotels from pipe-dream markets into functional ventures (Das et al. 2023). Only if humanity tackles technical, aesthetic, and ethical challenges in unison could we herald a new age of space tourism that reconciles luxury with sustainability, security, and scientific development.

4 Implications of the Spacecraft Hotel Framework

A proposed framework for building orbital hotels or "floating sanctuaries" is ambitious and transformative in its vision of humanity's future in space. The framework intends to evolve space tourism from a fringe project to a foundational element for off-planet living by merging cutting-edge engineering with high-end aesthetics, safety measures, and environmental responsibility (Di Virgilio and Das 2023a). However, its realisation has profound theoretical, practical, social, and sustainable implications. This analysis discusses these dimensions and the promise and pitfalls they embody.

4.1 Theoretical Implications: Revolutionizing Theories on Space Habitation

The framework translates foundational theories of space colonization like Gerard K. O'Neill's rotating habitats into designs that can be built on Earth (National Space Society 2023). Making the case for tethered modules and centrifugal gravity systems, it probes the possibilities of artificial gravity in close quarters; an essential aspect for long-term human sustainability off this globe (Di Virgilio and Das 2023b). This fits the multi-planetary vision of space tourism, but adds new variables, such as balancing the rotation speed with the guests' comfort (Majerova and Das 2023). A key feature of CLSS is also the advance of Biosphere 2's legacy (Allen et al. 2003), showcasing the evolution of closed systems from experimental disasters to reliable infrastructures.

Redefining Human-Centered Design

Standard architectural principles, based on Earth's orientation, gravity and hierarchies of space, were flipped upside down in zero-G environments. The framework encourages designers to reconsider "luxury" with flexible, alternative solutions such as interiors augmented by VR and adaptable furniture (Minh-Nhat et al. 2022a). These enhancements reach beyond the functional to discuss phenomenological theories of spatial perception (Soltani and Kirci 2019) to produce places that engage with human psychology. Incorporating color gradients to simulate directionality, for instance, recalls Gestalt principles of visual organization, suggesting that comfort in space is as much based on cognitive proximity as physical safety.

Systems Engineering Paradigms

The framework highlights the imperatives for holistic systems engineering, including that technical subsystem (e.g., radiation shielding, energy grids) must interoperate seamlessly. This corresponds with soft systems methodology, with the immediate feedback loops between design engineers and end-users (Checkland 2000). It also incorporates aspects of resilience theory, putting redundancy first in safety systems and scalability for life support and ensuring hotels can continue to operate in the face of unexpected perturbations essential in an unpredictable space environment (Minh-Nhat et al. 2022b).

4.2 Practical Implications: Technical Feasibility and Cost

Though the framework suggests phased timelines for implementation, several significant technological challenges remain (Mondal and Das 2023a). Artificial gravity systems require meticulous engineering to prevent motion sickness, and existing materials for radiation shielding (like polyethylene) are not transparent enough for the panoramic views preferred by most space tourists (Mondal and Das 2023b). Prototyping these systems will require significant investment (on the order of $2–5 billion per module), raising concerns about how such systems will be funded. Public-private partnerships similar to NASA's Commercial Crew Program might reduce cost but could also add time, if geopolitical tensions or budget cuts arise.

Regulatory and Logistical Challenges

However, no international legal regime governs orbital hotels today, posing uncertainty regarding safety certifications, accident liability and space traffic management (Mondal and Das 2023c). While the framework's dependency on ISS safety standards is practical, there will need to be some coordination with groups like the

United Nations Office for Outer Space Affairs (UNOOSA) as we dial these into larger tourist populations. Similarly, supply chain logistics are still untested, like how to ferry luxury materials into orbit. SpaceX's Starship, with its 100-ton payload capacity, would be able to fit the bill. However, if cost-effective for transport, the sheer number of launches will likely contribute to the space junk problem, making long-term operations complex.

Human Factors and Training

Staffing these hotels is a unique challenge. Crew members will require dual hospitality and emergency response expertise, like "space butlers" schooled in zero-G first aid and crisis management. Simulations on Antarctic research stations or underwater habitats could provide preparatory training, but the psychological toll of prolonged isolation on staff members and guests remains under-researched (Mondal et al. 2023a, b).

4.3 Social Implications: Elitism and Accessibility

The prohibitive prices of space tourism, initially estimated at $one million a night, will likely deepen global disparities, making orbital hotels playgrounds for the superrich. These risks create a "space aristocracy," worsening social stratification (Polkowska 2021). Nevertheless, suppose hotels use profits to reopen space, for example, by funding educational programs for low-income communities or subsidizing scientific research (Ngoc-Vinh et al. 2022). In that case, the framework's emphasis on sustainability may help mitigate criticism.

Cultural Shifts and Identity

Orbital hotels could enable new stories of culture globally, redefining humanity's relationship with space (Nguyen et al. 2022). The unique experience of viewing Earth from orbit (overview effect) may create an awareness or consciousness of global environmental issues with guests (Taylor 2021). By contrast, the commodification of space could mask its symbolic value as a "final frontier," which it threatens to become in the form of a luxury commodity (Tandon and Das 2023). Microgravity art and architecture could also pave the way to new schools of thought, rejecting our terrestrial ideas of shape and function.

Labor and Ethics

The framework's reliance on AI and automation may displace human workers, but the space hospitality industry's unique needs could leave room for niche job markets (Tien-Dung et al. 2022a). Ethical issues emerge about working conditions; for one, will workers live through more extended contracts with little contact with Earth? Setting equitable labor conditions, like maritime or Antarctic work regulations, will be crucial to curtail exploitation.

4.4 Sustainable Implications

There are many analogues of our environmental stewardship here on Earth. However, in space, we can innovate and enforce environmental values. Closed-loop systems and reliance on solar energy underpin the framework, which aligns with the circular economy principles, reducing resource extraction from the earth (Tien-Dung et al. 2022b). Pyrolysis reactors that convert waste to fuel could eliminate up to 80% of dependence on costly resupply missions (Hasan et al. 2025) and lay the path for future sustainable deep-space habitats. However, energy storage remains challenging (Vrana and Das 2023a). Radiation-resistant batteries or compact nuclear systems must be polished for uninterrupted operations (Vrana and Das 2023b).

Ethical Responsibility

Space tourism sustainability has ethical dimensions. As O'Sullivan (2022) claims, the moral failure of recreating the colonializing forms of extraction of Earth in space would be a disaster. The framework's emphasis on zero-waste systems and renewable energy represents a commitment to ethical innovation, but transparency will be the key to success (Yegen and Das 2023). Independent audits of sustainability claims and carbon footprints will be needed to establish public confidence.

The spacecraft hotel concept is an audacious melding of ambition and practicality, providing a blueprint for humanity's emergence as a spacefaring species. In theory, it connects decades of research with frontier innovation; in practice, it faces enormous technical and financial obstacles. Socially, it challenges inequality, offering the potential for unifying experiences that change the world. It is a sustainability trailblazer—but also needs to find the tightrope between growth and responsibility. As Antic (2020) argues, these are not simply tourist destinations, but prototypes for humanity's future, a future that will require us to reach for the stars while never losing our moral and ecological compass.

5 Conclusion

The concept of orbital hotels or "floating sanctuaries" indicates a watershed moment in humanity's quest to be a multi-planetary species. The framework condenses decades of theoretical investigation, technological innovation, and ethical meditation to codify a "feasible" path toward establishing luxurious, secure, and sustainable habitats in space. Though the challenges are substantial—encompassing engineering difficulties and societal inequities, the interdisciplinary framework offers a blueprint in aligning humanity's hopes with the pragmatic realities of living beyond Earth. So, as we find ourselves on the brink of this new frontier, it is worth taking a moment to consider the more significant lessons and responsibilities that this new challenge requires us to shoulder.

The framework elaborates on balancing technical innovation and strong human-centered design. Technologies such as artificial gravity, radiation shielding, and closed-loop life support are not just innovations with engineering implications, they are critical to securing human health in space and maintaining psychological health. As envisioned in the rotating habitats initially proposed, the essential factors of iterative innovation, developing and evaluating small-scale prototypes, would revolutionize Earth-centric societies and allow researchers to hone beneficial responses to issues in motion sickness and spatial disorientation. Likewise, the evolution of material science, including radiation-resistant smart glass and graphene-based textiles, demonstrates the possibility of merging function with luxury, creating a unique challenge for designers to alter how we view luxury in zero-G. Nevertheless, alongside these technical advancements, it is necessary to rigorously implement safety procedures, with redundant escape systems and AI medical infrastructure—granted some human elements can be minimized but not human cognition.

The framework, however, is not solely about engineering—it forces us to consider societal and ethical questions of profound gravity. Space tourism's steep price tag threatens to establish a "space aristocracy," in which orbital hotels are the sole territory of the ultra-elite. This exclusivity suggests such inequalities could carry over to the cosmos, subverting democratic ideals that often accompany space exploration discussions. However, the framework's focus on sustainability, through closed-loop systems, waste-to-resource technologies and harvesting solar energy, provides a counter-narrative. The orbital hotels through ecological stewardship could also set precedents on their planet, paving the path for responsible resource management. Additionally, collaborations with schools and universities and civic engagement campaigns could make space tourism more inclusive, allowing developing nations a piece of the cosmic pie while encouraging future explorers to see space as a shared legacy.

The cultural effects of orbital hotels would be just as revolutionary. The "overview effect," a designated transformation in consciousness claimed by astronauts who see the earth from space, might generate a refreshed sense of where stewardship is possible and, in turn, set the wheels in motion of global campaigns in the name of the planet. On the other hand, this commodification of space threatens to turn it into an upscale experience, severed from its symbolic role as a frontier of collective human accomplishment. Outcomes, such as independent audits of sustainability claims, equitable labor practices for staff, and international regulatory cooperation will be critical in ensuring that space tourism develops as a force for solidarity rather than splintering.

In this sense, orbital hotels have far more significant implications than tourism. Such habitats prototype long-term space colonization, testing technologies for future Martian bases or lunar settlements. Advances in hydroponic farming, 3D-printed habitats and strategies in psychological resilience, tested within the cramped quarters of orbital hotels, may become critical for maintaining life on other planets. Additionally, the government-private enterprise-academic institution partnership underpinning the framework provides a model for the type of global collaboration necessary to tackle existential challenges in space and on Earth.

Ultimately, the road to operational orbital hotels must also serve as a crucible for humanity's ethical maturation, every bit as much a test of our technical prowess. The interesting framework proposed in this chapter, can guide us through this journey and highlight that, while space tourism appears to be an exciting agenda, it must be complemented by ambition and responsibility to achieve a successful outcome. Our lesson from our terrestrial progress must be to look forward to this new era and against replicating the same exploitation and inequity that has diluted advancement on the planet's surface. Instead, may orbital hotels become the torchbearers of innovation, sustainability, and inclusivity, living proof of humanity's ability to reach the stars without becoming lost in their light. In this act, we pay homage not only to the pioneers of space exploration but also to all humankind and the dreamers of this planet.

References

Al-Issa N (2024) Redefining luxury: exploring the natural connection with sustainability beyond labels. Cogent Bus Manag 11(1). https://doi.org/10.1080/23311975.2024.2423054

Allen J, Nelson M, Alling A (2003) The legacy of biosphere 2 for the study of biospherics and closed ecological systems. Adv Space Res 31(7):1629–1639. https://doi.org/10.1016/s0273-1177(03)00103-0

Al-Rumaihi A, Shahbaz M, Mckay G, Mackey H, Al-Ansari T (2022) A review of pyrolysis technologies and feedstock: a blending approach for plastic and biomass towards optimum biochar yield. Renew Sust Energ Rev 167:112715. https://doi.org/10.1016/j.rser.2022.112715

Antic M (2020) Development of urban tourism within the urbanization and globalization framework. Glasnik Srpskog Geografskog Drustva 100(2):83–106. https://doi.org/10.2298/gsgd2002083a

Carbajales-Dale M, Murphy TW (2022) The environmental and moral implications of human space travel. Sci Total Environ 856:159222. https://doi.org/10.1016/j.scitotenv.2022.159222

Chandran A, Lamb H (2024) Orbital fails to take flight. Nat Phys 20(11):1691. https://doi.org/10.1038/s41567-024-02688-6

Checkland P (2000) Soft systems methodology: a thirty year retrospective. Syst Res Behav Sci 17(S1):S11–S58. https://doi.org/10.1002/1099-1743(200011)17:1

Cheng HY, Ng PK, Nathan RJ, Saptari A, Ng YJ, Yeow JA, Ng KY (2021) The conceptualisation and development of a Space-Saving multipurpose table for enhanced ergonomic performance. Inventions 6(4):67. https://doi.org/10.3390/inventions6040067

Chowdhury RP, Stegeman LA, Lund ML, Fry D, Madzunkov S, Bahadori AA (2023) Hybrid methods of radiation shielding against deep-space radiation. Life Sci Space Res 38:67–78. https://doi.org/10.1016/j.lssr.2023.04.004

Das S (2023a) Modern-day gaming and the role of FinTech applications in popularising various entertainment platforms across the world. In: Das S, Gochhait S (eds) Digital entertainment as next evolution in service sector. Palgrave Macmillan, Singapore. https://doi.org/10.1007/978-981-19-8121-0_11

Das S (2023b) Digital entertainment: the next evolution in service sector. Emerging Digital Solutions in Reshaping Different Industries Palgrave-Singapore. https://doi.org/10.1007/978-981-19-8121-0

Das S, Mondal SR, Struck F, Eissfeldt J, Beernink L, Sigouras V, Hoose S, Kopac L (2023) What can we see inside the crystal ball? Exploring opportunities and challenges for metaverse and circular economy. AJIS 12(6):221. https://doi.org/10.36941/ajis-2023-0165

Di Virgilio F, Das S (2023a) Digital solutions for social media: role of digital evolution on business enterprises. In: Das S, Gochhait S (eds) Digital entertainment as next evolution in service sector. Palgrave Macmillan, Singapore. https://doi.org/10.1007/978-981-19-8121-0_9

Di Virgilio F, Das S (2023b) Digitalization of enterprises from a new normal sustainable development perspective: opportunities and challenges. In: Mondal SR, Yegen C, Das S (eds) New normal in digital enterprises. Palgrave Macmillan, Singapore. https://doi.org/10.1007/978-981-19-8618-5_9

Ehlausanne (2025, January 28) Sustainability trends in hospitality: key benefits and strategies. eXceed Global Learning Pty Limited. https://insights.ehotelier.com/insights/2025/01/30/sustainability-trends-in-hospitality-key-benefits-and-strategies/

Evgeniou A (2024, August 30) The influence of space exploration on Sci-Fi literature and films. Unicaf – Scholarship Programme. https://www.unicaf.org/the-influence-of-space-exploration-on-sci-fi-literature-and-films/

Fowles J, O'Neill GK (1977) The high frontier: human colonies in space. Technol Cult 18(4):718. https://doi.org/10.2307/3103615

Frost E (2023, August 10) Space travel: a glimpse into the future of human exploration. Medium. https://medium.com/@knutlunder/space-travel-a-glimpse-into-the-future-of-human-exploration-b627e99b973b

Giaccardi E, Redström J (2020) Technology and more-than-human design. Design Des Issues 36(4):33–44. https://doi.org/10.1162/desi_a_00612

Gulino E (2019, September 4) The First-Ever commercial space hotel will have Low-Gravity rock climbing and an insane view of Earth. House Beautiful. https://www.housebeautiful.com/lifestyle/a28903483/first-commercial-space-hotel-von-braun-station/

Guzman A (2024, March 21) Counteracting bone and muscle loss in microgravity – NASA. NASA. https://www.nasa.gov/missions/station/iss-research/counteracting-bone-and-muscle-loss-in-microgravity/

Hasan M, Haque R, Jahirul M, Rasul M (2025) Pyrolysis of plastic waste for sustainable energy recovery: technological advancements and environmental impacts. Energy Convers Manag 326:119511. https://doi.org/10.1016/j.enconman.2025.119511

Horst F, Boscolo D, Durante M, Luoni F, Schuy C, Weber U (2022) Thick shielding against galactic cosmic radiation: a Monte Carlo study with focus on the role of secondary neutrons. Life Sci Space Res 33:58–68. https://doi.org/10.1016/j.lssr.2022.03.003

Howell E (2016, February 26) Bigelow aerospace: inflatable modules for ISS. Space.com. https://www.space.com/19311-bigelow-aerospace.html

Khafaji IAMA, Kamaran R (2019) The influence of spatial flexibility to improve sustainability of interior design by using smart technology (case study –future smart home in Iraq). Eur J Sust Devel 8(4):438. https://doi.org/10.14207/ejsd.2019.v8n4p438

Majerova J, Das S (2023) Sustainable development and digitalization in the supply chain for environmental protection: changes and trends. In: Mondal SR, Yegen C, Das S (eds) New normal in digital enterprises. Palgrave Macmillan, Singapore. https://doi.org/10.1007/978-981-19-8618-5_8
Mann I, Permoser JM (2021) Floating sanctuaries: the ethics of search and rescue at sea. SSRN Electron J https://doi.org/10.2139/ssrn.3913982
Minh-Nhat HO, Majerova J, Das S (2022a) Taking steps to help the phoenix rise from ashes: a roadmap for digital enterprises to develop new sustainable goals. In: Sustainable development and innovation of digital enterprises for living with COVID-19. Springer, Singapore, pp 35–51. https://doi.org/10.1007/978-981-19-2173-5_3
Minh-Nhat HO, Nguyen HL, Mondal SR (2022b) Digital transformation for new sustainable goals with human element for digital service enterprises: an exploration of factors. In: Sustainable development and innovation of digital enterprises for living with COVID-19. Springer, Singapore, pp 85–103. https://doi.org/10.1007/978-981-19-2173-5_6
Mofijur M, Hasan M, Sultana S, Kabir Z, Djavanroodi F, Ahmed SF, Jahirul M, Badruddin IA, Khan TY (2023) Advancements in algal membrane bioreactors: overcoming obstacles and harnessing potential for eliminating hazardous pollutants from wastewater. Chemosphere 336:139291. https://doi.org/10.1016/j.chemosphere.2023.139291
Mondal SR, Das S (2023a) Defining the new digitalized normal: implications of A.I., IoT, and blockchain for business growth. In: Mondal SR, Yegen C, Das S (eds) New normal in digital enterprises. Palgrave Macmillan, Singapore. https://doi.org/10.1007/978-981-19-8618-5_10
Mondal SR, Das S (2023b) Digital solution for industry 4.0 and 5.0 and social media. In: Das S, Gochhait S (eds) Digital entertainment as next evolution in service sector. Palgrave Macmillan, Singapore. https://doi.org/10.1007/978-981-19-8121-0_10
Mondal SR, Das S (2023c) Human-centric digitalization for new sustainable development: an explorative study in human resource management. In: Mondal SR, Yegen C, Das S (eds) New normal in digital enterprises. Palgrave Macmillan, Singapore. https://doi.org/10.1007/978-981-19-8618-5_12
Mondal S, Das S, Vrana VG (2023a) How to bell the cat? A theoretical review of generative artificial intelligence towards digital disruption in all walks of life. Technologies 11(2):44. https://doi.org/10.3390/technologies11020044
Mondal SR, Yegen C, Das S (2023b) New normal in digital enterprises – strategies for sustainable development. Palgrave-Singapore. https://doi.org/10.1007/978-981-19-8618-5 and Hardcover ISBN978-981-19-8617-8
Monzon I (2020, November 3) NASA instructor explains how ISS astronauts prepare for solar storms. International Business Times. https://www.ibtimes.com/nasa-instructor-explains-how-iss-astronauts-prepare-solar-storms-2892095
National Space Society (2023, November 24) The colonization of space – Gerard K. O'Neill, physics today, 1974. NSS. https://nss.org/the-colonization-of-space-gerard-k-o-neill-physics-today-1974/
Ngoc-Vinh N, Tien-Dung P, Das S (2022) Digitization of business; Need for recombination during COVID-19. In: Sustainable development and innovation of digital enterprises for living with COVID-19. Springer, Singapore, pp 1–17. https://doi.org/10.1007/978-981-19-2173-5_1
Nguyen TDM, Mondal SR, Das S (2022) Digital Entrepreneurial Transformation (DET) powered by new normal sustainable developmental goals (n-SDGs): elixir for growth of country's economy. In: Sustainable development and innovation of digital enterprises for living with COVID-19. Springer, Singapore, pp 69–84. https://doi.org/10.1007/978-981-19-2173-5_5
O'Sullivan D (2022) The crisis of policy failure or the moral crisis of an idea: colonial politics in contemporary Australia and New Zealand. Pol Sci 74(1):1–17. https://doi.org/10.1080/00323187.2022.2099915
Patel C (2024, February 6) Space tourism: from science fiction to reality – cameron patel – Medium. Medium. https://medium.com/@cameronpatel325/space-tourism-from-science-fiction-to-reality-171ce6c56c3f

Polkowska M (2021) Space tourism challenges. Rev Eur Comp Law 45(2):153–182. https://doi.org/10.31743/recl.12223

Prameela SE, Pollock TM, Raabe D, Meyers MA, Aitkaliyeva A, Chintersingh K, Cordero ZC, Graham-Brady L (2022) Materials for extreme environments. Nat Rev Mater 8(2):81–88. https://doi.org/10.1038/s41578-022-00496-z

PultarovaT(2021,July26)TheriseofspacetourismcouldaffectEarth'sclimateinunforeseenways,scientists worry. Space.com. https://www.space.com/environmental-impact-space-tourism-flights

Rajendran S, Domalachenpa T, Arora H, Li P, Sharma A, Rajauria G (2024) Hydroponics: exploring innovative sustainable technologies and applications across crop production, with emphasis on potato mini-tuber cultivation. Heliyon 10(5):e26823. https://doi.org/10.1016/j.heliyon.2024.e26823

Raven P, Wackernagel M (2020) Maintaining biodiversity will define our long-term success. Plant Divers 42(4):211–220. https://doi.org/10.1016/j.pld.2020.06.002

Saleh R, Brem A (2023) Creativity for sustainability: an integrative literature review. J Clean Prod 388:135848. https://doi.org/10.1016/j.jclepro.2023.135848

Sam AA, Mondal J, Ghosh P (2017) Effect of rotation on the flow behaviour in a high-speed cryogenic microturbine used in helium applications. Int J Refrig 81:111–122. https://doi.org/10.1016/j.ijrefrig.2017.05.026

Seedhouse E (2020) Open-loop vs. closed-loop life support systems. In Springer eBooks (pp 75–99). https://doi.org/10.1007/978-3-030-52859-1_3

Sharma S, Rawal R, Shah D (2023) Addressing the challenges of AI-based telemedicine: best practices and lessons learned. J Educ Health Promot 12(1). https://doi.org/10.4103/jehp.jehp_402_23

Shin J, Cho J, Lee S (2021) Tactile-color system for accessibility of color education: 2.5D UV printed supplementary material for visually impaired students. 2023 17th International Conference on Ubiquitous Information Management and Communication (IMCOM), 1–7. https://doi.org/10.1109/imcom51814.2021.9377433

Smith VL (2000) Space tourism: the 21st century "frontier". Tour Recreat Res 25(3):5–15. https://doi.org/10.1080/02508281.2000.11014920

Soltani NS, Kirci NN (2019) Phenomenology and space in architecture. Int J Architect Eng Technol 6:1–6. https://doi.org/10.15377/2409-9821.2019.06.1

Staff I (2021, April 27) Critical Knowledge InSight: Mitigation of micrometeoroids and orbital debris risk | APPEL Knowledge Services. https://appel.nasa.gov/2021/04/27/critical-knowledge-insight-mitigation-of-micrometeoroids-and-orbital-debris-risk/

Tandon S, Das S (2023) Digital revolution complements sustainable goal achievement during and after pandemic: a research agenda. In: Mondal SR, Yegen C, Das S (eds) New normal in digital enterprises. Palgrave Macmillan, Singapore. https://doi.org/10.1007/978-981-19-8618-5_11

Taylor D (2021, March 31) The future of space tourism: op-ed. Space.com. https://www.space.com/future-of-space-tourism-op-ed

Team SVV (2024, June 6) The Future of Space tourism: Orbital hotels and moon bases – a new era in travel – Space Voyage Ventures. Space Voyage Ventures. https://spacevoyageventures.com/the-future-of-space-tourism-orbital-hotels-and-moon-bases/

Tien-Dung PHAM, Majerova J, Das S (2022a) Phases of possible recovery of digital enterprises in new normal business for living with COVID-19 times: opportunities for a new era in sustainable development goals. In: Sustainable development and innovation of digital enterprises for living with COVID-19. Springer, Singapore, pp 19–33. https://doi.org/10.1007/978-981-19-2173-5_2

Tien-Dung P, Nguyen XD, Das S (2022b) Sustainable goal achievement by digital revolution during and after pandemic; how much one wins and losses: a bird's eye view for future planning. In: Sustainable development and innovation of digital enterprises for living with COVID-19. Springer, Singapore, pp 133–147. https://doi.org/10.1007/978-981-19-2173-5_8

Tom (2024, October 31) 2001: A Space Odyssey: technical realism and metaphysical vision. Tom's Blog. https://tomyonashiro.com/2024/10/31/2001-a-space-odyssey-technical-realism-and-metaphysical-vision/

Trang NTT, Vukorep I, Stopp H (2019) Floating Village Cua Van: promoting climate-adaptive ecotourism with principles of living spaces. In: Advances in Science, Technology & Innovation/Advances in science, technology & innovation, pp 13–28. https://doi.org/10.1007/978-3-030-10804-5_2

Traweek MS, Tatara JD (1998) Overview of the Environmental Control and Life Support System (ECLSS) testing at MSFC. PubMed 5(1):5–12. https://pubmed.ncbi.nlm.nih.gov/11540464

Tussyadiah IP, Wang D, Jung TH, Dieck MT (2017) Virtual reality, presence, and attitude change: empirical evidence from tourism. Tour Manag 66:140–154. https://doi.org/10.1016/j.tourman.2017.12.003

Vrana V, Das S (2023a) Dynamic restructuring of digital media and entertainment sector: role of urbanization, industrial innovation, and technological evolution. In: Das S, Gochhait S (eds) Digital entertainment as next evolution in service sector. Palgrave Macmillan, Singapore. https://doi.org/10.1007/978-981-19-8121-0_2

Vrana V, Das S (2023b) Role of enterprises and their digital transformation in post-pandemic sustainable development. In: Mondal SR, Yegen C, Das S (eds) New normal in digital enterprises. Palgrave Macmillan, Singapore. https://doi.org/10.1007/978-981-19-8618-5_7

Yegen C, Das S (2023) Digital solution for print and publishing media. In: Das S, Gochhait S (eds) Digital entertainment as next evolution in service sector. Palgrave Macmillan, Singapore. https://doi.org/10.1007/978-981-19-8121-0_12

Zach (2023, October 18) Modern cottage style: the ultimate guide – Hudson Restoration. Hudson Restoration Inc. https://hudsonrestoration.com/modern-cottage-style-the-ultimate-guide/

Subhra R. Mondal Senior Researcher and academician at Duy Tan University, specializing in brand management, AI, and smart tourism. Editor of multiple Scopus-indexed books, her work explores digital recruitment and Web 5.0. She bridges academia and industry through collaborative projects.

Subhankar Das Senior Researcher and academician at Duy Tan University, publishing extensively on AI, blockchain, and digital transformation. Editor of seven Scopus-indexed books and reviewer for top journals. Mentors global PhD candidates and drives innovation in ASEAN academic collaborations.

The manufacturer's authorised representative in the EU is Springer Nature Customer Service Centre GmbH, Europaplatz 3, 69115 Heidelberg, Germany. If you have any concerns regarding our products, please contact ProductSafety@springernature.com

Printed and bound by CPI Group (UK) Ltd, Croydon, CR0 4YY
01/10/2026
02225925-0018